The Uncertain Machine

E.B Cohen

First edition, 2026

Contents

The Uncertain Machine

A History of Quantum Computing — from Feynman's Dream to the Race for Q-Day

E.B Cohen

The Uncertain Machine: A History of Quantum Computing

This is a work of narrative science history, drawn from the public record and the standard published scholarship. Quantum computing is a field that moves in months, not decades; where a claim is recent, disputed, or made by a company about its own machine, the text says so and gives the date, rather than a false air of permanence.

First edition, 2026

A Note on the Mathematics, and the Metaphors

This is a book about one of the most mathematical subjects human beings have ever studied, and it contains almost no mathematics. That is a deliberate choice, and it deserves a word of explanation.

Quantum computing is built on linear algebra — vectors, matrices, complex numbers, the geometry of high-dimensional spaces. To *design* a quantum algorithm you need that machinery, and no popular account can substitute for it. But to *understand what quantum computers are, why they might be powerful, who built them, and what is at stake,* you do not need to solve a single equation. You need good metaphors, used honestly and then corrected where they mislead. That is what this book tries to provide.

And correction is necessary, because the most popular metaphor of all is wrong. You have surely heard that a quantum computer "tries all the possible answers at the same time." It is a seductive image, and it is false. A quantum computer that merely tried every answer in parallel would be useless, because when you looked at it to read the result, you would get a single random answer, no better than a guess. The real power lies in something subtler and stranger — **interference**, the same phenomenon that makes ripples on a pond reinforce each other in some places and cancel to stillness in others. A quantum algorithm is a carefully choreographed act of cancellation, arranged so that the wrong answers destroy one another and the right answer is left standing. Getting the reader to feel the difference between "trying everything at once" and "arranging for the wrong answers to cancel" is one of the central tasks of this book.

Where I use the words *qubit, superposition, entanglement, gate,* and *decoherence,* I define them in plain language as they arise, and they are all collected in the Glossary at the back. A reader who remembers nothing of school physics will not be left behind. A reader who knows the field will, I hope, find the story worth retelling.

Introduction: The Machine That Computes With Maybe

In a windowless laboratory, cooled to within a hair of the coldest temperature the universe permits, hangs a machine that looks like nothing so much as a golden chandelier turned upside down. Tiers of gleaming plates descend from a steel canister, laced together by a filigree of coiled copper and niobium tubing, the whole assembly shrinking stage by stage toward a chip at the bottom no larger than a fingernail. The gold is not decoration; it is there to carry heat away and signals in. The tubing carries a helium mixture that, stage by descending stage, chills the chip to about one hundredth of a degree above absolute zero — colder than the deep of interstellar space, colder than anywhere in the natural cosmos. All of this elaborate cryogenic sculpture exists to protect a few dozen loops of superconducting metal from the one thing that would destroy what they are doing: the warm, noisy, ordinary world.

For inside that chip, something is happening that has no counterpart in the entire history of machinery. The little loops of current are being held in a state that classical physics forbids and classical language can barely describe — a state of *both*, of *maybe*, of possibilities held in suspension and made to interfere with one another like waves. This is a computer. But it does not compute the way any computer before it has computed. It does not store its information as definite ones and zeros marching through logic gates. It stores information as something the physicist Erwin Schrödinger, who helped discover it and then spent decades uneasy about it, called a superposition — a blend of possibilities that becomes a single definite fact only at the instant you look.

To the people who built it, this machine is the most exciting object in modern science: a device that exploits the deepest and weirdest layer of physical reality to perform calculations that the largest conventional supercomputers on earth could not finish before the sun burns out. To its skeptics — and there are serious, informed skeptics — it is a magnificent physics experiment that has been oversold as an imminent revolution, a machine perpetually five years from usefulness, riding a wave of investment and hype that may yet break. Both camps are describing the same golden chandelier. The remarkable thing about quantum computing, in the middle years of the 2020s, is that both camps might be right.

This book is the story of how we got here: how a philosophical puzzle about the nature of reality became, over the course of a single human lifetime, a global technological race with billions of dollars and the security of the world's secrets riding on its outcome. It is a story that begins not with a machine but with a catastrophe — the failure of classical physics, at the turn of the twentieth century, to explain something as mundane as the color of the light coming out of a hot oven. That failure cracked open a new physics so strange that its own founders never fully made peace with it, a physics of the very small in which particles are also waves, in which a thing can be in two states at once, in which two objects light- years apart can be bound together so that measuring one instantly determines the other, and in which the act of looking changes what is looked at.

For fifty years that strangeness was the property of physicists, who used it to build lasers and transistors and atomic bombs but treated its philosophical weirdness as a distraction. Then, in the early 1980s, a few unusual minds — a Russian mathematician, a reclusive Oxford theorist, and above all the irrepressible American physicist Richard Feynman — asked a question that no one had quite framed before. If the world at bottom runs on quantum rules, and if simulating those rules on an ordinary computer is hopelessly, exponentially hard, then why not build a computer that runs on quantum rules itself? Why not fight fire with fire, and compute with the very strangeness that makes the quantum world so hard to predict?

It was a beautiful idea, and for a decade it remained little more than that — a curiosity at the edge of theoretical physics, of interest to almost no one. What turned it into a field, and then into a race, was a single piece of mathematics published in 1994 by a quiet number theorist named Peter Shor, who showed that a quantum computer, if one could ever be built, would be able to break the codes that protect essentially every secret transmitted over the internet — every bank transfer, every state secret, every password. Overnight, a philosophical curiosity became a matter of national security. The money arrived. The engineers arrived. And a thirty-year struggle began: not to invent quantum computing, which had been invented on paper, but to *build* it, to coax these ghostly quantum states into existence and hold them still long enough to compute, in defiance of a universe that seems designed to snuff them out.

That struggle is the spine of this book. It runs through a menagerie of competing machines — ions suspended in electric fields, atoms held in tweezers of light, circuits of superconducting metal, single particles of light, defects in diamond — each a different wager on how to build a qubit. It runs through the invention of quantum error correction, the almost miraculous discovery that you can protect a fundamentally fragile quantum state by spreading it across many physical parts, and through the long climb toward the threshold where, at last, in December 2024, a machine crossed a line that had been theorized for thirty years. It runs through the "supremacy wars," the bitter and instructive quarrels over whether a given machine had really beaten the classical competition. And it runs, finally, toward a reckoning that is no longer hypothetical: the coming day, which cryptographers call Q-Day, when a quantum computer large enough to run Shor's algorithm will render the world's current encryption obsolete — a day that governments are already, quietly, racing to prepare for.

I have tried throughout to hold two things in balance that the literature on this subject too often keeps apart: genuine wonder at what has been achieved, and hard-headed skepticism about what has merely been promised. Quantum computing is neither the imminent miracle of the press releases nor the elaborate hoax of the debunkers. It is something more interesting than either: a real and profound scientific achievement, still short of practical usefulness, advancing faster than the pessimists expected and slower than the enthusiasts hoped, its ultimate destination genuinely unknown. This is the story of the uncertain machine — uncertain in its physics, which runs on probabilities; uncertain in its engineering, which teeters on the edge of the possible; and uncertain in its future, which no one, honestly, can yet foresee.

It begins, as the whole of modern physics begins, with the puzzle of a glowing object and a piece of mathematics that should have worked and did not.

Part I — The World Turned Strange

Chapter 1: The Ultraviolet Catastrophe

On a December evening in 1900, in a lecture hall of the German Physical Society in Berlin, a cautious, conservative professor of forty-two stood up to describe a small mathematical trick he had found to fit some experimental curves. Max Planck did not think he was starting a revolution. He thought he was patching a leak. By his own later admission he regarded the thing he was about to propose as "an act of desperation," a formula pulled out of the air to make the numbers come right, which he fully expected would be explained away by sounder physics once someone cleverer got to work on it. Instead, the trick he described that evening cracked the foundations of physics and never let them close again. It is conventional, and correct, to date the birth of the quantum age to that lecture. The whole strange story of this book runs back to it.

The problem Planck was trying to solve was, on its face, almost comically ordinary. Heat any object — a poker, an iron bar, the filament of a lamp — and it glows. As it gets hotter the glow shifts from dull red to orange to yellow to a brilliant blue-white. This is not mysterious to anyone who has watched a fire. What was mysterious, at the end of the nineteenth century, was that no one could explain it. Physicists had a beautiful, complete, triumphant theory of heat and a beautiful, complete, triumphant theory of light and electromagnetism — the crowning achievements of the century — and when they put the two together to predict the color of the light coming out of a hot object, they got an answer that was not merely wrong but absurd. The theory predicted that a hot object should pour out an infinite amount of energy at the blue and ultraviolet end of the spectrum. Every glowing ember, every warm body, every star, according to the best physics of 1900, should be blasting the universe with a lethal, infinite flood of ultraviolet light. Since this manifestly does not happen — you can warm your hands at a fire without being vaporized — something was deeply, structurally wrong. The physicist Paul Ehrenfest would later give the failure its enduring and faintly apocalyptic name: the **ultraviolet catastrophe.**

To see why the catastrophe arose, picture the light rattling around inside a hot oven as a collection of waves, like the vibrations of a plucked string, each with its own frequency. Classical physics held a firm democratic principle about such vibrations: on average, every possible mode of vibration should get an equal share of the available heat energy. The trouble is that there is no limit to how high the frequencies can go. You can always fit in a wave that wiggles faster, and then a faster one still, without end. So if every one of that infinite number of high-frequency modes claims its equal share of energy, the total demand is infinite. The mathematics was clean; the conclusion was ruinous. Nature was somehow declining to give the high-frequency waves their "fair" share, and no one knew why.

Planck's desperate trick was to break the democracy. Suppose, he said — with no physical justification he could defend, purely to make the formula work — that the oven could not emit or absorb light energy in any old amount, in a smooth continuous trickle, but only in discrete lumps, whole packets, indivisible units. Suppose energy came in grains. And suppose, crucially, that the size of the grain depended on the frequency of the light: the higher the frequency, the bigger the minimum grain. He wrote this as the most famous little equation of the new physics, $E = hf$ — the energy E of a grain equals the frequency f multiplied by a tiny new constant h, which we now call **Planck's constant**, one of the fundamental numbers of the universe, so small (about 6.6 followed by thirty-three zeroes after the decimal point, before the six) that in everyday life the graininess of energy is utterly invisible.

Watch what this assumption does to the catastrophe. If energy comes in grains whose size grows with frequency, then to excite a very high-frequency mode of vibration you must pay for at least one whole, large grain of energy all at once. But at any given temperature there is only so much energy to go around, sloshing about at random. The high-frequency modes, whose entry fee is enormous, mostly cannot afford even a single grain, and so they sit dark and empty. The infinite demand collapses. The energy the object radiates rises with frequency, peaks, and then falls away to nothing at the ultraviolet end — exactly, to the last decimal, matching what experiments showed. The catastrophe was cured. The color of a hot object was explained. And the price of the cure was the idea that energy, at bottom, is not smooth but granular.

Planck himself flinched from what he had done. For years he tried to find a way to keep the grains as a mere calculating device while leaving the underlying physics continuous and classical, the way it was supposed to be. He could not. The graininess was real, and it was not a special feature of hot ovens; it was a feature of the world. The word for a single grain — a single indivisible packet of energy — was *quantum*, from the Latin for "how much," a bland accountant's word that would come to name the most counterintuitive science human beings have ever devised. What Planck had shown, without wanting to, was that at the finest scale the universe is not a continuum but a countable thing, that it deals in whole numbers, that there is a smallest coin.

It is worth pausing on how strange this is, because everything that follows in this book — every qubit, every superposition, every quantum computer — grows from the soil that Planck first turned. In the world of our senses, quantities are smooth. You can pour any amount of water, walk any distance, glow with any brightness. The idea that there might be a smallest possible amount of *something as abstract as energy*, that nature counts rather than pours, violates a bone-deep intuition. Yet it is true, and its truth is not a philosopher's abstraction but a measured, engineering-grade fact, the basis of the LED in your phone and the sensor in your camera. The first lesson of the quantum world, the one Planck stumbled into against his own instincts, is that reality is pixelated. The universe, examined closely enough, is made of discrete steps.

It is worth pausing on just how strange Planck's grains are, because the strangeness is the whole point of everything that follows. Think of the difference between a ramp and a staircase. Classical physics assumed the world was a ramp: energy could be added or removed in any amount, however small, in a smooth continuous slope. Planck was forced to conclude that, for light of a given color, the world is a staircase. You can stand on one step or the next, but never in between, and the higher the frequency of the light the taller each step becomes. To climb the ultraviolet staircase you must pay for a whole giant step at once, and at ordinary temperatures the object simply does not have that much energy to spend in one place — so the ultraviolet steps stay empty and the catastrophe evaporates. Another way to feel it: imagine a theatre that will only sell tickets in fixed blocks, and the blocks for the front rows cost a fortune. Most of the audience ends up in the cheap seats at the back. Nature, it turns out, seats its energy the same way.

How small is Planck's constant? Small beyond ordinary imagining. If you scaled a single grain of energy in visible light up until it weighed as much as a grain of sand, then on that same scale the energy you spend lifting a coffee cup would outweigh the Earth many times over. This is precisely why no one had noticed the graininess before: in every human-scale process the steps of the staircase are so fine that the staircase is indistinguishable from a ramp. The quantum world had been hiding in plain sight, its texture too fine for the eye.

Three things are commonly misunderstood about this moment. The first is that Planck "discovered the photon." He did not, and would have resisted the idea; his grains were a property of the walls of the oven exchanging energy with the light, not of light itself — that leap belonged to Einstein, and it is the subject of the next chapter. The second misconception is that Planck was a bold revolutionary. He was the opposite: a temperamental conservative who spent a decade trying to un-discover his own result, to find some way to fold the grains back into classical continuity. He never managed it, and late in life he described the futility of arguing against the new physics with a famous, rueful observation that science advances "one funeral at a time" — that old theories die not by persuasion but because their defenders eventually do. The third is that the ultraviolet catastrophe was some obscure technicality. It was not. It was the crack through which the entire twentieth century poured: lasers, transistors, LEDs, solar cells, MRI machines, and eventually the quantum computer all trace their lineage to the admission that energy comes in lumps.

You meet Planck's law every day without knowing it. The reason a dimmer bulb glows warm orange while a hotter one burns white is the blackbody curve he explained; the reason a blacksmith can judge the temperature of iron by its color is the same law read backwards. The most perfect blackbody ever measured is not in any laboratory but fills the whole sky: the cosmic microwave background, the cooled afterglow of the Big Bang, matches Planck's 1900 formula to a precision that still astonishes physicists — a curve sketched to fix an oven, found written across the entire universe.

It is worth pausing to notice how the ultraviolet catastrophe fits a pattern that recurs throughout the history of physics — and throughout this book. A theory that is beautiful, complete, and triumphant is applied to some perfectly ordinary phenomenon and returns not a small error but an absurdity, an infinity, a prediction so wildly wrong that it signals not a mistake in the calculation but a crack in the foundations. The infinity of the ultraviolet catastrophe was such a signal, and Planck's grains were the first patch over the crack. A century later, quantum computing would grow from the same soil: the impossibility of simulating quantum systems on classical machines is, in its way, another catastrophic infinity — the exponential explosion of amplitudes — and the response, again, is not to patch the old framework but to build a new kind of machine that speaks the grainy, quantized language nature actually uses. The lesson Planck taught, reluctantly, is that when the mathematics returns an absurdity, nature is usually trying to tell you something profound about its structure.

There is a human detail worth preserving, because it captures how genuinely reluctant the revolution's first author was. Planck was, by temperament and conviction, a deep conservative in physics, a man who revered the classical edifice and wanted nothing more than to defend it. He introduced his quantum of energy not as a bold new principle but as a desperate, temporary expedient — a calculational trick he fully expected would be explained away and absorbed back into continuous, classical physics once someone cleverer came along. For years afterward he labored to undo his own discovery, to find some route back to the smooth world he trusted. He never found it, because there was none. The graininess was not a trick; it was the truth. That the quantum age began with its founder trying to talk himself out of it is one of the great ironies of science, and a reminder that revolutions are often started by people who would much rather they hadn't happened.

Consider, finally, how completely Planck's reluctant grains have reshaped the material world you inhabit — for the abstraction has the most concrete possible consequences. The quantization of energy that cured the ultraviolet catastrophe is the same principle that makes the laser possible, and lasers read your data from optical discs, carry the internet's traffic through fiber, scan your groceries, and perform delicate surgery. It underlies the light-emitting diode, so that the quantized energy jumps of electrons in a semiconductor produce the precise colors glowing in every screen and lightbulb around you. It explains the photovoltaic effect that turns sunlight into electricity on rooftops, and the discrete atomic energy levels that make atomic clocks — and thus satellite navigation — possible. Nearly every distinctively modern technology traces some part of its lineage to the admission, forced on a reluctant professor in Berlin in December 1900, that energy comes in indivisible lumps. The strange grain of energy, invisible in everyday life because it is so fantastically small, turns out to run the visible world. And the story it began — the story of learning to see, and eventually to compute with, the granular quantum texture of reality — would run through the entire century that followed and into our own.

Chapter 2: Einstein's Reluctant Revolution

The year 1905 is remembered by physicists as the *annus mirabilis*, the miracle year, in which an obscure clerk in the Swiss patent office in Bern, examining applications for electrical gadgets by day and doing physics in the margins of his time, published four papers, any one of which would have made a career. One explained the jittery dance of pollen grains in water and in doing so proved, once and for all, that atoms are real. One laid out the special theory of relativity and overturned three centuries of assumptions about space and time. One, almost as an afterthought, contained the equation $E = mc^2$. And one — the paper Einstein himself, with characteristic mischief, called "very revolutionary," the only one of the four he described that way — concerned light, and it is the one that matters for our story.

The puzzle Einstein seized on was called the **photoelectric effect**, and like the glowing oven it was maddeningly simple to state. Shine light on a piece of metal, and under the right conditions the metal spits out electrons. This much was known. What made no sense, under the reigning theory that light was a continuous wave, was the *details*. If light were a wave, its energy would depend on its brightness. Shine a brighter light and you deliver more energy, so you ought to knock electrons out with more force. Shine a dim light and, though it might take a while for enough energy to accumulate, you ought eventually to pry electrons loose. That is the sensible, wavelike expectation. It is also flatly contradicted by what happens. In reality, whether the metal emits electrons at all depends not on the brightness of the light but on its *color* — its frequency. Red light, no matter how blindingly bright, might fail to eject a single electron from a given metal. Blue or ultraviolet light, even a feeble amount of it, would eject electrons instantly. And making the blue light brighter did not make each ejected electron faster; it only made *more* of them come out, each with the same energy as before.

None of this fits a wave. All of it fits grains. Einstein's leap — building directly on Planck's $E = hf$, but pushing it far past where Planck was willing to go — was to say that light itself is composed of discrete packets of energy, later named **photons**, each carrying an energy fixed by its frequency. Now the photoelectric effect becomes transparent. Ejecting an electron from a metal is like knocking a coconut from a tree: it takes a single blow above a threshold of force. A photon of red light is a small pebble; no matter how many of them you throw — no matter how bright the red light — each individual pebble is too weak to dislodge the electron, and throwing more weak pebbles accomplishes nothing. A photon of blue light is a proper stone, above the threshold, and a single one does the job. Brighten the blue light and you are simply throwing more stones, so more electrons fly off, but each is knocked loose by one stone and departs with the same energy. The frequency sets the energy per grain; the brightness sets the number of grains. The mystery dissolves the instant you accept that light is granular.

It is hard, from this distance, to feel how audacious this was. The wave theory of light was not a speculation; it was one of the great secured triumphs of nineteenth-century science. Thomas Young had shown a hundred years earlier that light passing through two slits produces an interference pattern of bright and dark bands — the unmistakable signature of waves, of crests reinforcing and cancelling, exactly like water waves in a harbor. James Clerk Maxwell had crowned the century by showing that light *is* an electromagnetic wave, a self-sustaining ripple of electric and magnetic fields, and had predicted its speed from first principles. To stand up in 1905 and say that this triumphantly established wave was, after all, a hail of particles was very nearly a provocation. Einstein knew it. He did not claim to have overthrown the wave theory; he claimed, more disturbingly, that light somehow needed to be *both* — a wave in some experiments, a stream of particles in others — and that no one yet understood how those two pictures could be true at once.

This was the first appearance of what would become the central and inescapable strangeness of quantum physics: **wave–particle duality**, the refusal of the fundamental constituents of the world to be cleanly one thing or the other. Light spreads and interferes like a wave, yet arrives and deposits its energy in lumps like a particle. Neither picture is a mere approximation to the other; both are needed, and which one you see depends on what you do to look. Over the next two decades physicists would discover, to their continuing discomfort, that this duality runs both ways: not only do waves behave like particles, but the things everyone had confidently called particles — electrons, atoms, whole molecules — behave, under the right conditions, like waves. The world at its foundations declines to sort itself into the tidy categories our minds prefer.

The scientific establishment resisted. Even Planck, who had lit the fuse, thought Einstein had gone too far; when Planck recommended the young man for the Prussian Academy of Sciences a few years later, he asked his colleagues to forgive Einstein's wild idea about light-quanta as the excess of a bold mind. The great experimentalist Robert Millikan spent a decade trying to prove the photon hypothesis wrong, performing meticulous experiments precisely to refute it — and instead confirmed Einstein's equation to high precision, a result he accepted with visible reluctance. In the end the evidence was overwhelming. It was for the photoelectric effect and the light-quantum, not for relativity, that Einstein received the Nobel Prize in 1921 — a fitting irony, since he would spend the second half of his life as the most distinguished and stubborn opponent of the full quantum theory his own revolution had helped to spawn.

To grasp why Einstein's step was so radical, consider what "light is a wave" had explained. For a century, every experiment — diffraction, interference, polarization — had confirmed that light spreads and overlaps like ripples on a pond. Thomas Young's double-slit experiment of 1801 was the crown jewel: shine light through two slits and you get not two bright bars but a striped pattern of light and dark, the unmistakable signature of waves adding and cancelling. To say, in 1905, that light was really a hail of particles was to seem to throw all of that away. Einstein did not throw it away; he did something far more unsettling. He said light was somehow both — that it travels as a spread-out wave but arrives, is absorbed, and is emitted as a compact grain. The wave tells you where the grains are likely to land; the grains are what actually land. This uneasy marriage, wave-particle duality, would become the beating heart of quantum mechanics, and no one has ever made it feel comfortable.

The evidence Einstein seized on was the photoelectric effect: shine light on a metal and it can knock electrons loose. The wave picture made a confident prediction — brighter light carries more energy, so brighter light should knock electrons out harder. Experiment said no. Dim blue light freed electrons with plenty of energy; intense red light, no matter how bright, freed none at all. Only the color — the frequency — set the energy of the ejected electrons; brightness merely set how many. This is exactly what you would expect if light arrived in grains whose individual punch depended on frequency, just as Planck's staircase demanded. One blue grain has enough energy to eject an electron; a red grain does not, and a billion red grains are still just a billion individual weaklings, none able to do the job. A useful analogy: you cannot knock a coconut down by throwing a million ping-pong balls at it, but a single well-aimed cricket ball will do it. Intensity is the number of throws; frequency is the mass of each ball.

There is a delicious irony buried in Einstein's career. When he won the Nobel Prize in 1921, it was awarded not for relativity — still thought too speculative by some of the committee — but "for his discovery of the law of the photoelectric effect." The most famous physicist in history was honored for the one idea he found most troubling. And troubled he remained. The very word "photon" was not his; it was coined in 1926 by the chemist Gilbert Lewis. Einstein spent his later decades in a long, losing argument against the full implications of the quantum theory he had helped to found — a story that will return in a later chapter with his most famous complaint about God and dice. He is often cast as the man who could not accept the future. It is fairer to say he was the man who saw, earlier and more clearly than anyone, just how strange that future would be, and could not bring himself to stop asking why.

You carry Einstein's grains in your pocket. The camera in your phone works because photons striking a silicon sensor knock loose electrons, one grain at a time, exactly the effect he explained; the light meter, the solar panel on a roof, the automatic door that senses your approach, all run on the photoelectric principle. Every digital photograph ever taken is a census of individual photons. The particle of light that Einstein reluctantly proposed is now counted, one by one, billions of times a second, in devices sold by the billion.

It is worth appreciating the peculiar courage of Einstein's 1905 leap, because it reveals something about how scientific revolutions actually happen — and why they are so hard. Planck had introduced his grains as a reluctant trick about how ovens exchange energy with light; he did not believe light itself was grainy. Einstein took the grains and said something far more radical and far more disturbing: that light *is* grains, everywhere and always, streaming through space as discrete packets. This flew in the face of a century of triumphant evidence that light was a wave — diffraction, interference, polarization, all beautifully explained by the wave theory and by Maxwell's magnificent equations. To assert that light was really particles seemed to throw all of that away, and for years Einstein stood almost alone in taking his own light-quanta seriously. Even the greatest physicists of the age resisted; when they nominated Einstein for membership in the Prussian Academy of Sciences in 1913, they praised his genius while explicitly excusing his "sometimes missing the target in his speculations, as for example in his theory of light quanta." The most respected minds in physics regarded his best idea as an embarrassing overreach.

The vindication, when it came, was slow and grudging, and its timing tells a story. Einstein proposed light quanta in 1905; the decisive experimental confirmation of his photoelectric predictions came a decade later, from the meticulous work of the American physicist Robert Millikan, who — remarkably — had set out to *disprove* Einstein's theory and instead confirmed it in precise detail, all the while doubting the interpretation. Not until the 1920s, with Arthur Compton's demonstration that light quanta bounce off electrons like billiard balls, carrying momentum as particles must, did the physics community fully accept that light genuinely has a particle nature. The word "photon" was coined only in 1926, by the chemist Gilbert Lewis, more than two decades after Einstein's paper. And Einstein's Nobel Prize, awarded in 1921, cited the photoelectric law rather than relativity — honoring the man for the one idea he found most troubling and that the world had taken longest to believe.

The deepest legacy of Einstein's reluctant revolution is the concept it forced upon physics: wave-particle duality, the maddening insistence that light — and, it would soon emerge, matter itself — is somehow both wave and particle at once, spreading like a wave yet arriving in lumps. This is not a comfortable idea, and no one has ever made it feel comfortable; it is simply how nature is. It would become the beating heart of quantum mechanics and, eventually, of quantum computing, where the wave-like spreading of possibilities and the particle-like discreteness of measurement outcomes are exactly the twin features a quantum computer exploits and endures. The superposition that lets a qubit explore many possibilities is the wave; the single definite answer you get when you measure is the particle. Every quantum computation lives in the tension Einstein first exposed, between the smooth wave of possibility and the sharp grain of actuality. Once physics accepted that light was both, the old comfortable question — is it *really* a wave or *really* a particle? — had to be abandoned and replaced by a stranger question that the next generation would spend the 1920s learning to ask: not what a thing is, but what happens when you look.

Chapter 3: The New Mechanics

For twenty years after Planck, quantum physics was less a theory than a collection of clever fixes — an "old quantum theory," physicists later called it, half-classical and half-new, held together with rules of thumb that worked without anyone knowing why. Niels Bohr, a young Dane working in Manchester, had used it in 1913 to explain the hydrogen atom, picturing the single electron as orbiting the nucleus like a planet but permitted only certain discrete orbits, jumping between them by emitting or absorbing whole quanta of light. The model predicted the exact colors that hydrogen emits, a stunning success, and it made Bohr famous. But it was a patchwork. It could not handle atoms with more than one electron. It gave no reason *why* the orbits were quantized. It was, as its own users admitted, a set of magic numbers in search of a theory. By the early 1920s the field had stalled in paradox, and a restless generation of very young physicists — most of them under thirty, some barely out of their teens — sensed that a real theory, a genuine mechanics of the quantum world, was waiting to be found. Between 1925 and 1927 they found it, twice, in two utterly different-looking forms that turned out to be the same thing.

The first form came from a twenty-three-year-old named Werner Heisenberg, working on the windswept North Sea island of Helgoland in the summer of 1925, where he had fled to escape a crippling attack of hay fever. Heisenberg made a radical methodological decision: he would refuse to talk about things that could not be observed. No one had ever seen an electron's orbit; the orbit was a classical picture smuggled into a non-classical world. What one *could* observe were the frequencies and brightnesses of the light an atom emits. So Heisenberg resolved to build a theory out of those observable quantities alone, arranging them in tables and finding rules for how the tables combined. The rules he found were strange. When he multiplied two of his tables together, the answer depended on the order: table A times table B did not equal table B times table A. For an ordinary multiplication this is nonsense — three times five equals five times three — and Heisenberg was so unsettled that he was unsure whether his scheme made sense at all. He sent it to his mentor Max Born, who recognized, with a jolt, that Heisenberg had unwittingly reinvented a piece of nineteenth-century mathematics called **matrix algebra**, in which exactly this order-dependence, this *non-commutativity*, is a known feature. Born and another young colleague, Pascual Jordan, worked the insight into a complete system: **matrix mechanics**, the first true quantum theory.

That non-commutativity — the fact that the order in which you do things to a quantum system matters — was not a technical quirk. It was the mathematical heart of the new physics, and it is worth flagging now because it echoes straight through to the quantum computer. In a quantum computer, operations are performed by "gates," and quantum gates, like Heisenberg's matrices, generally do not commute: apply gate A then gate B and you get one result; reverse the order and you get another. The whole art of quantum programming is the art of ordering non-commuting operations so that their combined effect steers the system's probabilities where you want them. The seed of that art was planted on Helgoland in 1925.

The second form of the theory arrived a few months later from a very different kind of physicist. Erwin Schrödinger was Austrian, urbane, thirty-eight — practically elderly by the standards of this revolution — and he found Heisenberg's abstract tables of numbers physically repellent, "a theory without a picture." He wanted a wave. His starting point was a suggestion made in 1924 by a French doctoral student, Prince Louis de Broglie, who had proposed, in a dissertation so bold his examiners consulted Einstein before passing it, that if light waves can behave like particles, then material particles ought to behave like waves — that an electron, or any object, has a wavelength inversely related to its momentum. For everyday objects this wavelength is so absurdly tiny as to be undetectable; for an electron it is comparable to the size of an atom, and therefore matters enormously. Schrödinger took de Broglie's matter-waves and asked the natural question: if the electron is a wave, what is the equation that governs its rippling? Over a productive holiday in the Swiss Alps, in the company of a lover whose identity history has not recorded, he wrote it down. The **Schrödinger equation** is to quantum mechanics what Newton's laws are to ordinary mechanics: the master rule that tells you how the state of a system changes from one moment to the next. It remains, a century later, the single most important equation in the field, and it is, in disguise, the equation that governs a quantum computer's every operation.

At first the two theories — Heisenberg's clattering matrices and Schrödinger's smooth waves — looked like rivals, and their creators regarded each other's work with something close to distaste. Heisenberg wrote privately that the more he thought about Schrödinger's waves the more repulsive he found them; Schrödinger wrote that he had been "discouraged, if not repelled," by Heisenberg's matrices. Within a year, to everyone's astonishment, Schrödinger and others proved that the two theories were mathematically identical — the same physics wearing two costumes, guaranteed to make the same predictions about every experiment. This was the moment the "old quantum theory" of patches and magic numbers gave way to **quantum mechanics** proper, a complete, consistent, breathtakingly accurate framework that has never once been contradicted by experiment in the hundred years since. Everything in this book — every qubit and gate and quantum algorithm — is an application of the mechanics those young men wrote down between 1925 and 1926.

The revolution of the mid-1920s arrived twice, in two disguises, and the fact that they turned out to be the same thing is one of the great surprises in the history of science. In 1925, on the wind-scoured North Sea island of Helgoland, where he had fled to escape a crippling attack of hayfever, the twenty-three-year-old Werner Heisenberg worked out a way to describe the atom using nothing but tables of numbers representing what could actually be observed — the frequencies and brightnesses of the light atoms emit —

and refusing to speak of electron "orbits" no one could see. He later recalled climbing a rock at dawn, dizzy with the sense that he was looking through the surface of things at a strange and beautiful interior. His tables obeyed a bizarre rule: the order of multiplication mattered, so that A times B need not equal B times A. Heisenberg had reinvented, without knowing it, the mathematics of matrices, and his mentor Max Born recognized it. This was matrix mechanics.

Barely a year later, the Austrian physicist Erwin Schrodinger — older, worldly, reportedly inspired during a Christmas holiday tryst in the Alps — published something that looked utterly different: a wave equation, a smooth continuous formula describing the electron as a spread-out wave of some kind. Physicists were enormously relieved, because waves were familiar and Heisenberg's clattering matrices were not. And then came the shock: Schrodinger and others proved that the two theories, however unalike they looked, made exactly the same predictions. They were one theory wearing two costumes. It was as though two explorers had mapped a new continent from opposite coasts and discovered, on meeting in the interior, that their maps matched perfectly.

But what was "waving"? Schrodinger hoped his wave was the electron itself, smeared out in space like a little cloud. Max Born supplied the answer that stuck, and it is the answer that makes quantum mechanics quantum: the wave is not a wave of stuff but a wave of probability. Its height at a point, squared, tells you the chance of finding the particle there if you look. The electron is not spread out like butter; it is spread out like a rumor — a distribution of possibilities that collapses to a single definite fact the instant a measurement is made. This is the famous "Born rule," and for it Born too received a Nobel Prize, though tellingly not until 1954, nearly three decades later, so slow was the community to digest what it meant.

A concrete way to feel the difference: in the old Bohr atom, an electron ran around the nucleus on a track like a planet, and you could imagine pointing to where it was. In the new mechanics there is no track and no "where" between measurements — only a probability cloud, denser here, thinner there, describing where the electron might be found. Ask the wrong question and nature gives no answer at all; ask "where is it?" by measuring, and nature picks a spot, apparently at random, weighted by the cloud. The determinism of Newton, in which the present exactly fixes the future, was gone. In its place stood a machinery that predicted only odds.

It is a common misconception that quantum mechanics is "fuzzy" or "imprecise." The opposite is true. The Schrodinger equation is as exact and deterministic as anything in physics: given the wavefunction now, it tells you the wavefunction later with perfect precision. What is probabilistic is only the final step, the measurement, where the cloud of possibility is forced to yield a single result. Everything up to that instant evolves like clockwork; only the last tick rolls the dice. This split — smooth deterministic evolution punctuated by abrupt random measurement — is the notorious "measurement problem," and physicists still argue about what it means. What no one argues about is whether the theory works. It is the most accurately tested theory in the history of science, verified in some cases to twelve decimal places.

The dual birth of quantum mechanics — Heisenberg's matrices in 1925 and Schrodinger's waves in 1926, looking utterly different yet proving to be the same theory — is worth dwelling on, because it illustrates a profound and recurring feature of deep physics: that the same underlying reality can wear radically different mathematical costumes. Heisenberg's approach was austere and forbidding, built from tables of numbers representing only what could be observed, refusing to speak of anything as unpicturable as an electron's path; physicists found it alienating. Schrodinger's approach was warm and familiar, describing the electron as a smooth wave obeying a differential equation of the kind physicists had used for centuries; they embraced it with relief. And then came the shock that the two, however unalike, made identical predictions in every case — that they were one theory seen from two sides. The lesson, which the reader should carry into the quantum computing chapters ahead, is that the mathematical formalism is a tool for capturing reality, not reality itself, and that the same quantum truth can be expressed in the language of matrices, of waves, of circuits, or of qubits, each illuminating a different facet.

The Born rule — Max Born's insight that the wave is a wave of *probability* — deserves emphasis as the precise moment quantum mechanics became irreducibly strange, and as the hinge on which everything in this book turns. Schrodinger himself hoped his wave was the electron, smeared out in space like a little cloud of charge. Born saw that it could not be, and supplied the interpretation that stuck: the wave's height at a point, squared, gives the *probability* of finding the particle there if you look. The electron is not spread like butter but spread like a rumor — a distribution of possibilities that collapses to a single definite fact the instant a measurement is made. With this, the clockwork determinism of Newton, in which the present exactly fixes the future, was gone, replaced by a machinery that predicts only odds. That this probabilistic interpretation took Born until 1954 to be honored with a Nobel Prize, nearly three decades after he proposed it, testifies to how slowly and reluctantly the community absorbed what it meant — much as Planck and Einstein had been slow to be believed before him.

It is worth correcting, firmly, the widespread notion that quantum mechanics is "fuzzy" or "vague," because the opposite is true and the distinction matters enormously for understanding quantum computers. The Schrodinger equation is as exact and deterministic as anything in physics: given the wavefunction now, it tells you the wavefunction at any later time with perfect, clockwork precision. What is probabilistic is only the final step — the measurement — where the smoothly evolving cloud of possibility is forced to yield a single result. Everything up to that instant evolves like clockwork; only the last tick rolls the dice. This split — deterministic evolution punctuated by abrupt random measurement — is the notorious "measurement problem," and it is precisely the structure a quantum computer exploits: a quantum program is a stretch of perfectly deterministic, precisely engineered evolution (the gates), sculpting the wave of amplitudes, followed by a single brutal, probabilistic measurement that collapses the result. The determinism of the evolution is what lets the programmer arrange interference with exquisite precision; the randomness of the measurement is the bottleneck the whole art of quantum algorithms is designed to overcome. Out of the new mechanics fell the two ideas this book cannot do without — superposition and its shadow, collapse — and it is to the hard limit these ideas imposed on knowledge itself that we turn next.

Chapter 4: God Does Not Play Dice

The meaning of Schrödinger's wave was supplied, within months, by the same Max Born who had recognized Heisenberg's matrices. In a short paper in 1926, tucked into a footnote added in proof, Born proposed the interpretation that would define quantum mechanics and unsettle physics forever. The wavefunction, he said, is not a physical wave and not the electron itself. It is a **wave of probability**. Where the wavefunction is large, the electron is likely to be found; where it is small, unlikely; where it is zero, never. More precisely — and this detail matters for everything that follows — the probability is given by the *square* of the wavefunction's magnitude. The wavefunction itself can take negative and even complex (imaginary) values; only when you square it do you get an honest, positive probability. This squaring, the **Born rule**, is the bridge between the ghostly mathematics of the quantum world and the concrete numbers that come out of a measurement, and a version of it governs exactly what happens when you finally read the answer out of a quantum computer.

Grasp what Born was really saying, because it is the pivot on which quantum strangeness turns. In Newton's world, if you know a particle's position and velocity now, you can calculate exactly where it will be forever after; the universe is a clockwork, in principle perfectly predictable, and any uncertainty is mere ignorance. Born's rule demolishes this. According to quantum mechanics, even if you know *everything* there is to know about a particle — its complete wavefunction, the most complete description physics permits — you still cannot predict where a measurement will find it. You can predict only the *odds.* Fire identically prepared electrons at a screen, one after another, each in precisely the same quantum state, and they will strike different spots, scattered at random, building up over thousands of trials into the smooth probability pattern the wavefunction predicts — but with no way, even in principle, to say where any individual electron will land. Randomness, in this view, is not a confession of ignorance to be cleaned up by better instruments. It is woven into the fabric of reality at the deepest level. The universe, at bottom, plays dice.

That last phrase belonged to the theory's most illustrious enemy. Albert Einstein, who had done as much as anyone to create quantum physics, recoiled from where it had gone. "Quantum mechanics is very impressive," he wrote to Born in December 1926, in a letter that has become the most famous single sentence in the history of the debate. "But an inner voice tells me that it is not yet the real thing. The theory says a lot, but does not really bring us any closer to the secret of the 'Old One.' I, at any rate, am convinced that *He* does not play dice." Einstein did not doubt that quantum mechanics *worked* — that its predictions matched experiment with unnerving accuracy. He doubted that it was *complete.* Surely, he insisted, beneath the probabilities there must be some deeper layer of reality, some hidden variables, that determines the actual outcome of each measurement, so that the randomness is only apparent, only our ignorance of the deeper machinery, exactly as the seemingly random tumble of a thrown die is really the determinate result of forces we simply cannot track. Bohr, leading the theory's defenders, insisted just as fiercely that there was no deeper layer, that the probabilities were the bottom of reality, and that asking what the electron "really" does when no one is measuring it was a meaningless question. Thus began the Bohr–Einstein debates, a decades-long intellectual duel between two of the greatest minds of the century, conducted in letters, at conferences, in thought experiments of escalating ingenuity — a duel that, remarkably, would eventually be settled by actual laboratory experiment, and settled against Einstein, in a way that directly underwrites the quantum computer.

Bundled with the probabilistic wavefunction came a second shock, delivered by Heisenberg in 1927: the **uncertainty principle**. It states that certain pairs of properties of a quantum object — most famously position and momentum — cannot both be sharply defined at the same time. The more precisely a particle's position is pinned down, the more wildly uncertain its momentum becomes, and vice versa, with the trade-off governed once again by Planck's tiny constant. This is not a statement about the clumsiness of our measuring instruments, a matter of the measurement disturbing the thing measured, though that is how Heisenberg first, imperfectly, explained it. It is a statement about what it is even *possible* for a quantum object to *be*. A particle with a perfectly definite position simply does not possess a definite momentum — not hidden, not unknown, but genuinely undefined, the way the "top edge" of a sphere is not hidden but nonexistent. Reality, at the quantum scale, does not come with all its properties simultaneously sharp. It is intrinsically fuzzy, and the fuzziness is lawful and precise.

Out of these ideas — the probabilistic wavefunction, the Born rule, the uncertainty principle, Bohr's insistence that the unobserved has no definite properties — congealed the standard understanding of quantum mechanics, assembled largely in Bohr's institute in the Danish capital and therefore called the **Copenhagen interpretation**. Its central and most disturbing tenet concerns what happens at the moment of measurement. Before you look, a quantum system exists in a **superposition** — a blend of all its possible states at once, the electron's wavefunction genuinely spread across many positions, the possibilities coexisting and, crucially, able to interfere with one another. At the instant of measurement, this superposition **collapses**: the spread-out wave of possibilities snaps, at random and instantaneously, into a single definite outcome, and all the other possibilities vanish. Look at the electron and it is suddenly *here*, at one spot, its former everywhere-ness gone. Why measurement should have this violent, special, discontinuous effect — what counts as a measurement, why looking should collapse a wave that otherwise evolves so smoothly — is the notorious **measurement problem**, and a century on there is still no consensus about it. But whatever its ultimate meaning, the two-part rhythm the Copenhagen picture describes — smooth superposition while unobserved, sudden random collapse when measured — is precisely the rhythm of a quantum computation. A quantum computer spends its working life in superposition, its qubits exploring a vast space of possibilities in parallel, and then, at the end, is measured, collapsing to a single string of ordinary bits that, if the algorithm was clever, is the answer you sought. Every quantum program is a wager placed inside a superposition and cashed out by a collapse.

Heisenberg's uncertainty principle is routinely misdescribed, so it is worth stating carefully, because the truth is stranger than the myth. The myth says: measuring a particle disturbs it, so you cannot know its position and momentum at once — a mere practical clumsiness of our instruments. The reality is deeper. Quantum mechanics says a particle does not possess a sharp position and a sharp momentum simultaneously, even in principle, even unmeasured. The two quantities are like the width and the pitch of a musical note: a note that lasts only an instant has no well-defined pitch, and a note of perfectly pure pitch must ring on forever. Position and momentum are related the same way, through the mathematics of waves, and Planck's constant sets the size of the unavoidable trade-off. Sharpen one and the other must blur; this is not a limit on knowing but a limit on being.

This is what horrified Einstein. If the deepest description of a particle is a cloud of probabilities that only becomes definite when measured, then the universe has genuine chance sewn into its foundations, and the outcome of an individual quantum event is not merely unknown but undecided until it happens. In a 1926 letter to Max Born, Einstein made the objection that would follow him to the grave: he was convinced that God "does not play dice." Born's camp replied, in effect, that it was not Einstein's place to tell God what to do. The exchange was affectionate — the two men admired each other deeply and corresponded for decades — but it marked the opening of the longest and most productive argument in modern physics, one that Einstein would keep sharpening until he found what he believed was its killer form, the subject of the chapters just ahead.

The uncertainty principle has consequences that sound like science fiction but are measured facts. Because a particle can never be perfectly still — perfect stillness would mean perfectly known position and momentum at once — even at the coldest imaginable temperature, absolute zero, quantum systems retain a residual jitter called zero-point energy. Empty space is not truly empty: it seethes with fleeting "virtual" particles borrowing energy for instants too short for the universe to notice, and this seething produces real, measurable effects, from the tiny Casimir force that pulls two mirrors together in a vacuum to the faint Lamb shift in the light of hydrogen. The uncertainty that unsettled Einstein is, quite literally, holding empty space open.

There is also a startling upside hidden in the principle, one our story will need. Because quantum quantities are undecided until measured, a quantum system can explore many possibilities at once in a way no classical system can — and any attempt to eavesdrop on a quantum state, to pin down what it is doing, unavoidably disturbs it. That double edge — you cannot copy or spy on a quantum state without leaving a mark — is a nuisance for anyone building a computer but a gift for anyone building a code. Decades later it would become the basis of quantum cryptography, a way of sending secrets whose security rests not on the difficulty of a mathematical puzzle but on the laws of physics themselves. Heisenberg's limit, which Einstein saw as a defect in the theory, turns out to be the very thing that lets two strangers share a key that no one can steal without being caught.

The common misconception worth dispelling is that "uncertainty" reflects human ignorance that better technology might someday cure. Generations of physicists have hunted for a hidden layer of clockwork beneath the quantum haze — "hidden variables" that would restore Einstein's dice-free determinism — and, as we will see, experiment has ruled out the most natural versions of that dream. The randomness appears to be woven in, not painted on. When a single radioactive atom decays, there is, so far as any experiment can tell, no prior fact about the exact instant it will happen; the moment is chosen, if that is the word, only when it arrives.

So by the late 1920s physics had a theory of unprecedented power and unprecedented weirdness: exact in its equations, probabilistic in its verdicts, and insisting that reality is undecided until observed. Einstein refused to accept that this was the final word. Rather than merely grumble, he set out to prove the theory incomplete by following its own rules to what he was sure would be an absurd conclusion — two particles so linked that measuring one would instantly fix the other across any distance. He intended it as a reductio ad absurdum. Nature, it would turn out, had other plans.

Chapter 5: Spooky Action

By 1935 Albert Einstein had been arguing with the quantum theory for nearly a decade, and losing. At the great Solvay conferences of 1927 and 1930 he had arrived each morning with an ingenious thought experiment designed to catch quantum mechanics in a contradiction — a box that weighed a single photon, a shutter timed to a hair — and each evening Bohr, sometimes after a sleepless night, had found the flaw and turned the trap against its maker. Einstein had failed to show that quantum mechanics was *wrong*. So he changed his line of attack. He would try to show, instead, that it was *incomplete* — that it could not be the whole story, because the whole story it told was too absurd to be believed. The weapon he forged for this purpose, in a paper written that year with two younger collaborators, Boris Podolsky and Nathan Rosen, is known ever after by their initials: the **EPR paradox**. It is one of the most cited papers in the history of physics, and its intended target survived it, transformed, to become a cornerstone of quantum technology.

EPR turns on a phenomenon Schrödinger, responding to the paper that same year, named **entanglement** — *Verschränkung* in his German — and which he called "not one but rather *the* characteristic trait of quantum mechanics, the one that enforces its entire departure from classical lines of thought." Here is the idea, stripped to its essence. Quantum mechanics permits two particles to be prepared together in a joint state so intimately linked that neither has a definite property of its own; only the *pair* has definite properties. Imagine two coins minted together in such a way that quantum mechanics guarantees they will always land opposite — one heads, one tails — yet insists that until you look, neither coin is heads or tails; each is in a superposition of both. Now separate them. Send one to a laboratory on Earth and carry the other to the Moon, or to a distant star. The entanglement holds: still neither coin has a definite face; still the theory says they will land opposite. Then look at the Earth coin. It shows heads — at random, with no prior fact of the matter deciding it. And in that instant, quantum mechanics says, the Moon coin, which a moment before was itself an even blend of both possibilities, becomes definitely tails. Not "revealed to have been tails all along" — *becomes* tails, its superposition collapsing the instant its distant partner is measured, faster than any signal traveling at the speed of light could carry the news.

To Einstein this was a *reductio ad absurdum*, a proof that something had to give. Either measuring the Earth coin somehow reaches across the void and physically changes the Moon coin instantaneously — "spooky action at a distance," he scoffed, a phrase dripping with contempt, since it seemed to violate the relativistic prohibition on faster-than-light influence that Einstein himself had established — or, far more reasonably, the coins had their outcomes fixed all along, written into some hidden variables at the moment they were minted, and quantum mechanics, in refusing to acknowledge those predetermined values, was simply an incomplete description of a reality that was, underneath, perfectly ordinary and local and definite. EPR posed this as a dilemma and left no doubt which horn they thought sane: reality is local; there is no spooky action; therefore the hidden values exist; therefore quantum mechanics is incomplete. It was Einstein's best and most enduring argument, and for thirty years it looked like an unanswerable philosophical standoff — a matter of taste and metaphysics, with no experiment in sight that could decide between Einstein's local hidden reality and Bohr's spooky quantum wholeness.

That is what makes entanglement so important to our story, and it is worth being clear about why. For three decades entanglement was treated as an embarrassment, a philosophical curiosity, the weird corner of the theory that polite physicists did not visit. "Shut up and calculate," ran the working motto of the postwar generation: use the equations, which never fail, and do not ask what they mean. Entanglement was filed under "meaning," and ignored. Yet it is precisely entanglement — this capacity of quantum particles to be correlated more tightly than any classical objects can be, to share a joint existence that is more than the sum of its parts — that gives a quantum computer much of its power. A quantum computer with its qubits merely in superposition, each doing its own thing, would be a limited device. It is when the qubits are *entangled*, woven into a joint state whose description grows exponentially with their number, that the machine becomes something no classical computer can efficiently mimic. The 2^n-dimensional space that a quantum computer explores — the space that holds a thousand trillion trillion possibilities in a mere hundred qubits — is a space of entangled states. Einstein's "spooky action," the phenomenon he introduced to discredit quantum mechanics, is the very engine room of the technology this book is about.

There is a further gift hidden in entanglement, one that would take until the 1980s and 1990s to unwrap. Because the outcomes of measuring entangled particles are correlated but individually random, entanglement lets you do things with information that are classically impossible. It underlies **quantum teleportation**, in which the exact quantum state of one particle is transferred to a distant partner without any particle making the journey — a protocol we will meet, and which is a workhorse of quantum communication. It underlies **superdense coding**, in which a shared entangled pair lets you pack two classical bits of message into the transmission of a single qubit. It underlies the most secure form of communication known to physics, in which the very act of eavesdropping disturbs the entangled system and betrays the spy. None of these were imagined by Einstein, Podolsky, and Rosen, who saw in entanglement only an absurdity. They had put their finger, in trying to bury quantum mechanics, on its most valuable feature.

The thought experiment Einstein published in 1935 with his young colleagues Boris Podolsky and Nathan Rosen — forever after the "EPR paper" — was meant to be a knockout blow, and to understand our whole story you must understand what he was trying to knock out. Take two particles, let them interact, then send them flying far apart — to opposite ends of a laboratory, or opposite ends of the galaxy. Quantum mechanics says they can be left in an entangled state, a single shared description in which neither particle has a definite property of its own, yet their properties are perfectly correlated. Measure one and you instantly know the other. Einstein's point was sharp: either the distant particle had a definite property all along that the theory failed to mention — making quantum mechanics incomplete — or measuring here somehow reaches out and fixes reality there, faster than light, which he dismissed with a phrase that has echoed ever since as "spooky action at a distance." Surely, he argued, the sensible conclusion was the first: the particles carry hidden instructions, and quantum mechanics is merely an incomplete sketch of a deeper, ordinary reality.

An everyday image helps, and also shows where the everyday image fails. Imagine a pair of gloves separated into two boxes and shipped to opposite ends of the Earth. Open one box, find the left glove, and you know instantly the other is the right — no spookiness, no signal, just correlation that was fixed when the gloves were packed. Einstein's intuition was that entangled particles are exactly like the gloves: their fates were sealed at the source, and the "instantaneous" knowledge is nothing more than reading a pre-written note. This is such a reasonable picture that almost everyone who hears it agrees with Einstein. The astonishing thing — the thing that would take thirty years and a quiet Irish physicist to bring into the open — is that the glove picture is not merely unproven but demonstrably wrong. Entangled particles are correlated more tightly than any pre-written notes could ever manage.

What makes entanglement so hard to swallow is that the correlation is not of one fixed property but of any property you choose to measure, along any axis, even one selected at the last instant after the particles have parted. The gloves can only answer the single question "left or right?" Entangled particles seem to answer whatever question you spring on them, and to answer in lockstep with their partner, as if each somehow knew which question the other would face. It is this ability to be correlated in every direction at once that no set of hidden instructions can reproduce — and that, decades on, becomes the raw fuel of quantum computing. Entanglement is not a curiosity at the edge of the theory; it is the resource that lets a quantum computer's parts act as an indivisible whole, so that the state of many qubits together holds vastly more than the sum of the qubits taken one at a time.

It is a persistent misconception, repeated in countless breathless articles, that entanglement lets you send messages faster than light. It does not, and the reason is subtle and beautiful. When you measure your particle you get a random result — you cannot choose it — and your distant partner also gets a random result; only when the two of you later compare notes, by ordinary slower-than-light communication, does the eerie correlation reveal itself. Each side alone sees pure noise. Nature guards its own speed limit with exquisite care: the spooky link is real, but it carries no usable signal on its own. This is why quantum teleportation, a real and demonstrated technique that uses entanglement to transmit a quantum state, still requires an ordinary phone call alongside it to complete the trick.

It is worth appreciating that Einstein, in trying to destroy quantum mechanics, gave it its most precious concept — a pattern of unintended consequence that runs like a thread through this whole history. He devised the EPR thought experiment in 1935 as a weapon, a reductio ad absurdum meant to expose quantum mechanics as incomplete by showing that it implied something he was certain must be false: that measuring one particle could instantly fix the state of its distant partner. He was so sure this "spooky action at a distance" was absurd that he took its apparent necessity as proof the theory must be hiding something — that the particles must carry definite properties all along, mere "hidden variables" that the incomplete theory failed to mention. He intended to bury entanglement. Instead he named it, defined it, and thrust it to the center of physics, where it would sit for decades as the field's deepest puzzle and, eventually, become the master resource of an entire technology. The man who could not accept entanglement was the man who taught the world to see it.

The glove analogy deserves to be pressed once more, because understanding exactly where it fails is understanding what makes entanglement revolutionary. A pair of gloves shipped in separate boxes to opposite ends of the Earth are perfectly correlated — open one, find the left, and you know instantly the other is the right — but there is nothing mysterious here: the gloves' fates were sealed when they were packed, and your knowledge merely reads a pre-written fact. Einstein's conviction was that entangled particles are exactly like the gloves, their correlations fixed at the source. The astonishing discovery, which would take Bell's theorem and Aspect's experiments to establish, is that this reasonable picture is not merely unproven but *demonstrably wrong*. Entangled particles are correlated more tightly than any pre-written instructions could ever manage — correlated no matter which of many incompatible questions you choose to ask, along any axis, even one selected at the last instant after the particles have parted. The gloves can answer only "left or right?" Entangled particles seem to answer whatever question you spring on them, in lockstep with their partner. It is this correlation-in-every-direction-at-once that no hidden instructions can reproduce, and that, decades on, becomes the raw fuel of quantum computing.

The subtle law that keeps entanglement from enabling faster-than-light communication deserves restating, because it is both beautiful and perpetually misunderstood. When you measure your half of an entangled pair, you get a random result you cannot choose; your distant partner also gets a random result; and only when the two of you later compare notes, over an ordinary channel limited by the speed of light, does the eerie correlation reveal itself. Each side alone sees nothing but noise. Nature permits the spooky link but forbids using it to send a message, guarding its own speed limit with exquisite care. This is why quantum teleportation, a real and demonstrated technique that uses entanglement to transmit a quantum state, still requires an ordinary phone call alongside it to complete the trick — without the classical message, the distant qubit remains random garbage. Einstein believed he had exposed a fatal flaw; in truth he had identified the single most important phenomenon in the field, and named it with a sneer that has echoed ever since. For thirty years his challenge sat as untestable philosophy, until a quiet Irish physicist found a way to turn it into a number an experiment could measure.

Chapter 6: Bell's Inequality

John Stewart Bell was born in Belfast in 1928 to a family of modest means, left school at sixteen for lack of money, and worked as a laboratory technician before scraping his way to a physics degree. He spent his career at CERN, the great European particle-physics laboratory near Geneva, designing accelerators — that was his day job. Quantum foundations, the questions Einstein and Bohr had fought over, were his private passion, pursued, as he put it, on the side, almost furtively, because the profession regarded such questions as unserious, the province of philosophers and cranks. Bell thought the profession was wrong. He believed the Bohr–Einstein debate was about something real, and he had a heretic's sympathy for Einstein — a hope that the great man might turn out to be right, that beneath the quantum probabilities there might lie a sensible, local, deterministic world. In 1964, on sabbatical in the United States, he sat down to formalize that hope. What he proved instead astonished him, and changed physics.

Bell's achievement was to take the vague, metaphysicalseeming dispute of EPR and extract from it a sharp, quantitative prediction that different theories of the world would answer differently. He asked: suppose Einstein is right. Suppose entangled particles really do carry hidden, predetermined values, fixed at the moment they are created, so that all the correlations we see are just the playing-out of instructions written in advance — no spooky action, everything local, everything definite. What would such a world *require*? Bell considered experiments in which two entangled particles fly apart and each is measured, not along a single fixed direction, but along directions the experimenters are free to choose independently, at the last instant, at each end. He showed, with a piece of reasoning so simple it can be followed with schoolroom arithmetic and so profound it took the world's physicists years to absorb, that *any* theory built on local hidden values must obey a certain limit on how strongly the results at the two ends can be correlated, across the range of measurement choices. This limit is **Bell's inequality**. It is a ceiling. No local, predetermined, Einstein-style world can push the correlations above it.

And then Bell computed what quantum mechanics predicts for the same experiment. Quantum mechanics, with its entanglement and its spooky wholeness, sails straight past the ceiling. For certain choices of measurement directions, the quantum correlations are *stronger* than any local hidden-variable theory can possibly produce. The two pictures of reality — Einstein's local determinism and the full quantum theory — no longer merely differed in flavor or philosophy. They made *different numerical predictions* about a doable experiment. One of them had to be wrong, and Nature would decide which.

This was a thunderclap, though a delayed one, for it took some years and some stubborn individuals before anyone managed to do the experiment. Measuring Bell's inequality is hard: you must create pairs of entangled particles, usually photons; separate them; and measure each along rapidly, randomly chosen directions, closing off every loophole by which a sneaky local mechanism might mimic the quantum result — making sure, for instance, that the choice of measurement at one end cannot possibly send a signal, even at light speed, to influence the other end before its measurement is done. Through the 1970s a few pioneers, notably John Clauser at Berkeley, built the first crude versions and found results favoring quantum mechanics, though with loopholes still open. The decisive early experiments came in 1982 from a French physicist, Alain Aspect, who used a clever fast-switching scheme to change the measurement directions while the photons were already in flight, slamming shut the most important loophole. His results were emphatic. Bell's inequality was violated. The correlations exceeded the local ceiling. Einstein's dream of a local hidden reality beneath the quantum probabilities was, as a matter of experimental fact, dead.

It is worth being careful about what this does and does not mean, because it is often garbled. The violation of Bell's inequality does *not* mean you can send messages faster than light; the individual results at each end are random, and only when the two lists of outcomes are later brought together and compared does the eerie correlation appear, so no usable signal outraces light and relativity is safe. What it means is deeper and stranger than a fast telegraph. It means that the world is not *locally real* in the way Einstein took for granted — that either the properties of particles are genuinely undefined until measured (reality is not "real" in the classical sense) or that distant events are bound together in a way that transcends any local mechanism (the world is not "local"), or both. The comfortable classical assumption that objects have definite properties sitting in them, independent of observation, and that nothing influences anything faster than light, cannot both be true. Nature had been asked Einstein's question in the plainest possible terms, and Nature had answered: the quantum weirdness is not a veil over an ordinary world. It goes all the way down.

For decades this was regarded as a beautiful but purely

foundational result, a settling of old scores with no practical consequence. That judgment was spectacularly wrong. The violation of Bell's inequality is now the bedrock certification of quantum technology. It is the guarantee that entanglement is a genuine physical resource and not a bookkeeping illusion — that when a quantum computer entangles its qubits it is doing something with no classical shortcut, and that when a quantum cryptographic system detects an eavesdropper it is relying on a correlation no classical spy can fake. Bell tests are now run as calibration checks in quantum laboratories; a "loophole-free" Bell violation, finally achieved in 2015 in Delft, is used to certify random-number generators whose unpredictability is guaranteed by physics itself. And in 2022, the Nobel Prize in Physics was awarded to John Clauser, Alain Aspect, and Anton Zeilinger — the experimentalists who turned Bell's inequality from a thought into a measurement — "for experiments with entangled photons, establishing the violation of Bell inequalities and pioneering quantum information science." The citation was explicit: the foundations had become a technology. John Bell himself did not live to see the prize; he died suddenly in 1990, at sixty-two. He had gone looking for a way to rescue Einstein's sensible world, and had instead proved, beyond appeal, how strange the real one is.

John Stewart Bell's achievement was to do something almost no one thought possible: to take the airy dispute between Einstein and Bohr and reduce it to arithmetic that an experiment could settle. Bell was an unlikely revolutionary — a working particle physicist from Belfast, the son of a family of modest means, who pursued the foundations of quantum mechanics as a private passion on the side, calling it his "hobby." In 1964 he asked a precise question: suppose Einstein was right, and each particle carries hidden instructions fixed at the source, with no spooky influence passing between them. What limit would that place on how strongly two distant measurements could agree, averaged over many different choices of what to measure? He derived an answer in the form of an inequality — a ceiling. If local hidden variables govern the world, a certain tally of correlations can never exceed a specific number. Quantum mechanics, he showed, predicts a tally that sails straight past that ceiling. The two pictures did not merely feel different; they made different numerical predictions. Nature could be asked to choose.

The genius of Bell's inequality is that it does not depend on the details of any particular hidden theory. It is a constraint that ANY local, common-sense account of reality must obey — any theory in which distant events cannot instantly influence each other and things possess definite properties before they are looked at. Violate the inequality and you must give up at least one of those two deeply cherished ideas: locality, that nothing outruns light, or realism, that properties exist before measurement. Bell handed physics a sharpened knife and pointed it at the foundations. Then the experimenters picked it up.

The measurements are done not with position and momentum but with a more convenient property: the polarization or spin of entangled photons, the angle along which each behaves like a tiny arrow. An experimenter on each side freely chooses an angle to measure, records a yes or no, and does this thousands upon thousands of times while the choices are switched at random. Tally the agreements and disagreements across the different angle settings and compute Bell's number. If it stays under the ceiling, Einstein wins and the world is made of gloves. If it breaks through, quantum mechanics wins and reality is stranger than any classical story allows. Through the 1970s and especially in Alain Aspect's landmark experiments in Paris in 1982 — which switched the measurement angles so fast that no signal traveling at light speed could carry word from one detector to the other in time — the verdict came in, again and again, decisively: the ceiling is broken. Bell's inequality is violated, exactly as quantum mechanics predicts and by exactly the predicted amount.

Let that land. The experiments do not merely fail to support Einstein's hidden variables; they rule out the entire family of local realistic theories, forever, regardless of what clever mechanism one might invent. The world is genuinely non-classical at its root. Either distant entangled events are linked in a way that transcends ordinary space, or particles have no definite properties until measured, or both — but the comfortable clockwork universe, in which everything has a definite value and nothing outruns light, is simply not the universe we live in. Few results in the history of thought have overturned so much with so little: a page of algebra and a table of photon counts.

The genius of Bell's move — turning an untestable philosophical dispute into a measurable number — is worth appreciating as one of the great intellectual feats of the twentieth century, because it accomplished something that had seemed impossible. For three decades the argument between Einstein and Bohr over the completeness of quantum mechanics had been regarded as pure metaphysics, a matter of taste about which reasonable physicists could disagree and then get back to work; there seemed no way to decide it by experiment. Bell found the way. He asked precisely what limit Einstein's picture — local hidden variables, definite properties fixed at the source with no spooky influence between distant particles — would place on how strongly two separated measurements could agree, averaged over many different choices of what to measure. He derived a specific numerical ceiling that any such local, common-sense theory must obey, and showed that quantum mechanics predicts a value that sails right past it. The two worldviews no longer merely felt different; they made different, measurable numerical predictions. Nature could be put on the witness stand and made to answer.

What makes Bell's inequality so powerful, and worth understanding precisely, is that it does not depend on the details of any particular hidden-variable theory. It is a constraint that *any* local, realistic account of the world must satisfy — any theory in which nothing outruns light and things possess definite properties before they are measured. To violate the inequality is to be forced to abandon at least one of these two deeply cherished assumptions: locality, that distant events cannot instantly influence one another, or realism, that properties exist before being looked at. This is why the experimental verdict, when it came, overturned so much with so little. Through the 1970s and especially in Alain Aspect's landmark Paris experiments of 1982 — which switched the measurement settings so rapidly that no signal traveling at light speed could pass between the detectors in time — the inequality was found violated, again and again, exactly as quantum mechanics predicted and by exactly the predicted amount. The comfortable clockwork universe, in which everything has a definite value and nothing outruns light, is simply not the universe we inhabit. A page of algebra and a table of photon counts had refuted a worldview as old as physics itself.

The story's coda — from disreputable hobby to Nobel Prize to industrial tool — is worth telling because it shows how the deepest foundational questions can become intensely practical. For decades this work was regarded as slightly disreputable, a philosophical indulgence unfit for a serious career; Bell pursued it on the side, calling it his hobby, and the early experimenters risked their reputations chasing what many colleagues dismissed as metaphysics. Then, in 2022 — long after Bell's death in 1990 — the Nobel Prize in Physics was awarded to John Clauser, Alain Aspect, and Anton Zeilinger precisely for these experiments on entangled photons and the violation of Bell inequalities. The hobby had become the foundation of a new industry. For Bell's inequality is not only a probe of metaphysics but a tool: the same tests that refuted Einstein now certify that a source is producing genuine entanglement, and that certification underwrites the security of quantum cryptography and the honesty of a quantum computer's most precious resource. Entanglement had been proven real, measurable, and stronger than any classical shadow — and the question was no longer whether the quantum world was strange, but whether that strangeness could be harnessed and made to compute.

Part II — Information Has a Physics

Chapter 7: The Universal Machine

In the spring of 1935, while Einstein and his colleagues were composing the EPR paper in Princeton and Bell was a boy of seven in Belfast, a shy, brilliant, socially awkward young Englishman lay in a meadow near Cambridge, thinking about a problem that had nothing, apparently, to do with physics. Alan Turing was twenty-two, a newly elected fellow of King's College, and he had become fascinated by a question the great German mathematician David Hilbert had posed about the ultimate limits of mathematics. Hilbert had asked whether there existed some definite mechanical procedure — some algorithm, though the word was not yet in use — that could, in principle, decide the truth or falsehood of any mathematical statement whatsoever. Was mathematics, at bottom, a machine that could be cranked? Turing set out to answer the question, and to answer it he had to do something no one had done before: he had to say, precisely, what a "mechanical procedure" even *is*. In pinning that down, he invented the theory of computation, and, on paper, the computer.

Turing's method was characteristically radical in its simplicity. He imagined stripping the act of calculation down to its barest possible essence. Picture, he said, a person doing a long computation with pencil and paper. What are they *really* doing? They read a symbol; based on that symbol and their current state of mind, they write a symbol, perhaps erase one, move their attention to a different spot on the page, and shift into a new state of mind; then they repeat. That is all. Everything a human "computer" does — and in 1935 a "computer" was a person, usually a woman, employed to do arithmetic — could be reduced to this mind-numbing loop of read, write, move, change state. Turing abstracted it into an imaginary device, now called a **Turing machine**: an endless tape divided into cells, a head that can read and write one symbol at a time and move left or right, and a little table of rules specifying, for each combination of current symbol and current internal state, what to write, which way to move, and which state to enter next. That is the entire machine. It is almost insultingly simple. And Turing proved that this trivial device, given the right table of rules, can carry out *any* computation that any human being, or any conceivable machine, could carry out. Anything that can be computed at all can be computed by a Turing machine.

The deepest stroke came next. Turing showed that you could design a single, special Turing machine — a **universal** Turing machine — whose rules let it read, from the beginning of its tape, a *description* of any other Turing machine, and then imitate that machine exactly. One machine that could become any machine, merely by being fed a different description. This is the theoretical birth certificate of the general- purpose computer, and it is one of the most consequential ideas of the twentieth century. Your laptop, your phone, the servers behind every website, are physical realizations of Turing's universal machine: one piece of fixed hardware that becomes a word processor, a game, a spreadsheet, or a video player depending only on the description — the software — you feed it. The separation of hardware from software, the very idea of a programmable machine, was Turing's, worked out years before any such machine was built, as a byproduct of a question in pure logic. (His answer to Hilbert's question, incidentally, was no: he proved that there are perfectly well-posed mathematical questions no mechanical procedure can ever decide, the most famous being whether a given program will eventually halt or run forever. Even the universal machine has limits it cannot cross. Those limits, too, will matter when we ask what quantum computers can and cannot do.)

Turing's abstraction contained a further idea that will run like a thread through this book: the idea that computation is **substrate-independent**. A Turing machine is defined by what it *does* — the pattern of reading, writing, and state-changing — not by what it is made of. You could build one out of gears and levers, out of electrical relays, out of vacuum tubes, out of transistors etched in silicon, out of water flowing through pipes, or, in a famous illustration, out of a sufficiently patient person pushing beads. All of these, if they implement the same rules, compute the same things. The logic floats free of the hardware. This principle — that a bit is a bit no matter how it is physically stored, and a computation is a computation no matter what performs it — was liberating and immensely productive. It let computer science develop as a mathematical discipline, indifferent to the messy details of physics. For half a century it seemed a bedrock truth.

And yet, as we will see, it contains a hidden assumption, and the whole drama of quantum computing lies in that assumption's failure. Turing, thinking in the 1930s, took for granted that the symbols on his tape were ordinary, definite, classical symbols — a cell held a 0 or a 1, one or the other, full stop. He could not have imagined a tape whose cells could hold a superposition of 0 and 1 at once, or two cells entangled so that their symbols had no independent existence. He was, without knowing it, describing a *classical* computer, and his sweeping claim that a Turing machine can compute anything computable was really a claim about what is computable *classically*. When, decades later, physicists asked what happens if you let the tape be quantum — if the head can put cells into superposition and entangle them — they found a genuinely more powerful machine, a **quantum Turing machine**, capable of computations no classical Turing machine can perform efficiently. Turing's universal machine was universal only within the classical world. There was a wider world he never saw.

It is worth dwelling on how radical Alan Turing's idea of 1936 was, because everything we mean by the word "computer" descends from it. Before Turing, a "computer" was a job title held by a person — usually a woman — who carried out calculations by hand, following fixed rules with pencil and paper. Turing's stroke of genius was to strip that activity down to its barest possible skeleton and ask what, in principle, it could ever accomplish. He imagined a machine of almost insulting simplicity: an endless paper tape divided into squares, a head that can read or write a single symbol on one square at a time, and a small table of rules telling it what to do next based only on what it currently sees and which of a few internal states it is in. That is all. No screen, no memory chips, no electricity even — just a tape, a head, and a rule table. And yet Turing proved that this austere device, given the right rule table, could carry out any calculation that any other computing machine could ever perform. It was universal.

The deepest twist came next. Turing realized that the rule table of any particular machine could itself be written down as symbols on the tape — a description of a machine, fed to a machine. So there exists a single Universal Turing Machine that, handed the description of any other machine along with that machine's input, will faithfully imitate it. This is the theoretical seed of everything: the reason one laptop can run a spreadsheet, a game, and a video call is that it is a physical approximation of Turing's universal imitator, a machine whose "hardware" does nothing but obey a program handed to it as data. Software, the very idea that instructions are a kind of data that can be stored, copied, and run, is Turing's child.

But Turing's 1936 paper is remembered as much for what it proved impossible as for what it made possible, and this is the part our story needs. He showed that some questions can never be answered by any program at all, no matter how clever or how long it runs. The most famous is the halting problem: there can be no general procedure that, examining any program and its input, always correctly decides whether that program will eventually stop or run forever. The proof is a jewel of self-reference — assume such a decider exists, then build a program that asks the decider about itself and does the opposite of whatever it predicts, and the whole thing detonates in contradiction. With this, Turing drew a hard boundary around computation: there are islands of the answerable in an ocean of the forever-uncomputable, and no machine, quantum or classical, can cross that shore. It is a common and important misconception that quantum computers will "solve everything faster" or break this barrier. They will not. A quantum computer can compute exactly the same set of functions a classical computer can — not one function more. Everything uncomputable to Turing's tape remains uncomputable to a quantum machine. What can differ, and this is the whole game, is not what can be computed but how quickly.

That distinction — possible versus practical — is the hinge of computer science, and it was sharpened in the decades after Turing into the theory of computational complexity. Some problems can be solved in a number of steps that grows gently with the size of the input: double the input, and the work merely doubles or squares. These are the "tractable" problems, the ones we call efficient, gathered into a class known as P. Others seem to demand an effort that explodes — grows exponentially — so that adding a few more items to the input multiplies the work a thousandfold, and a problem that takes seconds at small size would take longer than the age of the universe only slightly enlarged. A vivid example is the traveling salesman puzzle:

find the shortest route through a handful of cities and you can do it by hand, but let the list of cities grow and the number of possible routes balloons past the number of atoms in the galaxy. The universe is not big enough, nor old enough, to check them all one by one.

For decades it seemed that this wall of exponential difficulty was simply a fact of nature, an absolute limit set by the logic of step-by-step calculation itself. The Church-Turing thesis — the widely held belief that anything effectively computable is computable by a Turing machine — was even strengthened into a bolder claim: that any reasonable physical computer could be simulated by a Turing machine with only a modest, polynomial slowdown, so that no machine could ever be more than trivially faster than any other at the fundamental level. If that stronger thesis were true, the exponential wall would be permanent, and there would be no such thing as a quantum computer worth building. The entire subject of this book rests on the discovery that the stronger thesis is, quite possibly, false — that the laws of quantum physics may let a machine sidestep certain exponential walls that stop every classical computer cold.

A final reflection on Turing's legacy sets the stage for everything that follows, and it concerns the precise nature of the boundary he drew. Turing showed that computation has an outer limit — the uncomputable — that no machine can cross, and this limit is absolute: it binds classical and quantum computers alike, for a quantum machine can compute exactly the same set of functions a classical one can, not a single function more. But within the vast territory of the computable, Turing and his successors found a second, subtler frontier: the line between what can be computed *efficiently* and what, though computable in principle, would take longer than the age of the universe. This is the frontier of complexity, and it is on this inner frontier, not the outer one, that quantum computing does its work. A quantum computer breaks no law of computability; it merely, for a special few problems, moves the line between the practical and the hopeless — turning certain "not in a billion years" problems into "by this afternoon." Understanding that quantum computers shift the complexity frontier without touching the computability frontier is the single most important conceptual distinction for judging what they can and cannot do, and it immunizes the reader against the wilder claims that quantum machines will "solve everything."

It is worth making concrete just how absolute the exponential wall of complexity can seem, and why the mere possibility of scaling it was so revolutionary. Consider a problem whose difficulty doubles with each item added to the input — a not-uncommon situation. With ten items you face about a thousand possibilities, trivial to check; with fifty, about a quadrillion; with three hundred, more possibilities than there are atoms in the observable universe. No conceivable classical computer, however fast, could ever check them all, not if it ran for the entire lifetime of the cosmos on every atom in existence. For decades this exponential wall was regarded as an absolute fact of nature, a permanent limit set by the logic of step-by-step calculation itself, codified in the belief that any reasonable physical computer could be efficiently simulated by a Turing machine with only modest slowdown. If that belief held, no machine could ever be more than trivially faster than any other at the fundamental level, and there could be no quantum computer worth building. The entire field rests on the discovery that this belief is, quite possibly, false — that the laws of quantum physics may let a machine sidestep certain exponential walls that stop every classical computer cold.

That possibility, that physics itself might permit a shortcut through the exponential wilderness, transforms computation from a branch of pure mathematics into a branch of physics — which is precisely the insight that David Deutsch would later drive home. Turing had abstracted computation away from any physical machine to the pure logic of the tape and the rule table, and this abstraction was immensely powerful; but it quietly assumed that the physical world underneath was classical. Once you take seriously that the world is quantum, the question of what can be efficiently computed becomes a question about what the actual laws of physics permit, not about an idealized classical machine — and the answer, for a handful of precious problems, turns out to be different and better. But before anyone could imagine harnessing quantum physics to compute, a prior realization had to dawn: that computation is not an abstract game of symbols floating free of the world but a physical process, obeying physical law and paying physical costs in energy and heat. To see how deep that connection runs, we must follow a mischievous imaginary demon, conjured up to torment the science of heat a century before Turing was born.

Chapter 8: Maxwell's Demon and the Cost of Forgetting

If Turing showed that computation is substrate-independent — that the logic floats free of the hardware — the story of this book turns on the discovery that this freedom is not quite complete, that computation is also, unavoidably, a *physical* process, bound by the laws of thermodynamics, and that thinking hard about those bonds would lead, by a winding road, to the quantum computer. The person who dragged the physics of information into the light was a demon — an imaginary one, conjured in 1867 by the same James Clerk Maxwell who had unified light and electromagnetism, and set loose to torment physicists for the next century.

Maxwell's demon was a thought experiment aimed at the second law of thermodynamics, the law that says heat flows from hot to cold and never spontaneously the other way, that disorder — entropy — always increases, that you cannot un-mix the cream from the coffee. Maxwell imagined a tiny intelligent being guarding a trapdoor in a wall dividing a box of gas. Gas molecules fly about at random, some fast (hot), some slow (cold). The demon watches, and whenever a fast molecule approaches from the left it opens the door to let it through to the right, and whenever a slow one approaches from the right it lets it through to the left. Over time, without doing any obvious work — the door is frictionless, the demon merely *decides* — the fast molecules accumulate on one side and the slow ones on the other. One side grows hot, the other cold, order emerges from disorder, entropy *decreases*, and the second law of thermodynamics is violated by nothing more than intelligent sorting. You could run an engine off the temperature difference. You could, in effect, get useful work from nothing. The demon seemed to prove that mind, or information, could cheat the iron laws of physics.

For decades physicists chipped at the paradox without fully resolving it. The crucial move came in stages, and it reframed the whole question in terms of *information*. The demon, to do its sorting, must *measure* each molecule — must acquire information about which are fast and which are slow. And that information must be stored somewhere, in the demon's memory, whatever that memory is made of. Here was the opening. In 1961, a physicist at IBM named **Rolf Landauer** proved a result of startling depth, now called **Landauer's principle**: erasing information has an unavoidable thermodynamic cost. Specifically, to erase a single bit of information — to reset a memory cell to a blank state, wiping out whatever it held — you must dissipate at least a certain minimum amount of energy as heat into the environment, an amount fixed by the temperature and by Boltzmann's constant. Erasure is not free. Forgetting has a price, paid in heat and entropy.

Landauer's principle resolved Maxwell's demon at last, in an argument completed by his IBM colleague Charles Bennett in the 1970s and '80s. The demon can indeed sort molecules and locally decrease the gas's entropy — but only by recording information about each molecule in its finite memory. Sooner or later that memory fills up and must be erased to make room for more, and *that erasure*, by Landauer's principle, dumps at least as much entropy into the world as the sorting removed. The books balance. The second law is saved, not by any flaw in the sorting, but by the thermodynamic cost of the demon's inevitable forgetting. The paradox was killed by the recognition that information is physical — that a bit is not an abstract token but a physical state of some real system, and that manipulating bits is a physical process with physical consequences, obedient to thermodynamics.

That phrase — *information is physical* — became Landauer's motto, and it is the intellectual hinge on which this book turns. For most of the twentieth century, computer scientists had happily ignored the physics of their bits, treating computation as pure abstract logic in Turing's spirit. Landauer insisted that this was a convenient fiction with a hidden floor: every bit lives in some physical system, every logical operation is a physical operation, and the laws of physics therefore constrain what computers can do and how little energy they can do it with. Most of the time the physics in question was ordinary classical physics, and the constraints were remote from practice. But the deeper point was general, and it invited a dangerous question. If information is physical, and if the relevant physics is ultimately *quantum*, then what happens to computation when we take the quantum nature of the underlying physics seriously? What can you compute with bits that obey not the classical rules Turing assumed but the strange rules of superposition and entanglement? Landauer had built the bridge from computer science to physics. Others would cross it.

Maxwell's demon, conjured in 1867 by the great Scottish physicist James Clerk Maxwell, is one of those thought experiments that seems at first like a mere puzzle and turns out to be a doorway. Picture a box of gas divided in two by a wall with a tiny trapdoor. A gas is just a swarm of molecules flying about at a range of speeds — some fast, some slow, temperature being nothing but the average of all that motion. Now imagine a minuscule intelligent being stationed at the trapdoor, watching the molecules approach. Whenever a fast one comes from the right, it opens the door to let it into the left chamber; whenever a slow one comes from the left, it lets it into the right. It never does any work on the molecules — it only opens and closes a weightless door at the right moments. Slowly, sorting one molecule at a time, it gathers the fast molecules on one side and the slow on the other. The left side grows hot, the right grows cold, and this temperature difference could then drive an engine — all, apparently, for free.

The trouble is that this violates the second law of thermodynamics, the iron rule that in any closed system disorder — entropy — can only increase, that heat flows from hot to cold and never spontaneously the other way, that you cannot get useful work from nothing. The second law is arguably the most trustworthy statement in all of physics; the astronomer Arthur Eddington said that if your pet theory contradicts it, you have no hope but to collapse in deepest humiliation. Yet here was Maxwell's little demon, seemingly building order from chaos with nothing but cleverness and a frictionless door. For the better part of a century the demon haunted physics. Where, exactly, was the catch?

The first crack in the demon's scheme came in 1929 from Leo Szilard, who simplified it to a single molecule in a box and noticed something crucial: the demon must acquire information. It has to measure each molecule — is this one fast or slow, coming from the left or the right? — before it can decide whether to open the door. Szilard suspected that this act of measurement carried a hidden thermodynamic cost, an entropy price that would exactly cancel the demon's winnings and rescue the second law. He even worked out the fundamental price of one bit of information in terms of heat, a quantity that reappears everywhere in the science that followed. Information, Szilard glimpsed, is not free; it is physical, and it has a currency exchange with energy and entropy.

But the precise accounting stubbornly refused to balance until 1961, when Rolf Landauer at IBM located the true cost with a discovery so important it bears his name. Landauer's insight was that measuring information need not, in principle, cost anything irreducible — but ERASING it must. His principle states that whenever a bit of information is irreversibly destroyed — wiped, reset, forgotten — a minimum quantity of heat must be dumped into the surroundings, a tiny but unavoidable tax set by the temperature and by Boltzmann's constant. The demon's flaw was hiding not in the looking but in the forgetting. To keep sorting molecule after molecule, the demon's memory fills up, and to continue it must eventually erase old records to make room — and that erasure pours exactly enough entropy into the world to save the second law. The demon can cheat locally, for a while, but the bill always comes due when it clears its mind.

This is a genuinely astonishing conclusion, and it is worth savoring: the seemingly abstract act of erasing information — deleting a file, resetting a memory to zero — is a physical event that must generate heat, as surely as friction does. Landauer summed up his whole philosophy in a slogan that became the rallying cry of a new field: "Information is physical." Information is not an ethereal abstraction hovering above the material world; it is always encoded in the state of some physical system — the charge on a capacitor, the magnetization of a speck of iron, the spin of an electron — and it therefore obeys, and is constrained by, the laws of physics. You cannot reason about computation while ignoring thermodynamics, because computation is thermodynamics wearing a mask.

The persistence of Maxwell's demon as a serious problem — not a mere riddle but a genuine threat to the foundations of physics that took the better part of a century to resolve — is itself worth reflecting on, because it shows how a simple thought experiment can probe the deepest structure of nature. Maxwell himself proposed the demon almost playfully, to illustrate the *statistical* character of the second law of thermodynamics: heat flows from hot to cold not because it is forbidden to do otherwise but because the reverse is overwhelmingly improbable, and a being clever enough to sort individual molecules might, in principle, cheat the odds. What Maxwell could not have foreseen was that resolving his playful puzzle would require nothing less than a new understanding of the physical nature of information — that the demon's flaw would turn out to lie not in its physics but in its *memory*, in the ineliminable cost of the knowledge it must acquire and eventually discard.

The chain of minds that cracked the puzzle is a small monument to how science advances by relay, and each link deserves its name. Maxwell posed the demon in 1867. Leo Szilard, in 1929, simplified it to a single molecule and first glimpsed that the demon's act of *measurement* — acquiring information about each molecule — must carry a thermodynamic cost, and even computed the fundamental price of one bit. But Szilard's accounting did not quite balance, and the puzzle lingered until Rolf Landauer, at IBM in 1961, located the true cost with precision: measurement need not be costly, but *erasure* must be. Whenever a bit of information is irreversibly destroyed, a minimum quantity of heat must be dumped into the surroundings. The demon's memory fills as it sorts, and to keep working it must eventually erase old records — and that erasure pours exactly enough entropy into the world to save the second law. Charles Bennett completed the argument in the 1980s by showing precisely how the demon's memory could be reset only at the unavoidable Landauer cost. Puzzle solved, but only after a century and four minds.

The slogan Landauer distilled from all this — "information is physical" — is one of the most quietly revolutionary statements in twentieth-century science, and its reach extends far beyond demons and ovens. It asserts that information is never an ethereal abstraction floating above the material world but is always encoded in the state of some physical system — the charge on a capacitor, the magnetization of a speck of iron, the spin of an electron — and therefore obeys, and is constrained by, the laws of physics. You cannot reason about computation while ignoring thermodynamics, because computation *is* a physical process, consuming energy and generating heat with every irreversible step. This is why your laptop runs warm: every logic gate that forgets which inputs produced its output is erasing information and, by Landauer's principle, shedding heat, multiplied across billions of gates switching billions of times a second. But hidden in that same principle lay a tantalizing loophole — that a computation which never *erases* anything, which could in principle be run backward to recover its starting point, might cost, in the ideal limit, no energy at all. Pursuing that loophole would lead directly to the reversible logic that quantum mechanics, as it turned out, absolutely requires.

Chapter 9: Reversible Computing

Charles Bennett was the kind of scientist who follows an idea wherever it leads, across the boundaries that usually divide the disciplines, and the idea he followed in the early 1970s led from Maxwell's demon to a question that sounds almost paradoxical: can a computer run backward? Not literally reverse time, but compute in such a way that every step is undoable, so that from the final state you could always, in principle, reconstruct the initial one, throwing no information away. Landauer had shown that only the *erasure* of information carries a mandatory energy cost. It followed that a computer which never erased anything — a fully **reversible** computer — might compute with no fundamental thermodynamic price at all, however much heat real, imperfect hardware might waste in practice. The trouble was that ordinary computation seemed hopelessly irreversible. Every AND and OR gate, every step of a normal calculation, discards information, merging distinct inputs into identical outputs, and once merged they cannot be pulled apart. It looked as though forgetting were simply built into computing.

Bennett's contribution, in a landmark 1973 paper, was to prove that this appearance is false. He showed that *any* computation, no matter how many irreversible steps it seems to require, can be reorganized into a fully reversible one that reaches the same answer, at the modest cost of keeping some extra records along the way. The trick, in essence, is never to overwrite: instead of erasing intermediate results, you keep them, tucked away, so that the machine's history is always recoverable. When the computation is done and the answer read out, you can run the whole process backward to clean up the extra records reversibly, restoring the machine to its starting condition without ever performing a costly erasure. Reversible computation was not just theoretically conceivable; it was universal. Anything computable could be computed reversibly. Landauer's tantalizing loophole was real: there is no fundamental thermodynamic lower bound on the energy cost of computation itself, only on erasure, and erasure can be avoided.

To make reversible computing concrete, theorists devised **reversible logic gates** — logical operations that can be run in either direction because they never lose information. The most important of these is the **Toffoli gate**, invented by Tommaso Toffoli in 1980, a "controlled-controlled-NOT" that takes three bits in and puts three bits out: it flips the third bit if and only if the first two are both 1, and leaves them otherwise unchanged. Crucially, from its three output bits you can always reconstruct its three input bits — nothing is lost — and yet the Toffoli gate is powerful enough, all by itself, to build any logical circuit whatsoever. It is a universal reversible gate. Alongside it sits the simpler **controlled-NOT**, or CNOT, which flips a target bit depending on a control bit and is likewise reversible. These names should be filed away, because they are about to reappear in a new guise. When physicists came to build quantum computers, they found that the gates of a quantum computer must, as a matter of physical law, be reversible — and the Toffoli and CNOT gates, born in the study of classical reversible computing, became foundational quantum gates. The reversibility that Bennett pursued for thermodynamic reasons turned out to be *forced* on quantum computers by the deep structure of quantum mechanics.

Here is why. Recall from Part I that the evolution of a quantum system, between measurements, is governed by the Schrödinger equation, and that this evolution has a special mathematical character: it is **unitary**. Without the jargon, unitary evolution has a simple and profound feature — it is always reversible. The state of a quantum system at one moment determines its state at every earlier and later moment, running smoothly forward and backward, information conserved, nothing created or destroyed, right up until the discontinuous moment of measurement. This is not an option a quantum engineer can switch off; it is how the quantum world works. A quantum computer performs its computation by letting its qubits evolve under a sequence of unitary operations — the **quantum gates** — and because each such operation is unitary, each is automatically reversible. A quantum computer *cannot* have an irreversible AND gate of the ordinary kind, because such a gate would destroy information, and destroying information is exactly what unitary quantum evolution forbids. The abstract study of reversible classical computing, which might have remained a thermodynamic curiosity, turned out to be the indispensable preparation for quantum computing: it had already worked out how to do universal computation with reversible gates, precisely the constraint that quantum mechanics would impose.

The people who built this bridge sensed where it was heading. Bennett, at IBM, was steeped in both the physics of information and, increasingly, in quantum mechanics; he would go on, in 1984, to co-invent quantum cryptography, and in 1993 to help discover quantum teleportation. But the first person to put the pieces together into an actual model of a quantum-mechanical computer was a quieter figure, a physicist at Argonne National Laboratory named **Paul Benioff**. In 1980, drawing directly on the reversible-computing work, Benioff described how a Turing machine could be implemented as a quantum-mechanical system evolving under the Schrödinger equation — a computer whose parts were quantum objects obeying quantum laws. It was, in a sense, the first quantum computer, though a cautious one: Benioff's machine used quantum mechanics to implement *classical* computation reversibly, a quantum system pressed into doing the work of an ordinary Turing machine. He had shown that a computer could be built on quantum foundations. He had not yet shown that such a computer could do anything a classical computer could not. It reproduced classical computation in a quantum idiom, faithfully, without exploiting the one thing — superposition of many computational paths at once — that would make quantum computing powerful.

The idea of reversible computing sounds, on first hearing, like a contradiction. Computing feels inherently one-way: you feed in numbers, you get out an answer, and the inputs vanish into the result the way ingredients vanish into a cake. You cannot un-bake a cake, and it seems you cannot un-compute a sum — knowing the answer is 12 tells you nothing about whether you added 5 and 7 or 8 and 4. That loss of the inputs is exactly the information erasure that Landauer showed must cost heat. So the question Charles Bennett asked at IBM in 1973 was audacious: could you build a computer that never throws information away, that preserves at every step enough of its history to be run backward, and thereby, in principle, dodges Landauer's heat tax entirely?

Bennett proved the answer is yes, and the proof is elegant. The reason ordinary logic gates destroy information is that they have fewer outputs than inputs — an AND gate takes two bits in and gives one bit out, and the missing bit is forgotten, radiating its little quantum of heat. But you can redesign the gates so that nothing is ever lost. A reversible gate has as many outputs as inputs, and it is engineered so that from the outputs you can always uniquely reconstruct the inputs — the operation can be undone. Bennett showed that any computation whatsoever can be rebuilt out of such reversible steps, at the modest price of carrying along some extra bookkeeping bits that record what would otherwise have been discarded. The computation runs forward to produce its answer, you copy the answer out, and then — the clever part — you run the whole thing backward, cleanly un-computing all the intermediate scratch work and returning the machine to its starting state, having erased nothing and, in the ideal limit, spent no energy at all. Computation, it turned out, need not be a one-way street. It could be a round trip.

The reversible gate that became the workhorse of this idea is the Toffoli gate, invented by Tommaso Toffoli in 1980 — a "controlled-controlled-NOT" that flips a target bit only if two control bits are both 1, and leaves those controls untouched. Because the controls pass through unchanged, you can always tell what went in, and the gate is its own undo: apply it twice and you are back where you started. Toffoli proved this single gate is universal for reversible classical computing — with enough Toffoli gates and spare bits you can build any logical operation at all, all of it reversible, all of it in principle heat-free. That may sound like an engineering curiosity, but hold the thought, because it is the hinge on which our entire story turns.

Here is why it matters so enormously. Quantum mechanics, as we saw in the new mechanics of the 1920s, evolves by a law — the Schrodinger equation — that is perfectly reversible. Undisturbed quantum evolution never destroys information; run the equations backward and the past is recovered exactly. This means that any operation a quantum computer performs on its qubits, short of the final measurement, MUST be reversible, because the underlying physics permits nothing else. An ordinary irreversible AND gate simply has no quantum counterpart; it would violate the reversibility that quantum law demands. So the decades of thinking about reversible classical computation, which had looked like an esoteric footnote to thermodynamics, turned out to be the essential preparation for quantum computing. When physicists began to design quantum gates, they discovered they were already fluent in the grammar — the reversible logic Bennett and Toffoli had worked out was precisely the logic a quantum machine requires. The Toffoli gate itself lives on inside quantum computers as a fundamental building block.

There is a satisfying intellectual lineage here worth naming. Maxwell's demon posed a paradox about heat and information; Szilard and Landauer resolved it by proving that erasing information costs energy; Bennett and Toffoli responded by showing that computation could avoid erasure and so, in principle, avoid the cost — and in doing so they built, without quite intending to, the logical scaffolding on which quantum computation would stand. A chain of pure thought, stretching from a Victorian parlor puzzle about a trapdoor to the design of twenty-first-century quantum processors, each link forged by someone chasing a question that seemed, at the time, to have no practical use whatsoever. It is one of the great vindications of curiosity-driven science.

The counterintuitive heart of reversible computing rewards one more pass, because it overturns an assumption so natural we rarely notice we hold it. We instinctively feel that computation is inherently one-way, like cooking: you combine ingredients into a cake and cannot un-bake it, you add numbers into a sum and cannot recover which numbers you added. This felt inevitability is exactly the information erasure that Landauer proved must cost energy — and Bennett's radical insight was that the one-wayness is not inevitable at all, but a *choice* built into how we happen to design our logic. Redesign the gates so that nothing is ever thrown away — so that from the outputs you can always uniquely reconstruct the inputs — and computation becomes a round trip: run it forward to produce the answer, copy the answer out, then run it backward to cleanly un-compute all the intermediate scratch work, returning the machine to its starting state having erased nothing and, in the ideal limit, spent no energy at all. That any computation whatsoever can be rebuilt this way, out of reversible steps, is one of the quietly profound results of twentieth-century computer science, and it severed the assumed link between computing and irreversible heat.

The deep reason this esoteric-sounding result became the essential foundation of quantum computing deserves to be stated as plainly as possible, because it is the hinge connecting two apparently unrelated fields. Quantum mechanics evolves by a law — the Schrodinger equation — that is *perfectly reversible*: undisturbed quantum evolution never destroys information, and running the equations backward recovers the past exactly. This is not an optional feature but a fundamental property of quantum physics. It means that any operation a quantum computer performs on its qubits, short of the final measurement, *must* be reversible, because the underlying physics permits nothing else. An ordinary irreversible AND gate, which forgets which inputs produced its output, simply has no quantum counterpart; it would violate the reversibility that quantum law demands. So when physicists began designing quantum gates, they discovered they were already fluent in the required grammar — the reversible logic that Bennett and Toffoli had worked out for entirely different reasons, chasing a thermodynamic puzzle, turned out to be exactly the logic a quantum machine requires. Decades of seemingly impractical theorizing about reversible classical computation became, overnight, the indispensable preparation for the quantum age.

There is a lesson here about the value of curiosity-driven science that is worth pausing on, because this whole chain of ideas exemplifies it. Maxwell posed a playful paradox about a demon and a trapdoor in 1867; Szilard and Landauer, chasing its resolution, proved that erasing information costs energy; Bennett and Toffoli, responding to that, showed computation could avoid erasure and so, in principle, avoid the cost — and in doing so built, without quite intending to, the logical scaffolding on which quantum computation would later stand. Not one of these thinkers was trying to build a quantum computer; the idea barely existed for most of that history. Each was pursuing a question that seemed, at the time, to have no practical use whatsoever — the thermodynamics of information, the theoretical limits of energy-efficient computing.. Yet the chain of pure thought stretching from a Victorian parlor puzzle to the design of twenty-first-century quantum processors is unbroken, each link forged by someone following curiosity rather than application. It is one of the great vindications of fundamental research, and a reminder that the deepest practical revolutions often begin as abstract play. Reversibility, demanded by quantum law and prepared by decades of such play, would let a machine do something no classical computer can — hold and manipulate a superposition without collapsing it — and the field was now ready for the question that would light the fuse.

Chapter 10: The Simulation Problem

Every physicist who has ever tried to predict the behavior of more than a handful of quantum particles has run into the same wall, and the wall is made of exponents. It is worth understanding this wall in some detail, because it is not a mere technical nuisance. It is the deepest reason quantum computers exist, the problem they were invented to solve, and quite possibly the problem they will first solve usefully. The whole idea of a quantum computer was born as a response to it.

Recall a fact introduced back in Part I: each qubit — or, in nature, each quantum particle with two relevant states — doubles the size of the mathematical space needed to describe the system. One particle needs two numbers to describe its state. Two particles need four. Three need eight. Ten need a thousand and twenty- four. This is not because the particles are complicated individually; it is because quantum mechanics permits them to be *entangled*, woven into joint states that cannot be described by giving each particle's condition separately. To capture all the possible entangled combinations, you must track a number for every possible configuration of the whole system at once, and the number of configurations doubles with each particle added. The growth is **exponential**, and exponential growth is merciless. By the time you reach just three hundred particles — a speck of matter far too small to see — the number of complex numbers needed to write down the quantum state exceeds the number of atoms in the observable universe. There is not enough paper, not enough memory, not enough matter in the cosmos to store the description. No classical computer that could ever be built, however large, could hold it.

This is the **simulation problem**, and it is not hypothetical. It bites, hard, in real science. Chemists who want to predict how a molecule will behave — how a drug will bind to a protein, how a catalyst will speed a reaction, how a material will conduct — are asking questions whose answers lie in the quantum behavior of the molecule's electrons. But the electrons in even a modest molecule form an entangled quantum system whose exact description explodes exponentially. So chemists cannot solve their equations exactly for anything but the simplest cases. They are forced to approximate, with a toolbox of ingenious methods that work tolerably for many molecules and fail, sometimes badly, for the most interesting ones — the strongly correlated materials, the transition-metal catalysts, the reaction that fixes nitrogen from the air into fertilizer and quietly consumes something like two percent of the world's energy. Nature computes the answers effortlessly, of course: the molecule simply *is* what it is, the electrons simply *do* what they do, in real time. But we, trying to predict it from the outside with our classical machines, drown in the exponential.

The physicist who first stated the moral of this with full clarity, and drew the radical conclusion, was Richard Feynman, and the occasion has become legendary in the field: a talk at the First Conference on the Physics of Computation, held at MIT in May 1981, later published under the title "Simulating Physics with Computers." Feynman was then in his sixties, one of the most celebrated physicists alive, a Nobel laureate for his work on quantum electrodynamics, famous equally for his scientific brilliance, his showmanship, his bongo drums and safe-cracking and irreverence. He had a genius for cutting through formalism to the physical heart of a matter, and he cut to the heart of this one. The question he posed was simple: can a classical computer simulate physics — real, quantum physics — efficiently? And his answer was a flat, provocative no. "Nature isn't classical, dammit," he told the audience in a line that has become the field's unofficial motto, "and if you want to make a simulation of nature, you'd better make it quantum mechanical, and by golly it's a wonderful problem, because it doesn't look so easy."

The logic was inescapable and, once stated, obvious in retrospect. If simulating a quantum system on a classical computer requires exponentially exploding resources, then the mismatch is fundamental: you are trying to capture a quantum reality with classical machinery, and the two do not fit. But suppose, Feynman said, you built your simulator out of quantum parts. Suppose the computer itself were a controllable quantum system, its own particles entangling and superposing according to quantum rules. Then the simulator would have exactly the same kind of exponentially large state space as the thing being simulated — the exponential would be on both sides of the equation, matching, cancelling out. A quantum system would simulate a quantum system naturally, efficiently, without the classical blowup. You would fight fire with fire, model the quantum world with a quantum machine. Feynman sketched the idea of a "universal quantum simulator," a programmable quantum device that could be tuned to mimic any other quantum system, and he conjectured — he did not prove it; that would come later, from Seth Lloyd in 1996 — that such a machine could do the job with resources that grew only modestly, not exponentially, with the size of the system.

This was the conceptual birth of the quantum computer as a *useful* device rather than a philosophical curiosity. Benioff had shown a quantum computer could exist; Feynman showed why you might actually *want* one — because it could do something no classical computer could, namely simulate nature herself. And notice the beautiful inversion at the core of Feynman's insight. For fifty years, the exponential explosion of quantum possibilities had been regarded as a curse, the thing that made quantum systems so infuriatingly hard to predict, the reason chemistry and materials science were stuck with approximations. Feynman turned the curse into a resource. The very hugeness of quantum state space — the thing that defeats classical simulation — is exactly what a quantum computer harnesses. When people say a quantum computer with a few hundred qubits can, in some sense, "hold" more numbers than there are atoms in the universe, this is what they mean, and it is not marketing: it is the same exponential that Feynman recognized, put to work instead of merely endured.

The problem that would give birth to quantum computing is easy to state and terrifying to confront: nature, at the quantum level, is extraordinarily hard to simulate on an ordinary computer. Not merely tedious — hard in a way that grows exponentially worse with every particle you add, until the most powerful supercomputer ever built is defeated by a system you could hold on the tip of a pin. To feel the horror, you must understand where the explosion comes from, and it comes from superposition and entanglement, the very features that made the quantum world so strange in the opening chapters of this book.

Consider a single quantum particle that can be in two states

— call them 0 and 1, spin-up and spin-down. To describe it, a classical computer must store two numbers, the amplitudes for each possibility. Add a second particle and, because the two can be entangled, you cannot describe them separately; you must track every combination — 00, 01, 10, 11 — so now you need four numbers. A third particle brings eight; a fourth, sixteen. The pattern is merciless: each additional particle doubles the number of amplitudes you must store. This is the exponential wall in its purest form. Ten particles need about a thousand numbers — trivial. Thirty need a billion — a strain. Fifty need a quadrillion. At around three hundred entangled particles, the number of amplitudes required to write down the state exceeds the number of atoms in the observable universe. There is no computer, and there could never be a classical computer, large enough to hold the description of a few hundred humble quantum particles. The information is really there, woven into the entanglement, but it is beyond any classical machine's power to record, let alone to compute with.

This is not an academic embarrassment; it is a practical catastrophe for chemistry and materials science. The behavior of a molecule is dictated by its electrons, and electrons are quantum particles that entangle and superpose exactly as described. To predict, from first principles, how a drug molecule will bind to a protein, or how a proposed material will conduct electricity, or how nitrogen is fixed into fertilizer by a natural enzyme that does the job at room temperature while our industrial version demands crushing heat and pressure — for all of these you must solve the quantum equations of many entangled electrons, and the exponential wall stops you cold. Chemists have spent a century inventing ingenious approximations, clever shortcuts that trade accuracy for tractability, and these approximations run the modern pharmaceutical and materials industries. But they are approximations, and for the hardest, most entangled molecules they fail, sometimes badly, and there is no way to make them succeed by simply buying a bigger classical computer. The wall does not move.

Here is the thought that changed everything, and it is beautiful in its simplicity. If a quantum system is too hard for a classical computer to simulate because it holds exponentially much information, then perhaps the right tool to simulate a quantum system is another quantum system — one you can control. A machine built out of quantum parts would naturally, effortlessly, hold that same exponential wealth of amplitudes, not by storing them in memory but by BEING a physical embodiment of them. You would not compute the molecule's behavior number by number; you would build a controllable quantum system that dances to the same quantum music, let it evolve, and read off the answer. The simulator would speak nature's own language. Where the classical computer must translate quantum reality into an exponentially bloated description and choke, the quantum simulator would simply mirror it, particle for particle, entanglement for entanglement.

This is the seed of the whole field, and it inverts the usual relationship between difficulty and opportunity. The exponential wall, which had looked like nothing but an obstacle, was revealed to be also a resource. The very fact that a modest quantum system holds an amount of information no classical machine can match means that a modest quantum machine, properly steered, could perform feats of calculation no classical machine can match. The strangeness that had troubled Einstein and puzzled a generation was, it now appeared, a form of untapped computational power lying latent in the structure of matter itself. All that was needed was to learn to harness it — to take a system of quantum particles and make it not just exist but compute.

It is a common misconception that this means a quantum computer "tries all answers at once" and simply reads off the right one, the way a magic parallel machine might. As later chapters will make painfully clear, it is not so simple: the exponential information is there, but measurement collapses it to a single random outcome, and the entire art of quantum algorithms lies in the fiendishly clever business of arranging for the wrong answers to cancel and the right one to emerge. For now, though, the crucial point is the vision itself. Someone had to be the first to say, in public and in plain terms, that because nature is quantum mechanical, our machines for understanding nature ought to be quantum mechanical too — and to challenge the world's physicists to build them. Two people, working independently and from different directions, arrived at that frontier at almost the same moment. The first was a young physicist quietly proving that such a machine was even logically possible.

Part III — The Dream

Chapter 11: Benioff's Quantum Turing Machine

Before the quantum computer had a purpose, it had a proof of concept, and the person who supplied it worked in near-obscurity and stayed there for most of his life. Paul Benioff does not appear on magazine covers, and even within physics he is a name known mostly to specialists, yet he has a strong claim to have built the first quantum computer — on paper, in mathematics — and the field's own histories generally credit him as the founder of the theoretical model. That he is not more famous is partly a matter of temperament, partly of timing, and partly of the modest thing his machine actually did. His story is a useful corrective to the myth that revolutions arrive in a single flash. The quantum computer was assembled, idea by idea, by a succession of people who each saw a little further, and Benioff saw first.

Benioff was an American physicist, born in 1930, who spent his career at Argonne National Laboratory outside Chicago, working on the mathematical foundations of quantum mechanics — the kind of careful, rigorous, unglamorous work that rarely makes headlines but occasionally underpins everything that follows. In the late 1970s he became interested in the physics of computation, the same current of ideas that Landauer and Bennett were developing at IBM. He absorbed their central lesson — that computation is a physical process, and that a computer can, in principle, be made reversible and so dissipationless — and he asked a sharper question than anyone had asked before. Landauer and Bennett had analyzed computers using thermodynamics and classical physics. Benioff wanted to know: could you describe a computer entirely within *quantum mechanics*? Could you write down a genuine quantum-mechanical system — a collection of particles evolving under the Schrödinger equation, with a proper quantum Hamiltonian governing their interactions — whose evolution *was* a computation, step by faithful step?

In a series of papers beginning in 1980, Benioff showed that you could. He constructed a detailed quantum-mechanical model of a Turing machine: a quantum system whose states encoded the tape, the head position, and the internal state of Turing's abstract device, and whose evolution under the laws of quantum mechanics carried out, exactly, the read-write-move-change-state cycle of a classical computation. It was a tour de force of mathematical physics. It demonstrated, rigorously and for the first time, that the laws of quantum mechanics are fully compatible with universal computation — that there is nothing about the quantum world that prevents it from computing, that a computer is a perfectly legitimate quantum system. Turing had described computation abstractly; Benioff had shown it could be embodied in the most fundamental physics we have.

And yet Benioff's machine had a curious, self-imposed limitation, one that both marks the boundary of his achievement and sets the stage for the leap that followed. His quantum computer was designed to reproduce *classical* computation. At each step, although the machine was quantum-mechanical in its guts, it was arranged so that the tape held definite classical symbols — 0s and 1s — not superpositions. The quantum mechanics was there to implement the machine's workings reversibly and legitimately, but the computation it performed was exactly the computation an ordinary Turing machine would perform, no more. Benioff's device used quantum mechanics as a faithful engine for classical logic. It did not put its data into superposition; it did not entangle its bits to explore many computational paths at once; it did not exploit the exponential vastness of quantum state space that Feynman would soon identify as the whole point. It proved that a quantum computer could *exist* and could match a classical one. It did not yet show that a quantum computer could do anything a classical computer could not.

This distinction is the crux of the early history, and it is easy to blur, so it is worth stating plainly. There are two very different claims one might make about quantum computers. The weaker claim is: a computer can be built out of quantum-mechanical parts, obeying quantum laws, and it will compute correctly. The stronger claim is: such a computer, by exploiting distinctively quantum phenomena — superposition, interference, entanglement — can compute some things dramatically *faster* than any classical computer, performing feats that are not merely quantum-embodied versions of classical feats but genuinely beyond classical reach. Benioff had nailed down the weaker claim. The stronger claim — the one that makes quantum computing a revolution rather than a footnote — was still open. It hovered in the air around 1981, implicit in Feynman's argument about simulation, but not yet formulated as a general principle or proved for any concrete problem.

Benioff himself was aware of the frontier and would continue to contribute to the field for decades, a respected elder who never sought the limelight. He died in 2022, having lived to see the machines he first sketched grow into a global enterprise, though the public credit flowed mostly to others. His career is a reminder that priority in science is a tangled thing: the person who takes the first rigorous step is not always the one who takes the most celebrated one, and the "inventor" of quantum computing, if the title means anything, must be shared among several people who each contributed a distinct and necessary insight — Benioff the rigorous existence proof, Feynman the compelling purpose, Deutsch the general theory, Shor the killer application. History tends to compress this into a single heroic myth. The truth is a relay race, and Benioff ran the first leg.

History tends to hand the credit for quantum computing to Richard Feynman, whose charisma and fame made his call to arms unforgettable. But a year before Feynman's celebrated lecture, a quieter physicist named Paul Benioff had already done something essential and less glamorous: he had proved that a quantum computer was even possible in principle — that the laws of quantum mechanics permit a machine that computes, without contradiction. It is the kind of foundational spadework that rarely makes the legends but without which the legends have nothing to stand on.

Benioff, working at Argonne National Laboratory near Chicago, asked a precise and modest-sounding question in 1980: can Turing's universal machine — the tape, the head, the rule table — be described entirely within quantum mechanics? Could you write down a genuine quantum-mechanical system, obeying the Schrodinger equation, whose evolution over time exactly mimicked the step-by-step operation of a Turing machine? This mattered because there was a real worry that the answer might be no. Quantum evolution, recall, must be reversible and must preserve the delicate superpositions of the wavefunction, whereas ordinary computation, as we have seen, is full of irreversible, information-destroying steps. It was not obvious that the rigid, discrete, logical world of Turing's tape could be housed inside the smooth, reversible, continuous world of the Schrodinger equation without the two grinding against each other. Perhaps computation and quantum mechanics were fundamentally incompatible.

Benioff showed they were not. Drawing directly on the reversible-computing insights of Bennett and others, he constructed a detailed mathematical model of a Turing machine built from quantum parts — quantum spins standing in for the symbols on the tape, a quantum Hamiltonian orchestrating the reading, writing, and shifting — and demonstrated that its quantum evolution reproduced, faithfully, the behavior of the classical machine. The reversibility that Bennett had shown was possible for classical computation turned out to be exactly what allowed computation to be embedded in quantum law. Benioff's machine did not yet exploit the strange powers of the quantum world — it was, in effect, a classical computer laboriously reconstructed out of quantum bricks, doing nothing a classical computer could not — but that was precisely the point of the exercise. He had established the beachhead. He had proven that there was no hidden law of physics forbidding a quantum computer, no contradiction lurking in the idea, no reason of principle why the delicate machinery of quantum mechanics could not be marshaled to carry out a calculation.

To appreciate why this "merely possible" result was so important, consider how often grand visions in science founder not on engineering difficulty but on impossibility proofs — on someone showing that the dreamed-of machine would violate a conservation law or a logical principle and therefore can never be built by anyone, ever. Benioff's work slammed that door shut before it could open. Whatever the practical obstacles — and they would prove monumental — no one could say a quantum computer was forbidden. The rest would be a matter of ingenuity and patience, not of fighting the fundamental laws. He had shown the goal lay inside the map of the physically allowed.

Benioff's contribution is a small case study in how science actually advances, as opposed to how the stories are later told. The popular narrative craves a single eureka moment and a single hero, and Feynman, with his bongo drums and his safe-cracking and his Nobel Prize, made an irresistible one. But the real frontier is usually pushed forward by several people converging on the same territory from different paths, and by careful, unshowy work that clears the ground the visionaries then claim. Benioff and Feynman arrived at the quantum-computing frontier at almost the same moment and from opposite directions — Benioff asking whether such a machine was consistent with quantum law, Feynman asking what such a machine might be good for. Both were right, and both were needed. In the fullness of time, Benioff would be recognized as a father of the field, his name set beside the more famous ones, though he never sought the spotlight and rarely got it.

The importance of an "existence proof" in science is easy to underestimate, and Benioff's achievement is a perfect illustration of why such proofs matter so much. Grand visions in science founder more often on impossibility than on mere difficulty — on someone demonstrating that the dreamed-of machine would violate a conservation law, or a logical principle, or the second law of thermodynamics, and therefore can never be built by anyone, ever, no matter how ingenious. Before Benioff, there was a real and reasonable worry that a quantum computer might be impossible in exactly this fundamental sense — that the rigid, discrete, information-destroying logic of computation might be flatly incompatible with the smooth, reversible, superposition-preserving evolution that quantum mechanics demands. Had that incompatibility been real, the entire field would have been stillborn, and all the algorithms and hardware that followed would have been chasing a phantom. Benioff's proof that a quantum-mechanical model of a computer is consistent — that the Schrodinger equation can house a Turing machine without contradiction — slammed that door of impossibility shut, and in doing so cleared the ground on which everything else could be built.

The technical heart of Benioff's result rewards a second look, because it reveals how deeply the earlier thinking about reversible computation had prepared the way. The obstacle he had to overcome was precisely the tension we met in the discussion of Maxwell's demon and reversible computing: ordinary computation is full of irreversible, information-destroying steps, whereas undisturbed quantum evolution must be perfectly reversible, never destroying information. Benioff resolved the tension by drawing directly on the reversible-computing insights of Bennett and others, constructing his quantum Turing machine from reversible operations — quantum spins standing in for the symbols on the tape, a carefully designed quantum evolution orchestrating the reading, writing, and shifting, all of it reversible and therefore compatible with quantum law. The decades of seemingly esoteric work on reversible classical computation, which had looked like a footnote to thermodynamics, turned out to be exactly the preparation needed to embed computation in quantum mechanics. It is a striking example of how a line of pure, curiosity-driven inquiry can lie dormant for years before suddenly becoming indispensable.

There is a lesson here about how scientific credit is assigned, and it is worth stating plainly because it recurs throughout this history. The popular narrative of discovery craves a single eureka moment and a single hero, and in quantum computing that role fell to the charismatic Feynman, whose fame and flair made his call to arms unforgettable. But the real frontier is usually pushed forward by several people converging on the same territory from different directions, and by careful, unshowy work that clears the ground the visionaries then claim. Benioff and Feynman arrived at the quantum-computing frontier at almost the same moment — Benioff asking whether such a machine was *consistent* with quantum law, Feynman asking what such a machine might be *good for* — and both were right, and both were needed. That the quieter question, and the quieter man, are so much less remembered says more about how we tell the stories of science than about the relative importance of the contributions. Benioff proved the machine possible; Feynman would now proclaim it powerful.

Chapter 12: Feynman at MIT

Richard Feynman was, by the spring of 1981, the closest thing physics had to a folk hero. He had shared the Nobel Prize in 1965 for quantum electrodynamics, the fantastically precise theory of how light and matter interact; he had invented the diagrams that bear his name and that every particle physicist still draws; he had helped build the atomic bomb at Los Alamos as a young man and cracked the safes of his fellow scientists for sport; he lectured with a Brooklyn-inflected showmanship that packed halls and made undergraduates feel that physics was the most exciting game in the world. He would, three years later, become a household name to the wider public when he sat on the commission investigating the Challenger disaster and, in a televised hearing, dropped a piece of the shuttle's O-ring rubber into a glass of ice water to show, with devastating simplicity, how it lost its resilience in the cold. Feynman distrusted pomposity, formalism, and authority, and he had an unerring instinct for the physical essence of a problem. When he turned that instinct on the question of simulating physics, at the MIT conference in May 1981, he produced the argument that, more than any other single moment, gave the quantum computer its reason to exist.

We have already met the core of what he said in the previous part: that nature is quantum-mechanical, that simulating quantum systems on classical computers runs into an exponential wall, and that the way out is to build the simulator itself out of quantum parts. But it is worth dwelling on the man and the moment, because the *manner* of Feynman's insight is as instructive as its content, and because his talk framed the field's central ideas with a clarity no one has improved upon. Feynman did not approach the problem as a computer scientist or an engineer. He approached it as a physicist asking what the laws of nature permit. And he had a physicist's suspicion of anything that smelled like a free lunch. When he considered whether an ordinary, classical computer could simulate quantum mechanics — including quantum mechanics' most disturbing feature, the nonlocal correlations of entangled particles that violate Bell's inequality — he saw immediately that it could not do so efficiently, and he saw *why*. A classical computer, updating definite bits by local rules, simply cannot reproduce the spooky, stronger-than-classical correlations that Bell had identified and Aspect had just measured, except by the exponentially expensive brute force of tracking the entire joint state. The Bell correlations are, in a sense, a signature that the quantum world contains more than any efficient classical simulation can hold.

Feynman's proposed solution had two flavors, and the distinction between them maps onto a division that persists in the field to this day. The first flavor was the **quantum simulator** in the narrow sense: a special-purpose quantum device, built to mimic one particular physical system of interest by arranging for its own particles to obey the same equations. If you want to understand a complicated magnetic material, you build a cleaner, more controllable quantum system that behaves by the same rules and study *it* instead — an analog model, a quantum wind tunnel. This idea has borne fruit relatively early and is a going concern today; "analog quantum simulators" made of ultracold atoms are used right now to study problems in condensed-matter physics that classical computers cannot touch. The second, more ambitious flavor was the **universal quantum computer**: not a device hand-tuned to imitate one system, but a general-purpose, programmable quantum machine that could be instructed, by software, to simulate *any* quantum system at all — and, perhaps, to do much else besides. Feynman conjectured that such a universal quantum simulator was possible but did not prove it, and he was notably cautious about whether a quantum computer could do anything *useful beyond* physics simulation. That further question — the one that would eventually make quantum computing a matter of banks and spy agencies — he left almost entirely alone.

There is a revealing humility in Feynman's talk that later popular accounts often sand away. He was not announcing a technology; he was posing a problem and sketching the shape of an answer. "It doesn't look so easy," he said of building such a machine, and he meant it. He did not know how to construct the qubits, how to protect them from the noise of the world, how to make them interact in a controlled way — the very engineering problems that would consume the next forty years and are not fully solved yet. He was operating at the level of principle: showing that the laws of physics did not forbid a quantum computer and that such a machine, if built, would have a natural and important job to do. This is how deep ideas usually enter the world — as a clearly posed possibility, not a finished product. The genius lay in seeing that the possibility was there at all, and in seeing it through the specific lens of the simulation problem, which remains, four decades on, one of the two or three applications for which quantum computers are most likely to prove their worth.

The talk that lit the fuse was delivered in May 1981 at a conference on the physics of computation held at the Endicott House, a mansion owned by MIT, and it was given by the most famous physicist alive. Richard Feynman was then sixty-two, a Nobel laureate for his work on quantum electrodynamics, a legend for his role in the Manhattan Project and his gift for explaining the deepest ideas in plain, mischievous language. His keynote, later published as "Simulating Physics with Computers," did not present a finished theory or a working device. It did something more powerful: it framed the right question so vividly that a generation of physicists could not stop thinking about it.

Feynman's argument marched straight into the exponential wall. He wanted to know: can a classical computer efficiently simulate the physical world? For most of ordinary physics the answer is a comfortable yes. But when he turned to quantum systems, he ran headlong into the doubling catastrophe — the amplitudes that multiply beyond all counting as particles are added. With his characteristic bluntness he declared that nature simply could not be simulated easily on a classical machine, and delivered the line that has been quoted ever since, a small masterpiece of exasperated clarity: "Nature isn't classical, dammit, and if you want to make a simulation of nature, you'd better make it quantum mechanical, and by golly it's a wonderful problem, because it doesn't look so easy." In two sentences he had diagnosed the disease and prescribed the cure. If the world is quantum, and the world is hard to simulate because it is quantum, then build your simulator out of the same quantum stuff and let like mirror like.

What made Feynman's intervention electric was not the logic alone — Benioff and others were circling the same ideas — but who was saying it and how. When the most respected physicist of his age stands up and tells you that the accepted picture of computation is inadequate to nature, and that a fundamentally new kind of machine is needed, people listen in a way they never would to a technical paper by an unknown. Feynman conferred legitimacy. He made quantum computing a respectable thing for a serious physicist to think about, rescuing it from the fringe where speculative machines usually languish. He proposed what he called a "universal quantum simulator" — a controllable quantum system that could be tuned to imitate any other quantum system — and while he was hazy on the details, and even doubted whether a fully universal version was possible, he had planted the flag on the mountain and dared others to climb.

It is worth appreciating the man's peculiar genius for this role. Feynman distrusted grand formalism and loved to reason from physical pictures; he had a near-magical ability to see to the heart of a problem and state it in words a bright undergraduate could follow. He was also, by temperament, a builder and a tinkerer — he had spent the war years at Los Alamos cracking safes and repairing calculating machines for fun — so the idea of a physical machine that harnessed quantum weirdness appealed to him far more than any abstract theorem. And he had earned the right to be listened to on quantum mechanics if anyone ever had: his path-integral formulation, in which a particle explores all possible histories at once and the histories interfere, is itself a way of thinking about the quantum world that maps beautifully onto how quantum computation actually works. When Feynman spoke about superposition and interference, he was speaking his native language.

A common misconception is that Feynman "invented the quantum computer" at Endicott House. He did not build one, did not give a blueprint, did not write down an algorithm, and remained uncertain about how far the idea could be pushed. What he did was more like what a great composer does when he states a theme: he made the problem beautiful and urgent and impossible to ignore. The device he had most clearly in mind was a special-purpose quantum simulator, not the general-purpose, programmable quantum computer that would later emerge; the leap from one to the other still lay ahead. But every history of the field begins here, at that lecture, because here the vision passed from the private notebooks of a few into the public imagination of physics.

It is worth understanding why Feynman, of all people, was the ideal figure to launch quantum computing into the mainstream — for the messenger mattered nearly as much as the message. By 1981 Feynman was not merely a Nobel laureate but a genuine legend, famous far beyond physics for his role in the Manhattan Project, his safe-cracking and bongo-playing, his gift for explaining the deepest ideas in plain, mischievous language. When such a man stands up and declares that the accepted picture of computation is inadequate to nature and that a fundamentally new kind of machine is needed, people listen in a way they never would to a technical paper by an unknown. Feynman conferred *legitimacy*: he made quantum computing a respectable thing for a serious physicist to think about, rescuing it from the fringe where speculative machines usually languish. Legitimacy is a scarce and precious resource in science, and by spending his enormous reputation on the idea, Feynman did something no unknown could have done — he made the field safe to enter.

Feynman's particular genius for this role went deeper than fame, into the very structure of his physical thinking, and it repays attention because it connects to how quantum computers actually work. He distrusted grand formalism and reasoned from vivid physical pictures; he had a near-magical ability to see to the heart of a problem and state it in words a bright undergraduate could follow. And he had earned the right to speak on quantum mechanics if anyone ever had, for his own path-integral formulation — in which a particle explores all possible histories at once and the histories interfere, the ones that reinforce surviving and the ones that cancel vanishing — is itself a way of thinking about the quantum world that maps beautifully onto how quantum computation operates. When Feynman spoke of superposition and interference, he was speaking his native language, the same language of summed, interfering possibilities that he had used to reformulate quantum electrodynamics and win his Nobel Prize. The man who taught physics to think of nature as a sum over interfering histories was uniquely equipped to imagine a machine that computes by making possibilities interfere.

It is worth correcting, precisely, the myth that Feynman "invented the quantum computer" at Endicott House, because the truth is more interesting and more honest about how ideas mature. He did not build a machine, did not give a blueprint, did not write down an algorithm, and remained genuinely uncertain about how far the idea could be pushed — he doubted whether a fully universal quantum simulator was even possible. The device he had most clearly in mind was a special-purpose quantum *simulator*, a controllable quantum system tuned to imitate another quantum system, not the general-purpose, programmable quantum computer defined by its own logic that would later emerge from David Deutsch. What Feynman did was more like what a great composer does in stating a theme: he made the problem beautiful, urgent, and impossible to ignore, framing the right question so vividly that a generation could not stop thinking about it. Every history of the field begins at that 1981 lecture, not because Feynman finished the work but because he started it in the public imagination — and the crucial abstraction still to come, the quantum computer as a fully general programmable machine in its own right, would require a mind willing to take the strangeness of the quantum world with a literalness that even Feynman did not.

Chapter 13: The Oxford Heretic

David Deutsch is one of the most unusual figures in modern science, and to meet him is to be reminded that the popular image of the scientist as a bustling laboratory manager or a jet-setting conference-goer is only one possibility. Deutsch, born in Haifa in 1953 and raised in England, has spent most of his adult life working from his home in Oxford, nocturnal, reclusive, rarely teaching, seldom traveling, publishing few papers but each of them weighty, sustained by a loose affiliation with the University of Oxford's physics department and by an income that has never much interested him. He is a theorist of the purest kind, caring nothing for the machinery of academic careers and everything for ideas, and the idea that has driven him since youth is a conviction that most physicists find eccentric to the point of embarrassment: that the "many-worlds" interpretation of quantum mechanics is not one interpretation among several but the plain, literal truth about reality, and that the universe we see is a vanishingly thin slice of an unimaginably vast "multiverse" of parallel realities. Almost everything Deutsch has done in physics, including his founding contribution to quantum computing, flows from that conviction. It is a striking case of a technology born from metaphysics.

In 1985 Deutsch published a paper in the *Proceedings of the Royal Society* with the forbidding title "Quantum theory, the Church–Turing principle and the universal quantum computer." It is the founding document of quantum computing as a rigorous discipline, and it did for the quantum computer what Turing had done for the classical one: it defined the thing precisely and proved that a universal version exists. Deutsch's central move was to take Turing's implicit assumption — that the tape holds definite classical symbols — and lift it. Let the tape be quantum. Let its cells hold not just 0 or 1 but superpositions of 0 and 1; let the cells become entangled; let the machine's head manipulate these quantum symbols by unitary operations. The result is the **universal quantum Turing machine**, a precise mathematical model of a programmable quantum computer, the quantum analog of Turing's universal machine. Deutsch proved that such a machine exists — that there is a single quantum computer which, given the right program, can simulate any other quantum computer, and indeed, he argued, any finite physical system whatsoever. Where Benioff had built a quantum machine that mimicked classical computation, Deutsch built one that embraced superposition and entanglement as computational resources, and asked what such a machine could do that no classical machine could.

He gave the first concrete answer in the same body of work, with a toy problem now called **Deutsch's algorithm**, which we will examine in the next part. It is a modest thing — it determines a single yes-or-no property of a mystery function while consulting the function only once, where a classical computer would need to consult it twice — but its significance was enormous out of all proportion to its triviality. It was the first demonstration, on a specific well-posed problem, that a quantum computer can do something with fewer steps than any classical computer, by exploiting superposition and interference. It was proof of principle that the *stronger* claim about quantum computers — not merely that they can compute, but that they can compute *better* — is true. The floodgates that Shor would later throw open with his factoring algorithm were first cracked ajar by Deutsch's little coin-weighing puzzle.

Deutsch framed all of this in terms of a principle he stated with characteristic boldness, a strengthening of the old Church–Turing thesis. The original thesis, from the 1930s, held that any function that can be computed by any effective procedure can be computed by a Turing machine — a claim about mathematics and logic. Deutsch proposed a physical version: that any finitely realizable physical system can be perfectly simulated by a universal computing machine — and that the correct such machine is the universal *quantum* computer, not the classical one, because the universe is at bottom quantum-mechanical. In Deutsch's view, the classical Turing machine was never the true measure of what is computable in *our* universe; it was an approximation that happened to work for the special, tame, non-quantum problems people had considered. The real, physical limits of computation are set by the real, physical laws of the world, and those laws are quantum. Computation, Deutsch insisted, is not a branch of mathematics floating free of physics; it is a branch of physics, and its ultimate laws are quantum laws. This is a profound reorientation, and it is why Deutsch is often called the father of quantum computing, even though Benioff's model came first and Feynman's motivation was more influential in the short term. Deutsch gave the field its foundation and its manifesto.

What makes the story stranger, and more human, is *why* Deutsch was led to all this. He did not set out to invent a new kind of computer or to threaten cryptography or to speed up chemistry. He set out to prove that the multiverse is real. In his understanding, when a quantum computer places a qubit into a superposition and computes with it, what is *physically happening* is that the computation is being carried out across many parallel universes at once, each universe handling one branch of the superposition, the whole ensemble cooperating to produce an answer that no single universe could reach alone. To Deutsch, a quantum computer that factors a large number by exploring exponentially many possibilities is not a metaphor and not a mere mathematical convenience: it is a working demonstration that those exponentially many possibilities are physically *real*, playing out in parallel worlds. "When Shor's algorithm has factorized a number using 10^500 or so times the computational resources than can be seen to be present," he has written, "where was the number factorized? There are only about 10^80 atoms in the entire visible universe... So if the visible universe were the extent of physical reality, physical reality would not even remotely contain the resources required to factorize such a large number. Who did factorize it, then? How, and where, was the computation performed?" For Deutsch, the only answer that makes sense is the multiverse.

David Deutsch is the man who turned quantum computing from a suggestive vision into a precise science, and he did it by being, in the most productive sense, a heretic. Working in Oxford in the early 1980s — often literally from home, keeping nocturnal hours, indifferent to the ordinary machinery of an academic career — Deutsch took Feynman's and Benioff's scattered intuitions and asked the deepest possible version of the question. Turing had defined a universal classical computer by abstracting away from any particular physical machine to the pure logic of computation. Deutsch's radical move was to insist that this was a mistake — that computation is not a branch of mathematics floating free of the world, but a branch of physics, and that the correct universal computer must be defined by the actual laws of physics, which are quantum laws. In his landmark 1985 paper he set out to build the quantum analogue of Turing's universal machine: a single, fully general, programmable quantum computer capable, in principle, of simulating any physical process whatsoever.

This was the conceptual leap that Benioff and Feynman had not quite made. Benioff had built a quantum machine that imitated a classical one; Feynman had imagined a quantum simulator that mimicked particular quantum systems. Deutsch defined the universal quantum computer — a device with its own logic, its own gates, its own programs, that was not merely mimicking anything but computing in its own right, and that contained the classical computer as a mere special case of its far larger powers. He formulated a quantum version of the Church-Turing thesis, elevating it from a statement about abstract computability to a physical principle about what machines the universe permits. With Deutsch, the quantum computer became a well-defined mathematical object that theorists could reason about rigorously, prove theorems about, and design algorithms for. The field acquired a spine.

And then Deutsch did something to prove his machine was more than a philosophical curiosity: he invented the first quantum algorithm, a small, almost toy-like problem that a quantum computer could provably solve with fewer steps than any classical computer. It was modest — it answered a slightly contrived question about a function using a single query where a classical machine might need two — but its significance was enormous out of all proportion to its size. It was the first concrete proof that quantum interference could be marshaled to compute something faster, an existence proof that the strange resource was real and usable. That little algorithm, later sharpened and generalized with Richard Jozsa, was the seed from which Shor's and Grover's world-shaking algorithms would grow. It showed the doubters that "quantum speedup" was not hand-waving but a demonstrable mathematical fact.

To understand Deutsch, though, you must understand what drove him, because it is one of the strangest and most revealing facts in this whole history. Deutsch did not regard the many-worlds interpretation of quantum mechanics as an eccentric option to be entertained politely. He regarded it as simply true — as the plain, literal description of reality — and he regarded the quantum computer as its proof.

When a quantum computer factors a number too large for all the atoms in the universe to factor by classical brute force, Deutsch asked, where was the computation performed? His answer, delivered without a trace of irony, is that it was performed across enormous numbers of parallel universes, the branches of the quantum multiverse working in concert, their results brought together by interference. To Deutsch, a working quantum computer is an experimental demonstration that these other worlds are real, and he has said, provocatively, that anyone who denies the multiverse must explain how the machine gets its answers. It is a view most physicists find too much to swallow — but it was this conviction, this willingness to take the mathematics with deadly literal seriousness where others flinched, that gave Deutsch the nerve to define the machine in the first place.

This is a recurring pattern in the history of ideas worth pausing on: the person who makes the decisive breakthrough is often the one holding a belief the mainstream considers extreme, precisely because that belief lets them take seriously a possibility everyone else dismisses. Deutsch's many-worlds faith was, for him, not a distraction from the physics but the engine of it. He looked at a superposition and saw not a mathematical convenience or an epistemic haze but a real, branching reality, and he asked what you could DO with all those parallel branches. The answer was: compute. Whether or not one accepts his metaphysics — and the reader is under no obligation to — the machine he defined works exactly as he said it would, and it is the machine that fills the rest of this book.

It is worth dwelling on the precise nature of Deutsch's conceptual leap, because it is easy to miss how radical it was and how much it accomplished. Benioff had built a quantum machine that laboriously imitated a classical one, doing nothing a classical computer could not. Feynman had imagined a quantum simulator that mimicked particular quantum systems. Both still thought of a quantum machine as fundamentally a *mimic* — a device that imitates something else. Deutsch's insight was to define the quantum computer as a fully general, programmable machine *in its own right* — not imitating anything, but computing according to its own logic, its own gates, its own programs, and containing the classical computer merely as a special, impoverished case of its far larger powers. This is exactly the abstraction Turing had performed for classical computation, defining a universal machine by its logic rather than by any particular physical embodiment; Deutsch did it for the quantum world, and in doing so gave the field a rigorous mathematical object that theorists could reason about, prove theorems about, and design algorithms for. Before Deutsch, quantum computing was a collection of suggestive intuitions; after him, it was a science with a well-defined central object. He gave the field its spine.

The connection between Deutsch's metaphysics and his physics deserves careful emphasis, because it is one of the most revealing case studies in the whole history of science of how a "wrong" or fringe belief can be the very engine of a correct and profound discovery. Deutsch did not merely tolerate the many-worlds interpretation of quantum mechanics; he regarded it as simply true, as the plain literal description of reality, and he regarded the quantum computer as its proof. When a quantum computer performs a calculation too vast for all the atoms in a single universe, Deutsch asks, where was the computation done? — and answers, without irony, that it was distributed across enormous numbers of parallel universes working in concert, their results brought together by interference. Most physicists find this too much to swallow. But it was precisely this willingness to take the mathematics with deadly literal seriousness, where others flinched and reached for comforting reinterpretations, that gave Deutsch the nerve to define the universal quantum computer in the first place. He looked at a superposition and saw not a mathematical convenience or an epistemic haze but a real, branching reality teeming with parallel computation, and he asked what you could *do* with all those branches. The answer was: compute. The metaphysics that most found absurd was, for Deutsch, the engine of the physics.

There is a broader pattern here about scientific creativity that is worth naming explicitly, because it recurs in many fields. The person who makes the decisive breakthrough is often the one holding a belief the mainstream considers extreme, precisely because that belief lets them take seriously a possibility everyone else dismisses. A more cautious, more "sensible" thinker, unwilling to believe in parallel universes, might never have been driven to ask what computational power those universes could supply. Deutsch's heresy was productive: whether or not one accepts his multiverse — and the reader is under no obligation to, for the mathematics of quantum computing is entirely neutral about which interpretation you favor — the machine he defined works exactly as he said it would, and it is the machine that fills the rest of this book. This raises a question the next chapter must confront head-on: the seductive image of quantum computing as harnessing "parallel universes" is, for all its romance and for all Deutsch's sincere belief in it, a genuinely misleading way to understand how quantum algorithms actually achieve their speedups. To replace that seductive picture with an accurate one, we must first look squarely at the audacious interpretation Deutsch embraced — the idea that every quantum possibility is realized in a world of its own.

Chapter 14: Many Worlds

The interpretation of quantum mechanics that David Deutsch stakes his life's work on was born in the frustration of a graduate student. In 1957, a young man named Hugh Everett III, working toward his doctorate at Princeton under John Wheeler, grew dissatisfied with the standard Copenhagen story of quantum mechanics — the story we met in Part I, in which a quantum system evolves smoothly in superposition until a measurement causes it to "collapse," at random, into a single definite outcome. Everett found the collapse intolerable. It was, he thought, a clumsy, arbitrary, unexplained addition to the theory, a special rule invoked only for that mysterious event called "measurement," which the theory never defined. Why should measurement be special? A measuring device is just a large physical system made of atoms, obeying the same quantum laws as everything else. If those laws say quantum systems evolve smoothly in superposition, then surely the atoms of the measuring device, and of the physicist, and of the whole laboratory, should evolve smoothly too. Why insert a magical collapse?

Everett's radical proposal was to take the theory at its word and simply *delete the collapse.* Quantum systems evolve smoothly in superposition, always, without exception, and so does everything they interact with. But then what happens when you measure, say, an electron that is in a superposition of "spin up" and "spin down"? The Copenhagen story says the superposition collapses and you see one result. Everett's story says the superposition *does not* collapse. Instead, the electron's superposition spreads to *you*: the atoms of your measuring device, and then of your eyes, and then of your brain, become entangled with the electron, so that the whole system evolves into a superposition of two branches — one branch in which the device read "up" and you saw "up," and another branch in which the device read "down" and you saw "down." Both branches are real. Both are equally you. The world has, in effect, split into two, one for each outcome, and each copy of you, in each branch, is utterly convinced that a definite result occurred and that the other possibility simply didn't happen. There is no collapse because there is no need for one; there is only the endless branching of a universe that, at every quantum event, divides into all its possibilities, each going its own way, forever inaccessible to the others. This is the **many-worlds interpretation**, and the "worlds" are these ceaselessly proliferating branches — an unimaginably vast, ever-growing tree of parallel realities, the **multiverse**.

It sounds like science fiction, and it is easy to dismiss, but it is worth understanding why serious physicists take it seriously, because the reasons bear directly on what a quantum computer is. The great attraction of many-worlds is its economy of *principle*, if not of *worlds*. It requires no special collapse rule, no mysterious boundary between the quantum and classical realms, no privileged role for the observer. It simply says: the Schrödinger equation is the whole truth, it applies to everything without exception, and its universal application implies branching. Everything else — the appearance of definite outcomes, the randomness, the Born rule for probabilities — is supposed to emerge from the branching structure as experienced from inside a branch. There are real technical difficulties with making this fully work, especially in explaining *why* the probabilities come out as the Born rule dictates, and many-worlds is by no means the consensus view; physicists remain genuinely divided, with Copenhagen, many-worlds, and a handful of other interpretations all retaining serious adherents, and no experiment yet able to distinguish among them. But many-worlds is not a fringe position. In surveys of physicists it commands substantial support, and among those who work on quantum computing and quantum foundations it is especially popular, precisely because it takes the mathematics literally.

For Deutsch, and for a certain cast of mind, quantum computing turns the interpretation debate from idle philosophy into something closer to evidence. Here is the argument in its starkest form. A quantum computer running Shor's algorithm to factor a large number appears, in its internal workings, to evaluate a function on exponentially many inputs at once and to weave the results together by interference. Where, Deutsch asks, does this staggering amount of computation physically *happen*? In the many-worlds picture, the answer is clean and literal: the computation is distributed across exponentially many branches of the multiverse, each branch carrying out one strand of the work, the branches cooperating through interference to concentrate the answer.

The quantum computer is a device for harnessing the parallel labor of parallel worlds. On this view, a working quantum computer that factors a number too large for the whole visible universe to have factored by brute force is not just a useful machine; it is close to a *proof* that the other worlds are out there, pulling their weight. "Quantum computation," Deutsch has written, "will be the first technology that allows useful tasks to be performed in collaboration between parallel universes."

Not everyone is convinced, and the disagreement is illuminating rather than embarrassing. A Copenhagen-minded physicist will say that all this talk of parallel worlds is unnecessary storytelling laid over mathematics that stands perfectly well on its own; the qubit is in a superposition, the amplitudes interfere, the answer comes out, and there is no need to reify the intermediate possibilities as full-blown universes. A quantum computer, on this view, exploits the wave nature of quantum amplitudes, no more mysterious in principle than the interference of light in a hologram, and requires no metaphysical extravagance. Both camps agree completely on what the machine *does* and on every prediction of the theory; they disagree only on the story of what is "really" happening underneath — and, at least so far, no experiment can settle the story. This is the peculiar situation of quantum computing: a technology that works, whose workings can be described in completely standard and uncontroversial mathematics, and yet whose deepest meaning remains as contested as it was when Bohr and Einstein first quarreled. You can build the machine without settling the argument. But the machine keeps the argument alive, and gives it teeth, by making the "exponentially many possibilities" feel less like a figure of speech and more like a physical fact demanding explanation.

The many-worlds interpretation was the doctoral thesis of a young physicist named Hugh Everett III, submitted at Princeton in 1957, and it was born from a simple, stubborn refusal. Recall the measurement problem: quantum mechanics describes systems evolving smoothly as superpositions, but the moment a measurement is made, the superposition seems to "collapse" abruptly to a single definite outcome, chosen at random. This collapse had always been an awkward add-on, a second and mysterious law grafted onto the elegant Schrodinger equation, invoked only when a measurement occurs — but what counts as a measurement? Where is the line between the quantum world that superposes and the classical world of definite facts? The standard "Copenhagen" answer, associated with Niels Bohr, drew that line pragmatically but never explained it, and it troubled thoughtful people deeply. Everett's refusal was this: he simply declined to add the collapse at all.

Take the Schrodinger equation at its word, Everett said, all the way up. If it governs atoms, it governs the measuring apparatus, and the laboratory, and the physicist, and the entire universe. There is no special collapse and no magic line. So what happens when a physicist measures a particle that is in a superposition of two outcomes? The equation says the physicist becomes entangled with the particle and enters a superposition too — a superposition of "physicist who saw outcome A" and "physicist who saw outcome B." Both branches are equally real; the wavefunction simply splits, and each version of the physicist, in each branch, sees a single definite result and believes the other possibility never happened. The universe does not choose; it branches, endlessly, at every quantum event, into a vast and ever-proliferating tree of parallel realities. What we experience as the random collapse to one outcome is merely the fact that we, too, split, and each copy of us can see only its own branch.

It is a breathtaking idea, and its cost is obvious: to avoid the awkwardness of collapse, Everett had to accept a near-infinity of unobservable parallel worlds, each as real as our own, spawned ceaselessly. Most physicists in 1957 found the price absurd, and Everett's thesis was received coldly. Bohr's camp was dismissive; the young man's advisor, John Wheeler, though sympathetic, urged caution and diplomatic softening. Discouraged by the frosty reception, Everett left academic physics entirely, going to work as a defense analyst on nuclear war strategy for the Pentagon, and never published on the subject again. He died in 1982 at fifty-one, a heavy smoker and drinker who asked that his ashes be thrown out with the trash — and, by some accounts, was largely unaware that his despised idea was, in the years just before his death, beginning its long climb toward respectability. It is one of the more poignant stories in modern physics: a man who may have glimpsed the true architecture of reality, dismissed in his lifetime, vindicated only after.

The revival came in the 1970s and after, championed by physicists like Bryce DeWitt, who gave the interpretation its evocative name, and embraced with particular fervor by David Deutsch, for whom, as we have seen, the many-worlds picture was not just tenable but obviously correct and demonstrated by the very possibility of quantum computation. In Deutsch's telling, the reason a quantum computer can perform a calculation that would overwhelm all the matter in a single universe is that it is not confined to a single universe: it distributes the work across the branches of the multiverse and gathers the results through interference. The "parallelism" of a quantum computer, on this view, is quite literally the parallelism of parallel worlds.

Here, though, we must be careful, because this is where seduction shades into error. The many-worlds interpretation is a genuine and serious contender — perhaps a growing plurality of physicists working on quantum foundations find it the least bad option — but it is an INTERPRETATION, one of several, and every interpretation makes precisely the same experimental predictions. Copenhagen, many-worlds, pilot-wave theories, and various others all agree, to the last decimal, on what any experiment will show, including any quantum computation. Nothing a quantum computer does can distinguish between them. So while it is delicious to imagine your laptop's quantum successor borrowing computational muscle from a trillion sibling universes, that image is a choice of words, not a fact established by the machine. The machine works identically whether or not the other worlds are "really there."

The poignancy of Everett's personal story is worth preserving, because it is a cautionary tale about how the scientific community treats ideas ahead of their time. Hugh Everett proposed his interpretation as a graduate student at Princeton in 1957, and it was received with a coldness that shaped the rest of his life. Bohr's influential camp dismissed it; Everett's own advisor, John Wheeler, though sympathetic, urged diplomatic softening and cautious presentation, and the young man's radical vision was blunted in publication. Discouraged by the frosty reception, Everett left academic physics entirely, going to work as a defense analyst on nuclear-war strategy for the Pentagon, and he never published on quantum mechanics again. He died in 1982 at only fifty-one, a heavy smoker and drinker, reportedly largely unaware that his despised idea was, in the years just before his death, beginning its long climb toward respectability. He asked that his ashes be thrown out with the trash. It is one of the more melancholy episodes in modern physics: a man who may have glimpsed the true architecture of reality, dismissed in his lifetime, vindicated only after his death, when physicists like Bryce DeWitt gave his interpretation its evocative name and David Deutsch made it the foundation of his thinking about quantum computation.

The crucial point for the reader — the one that guards against a persistent and seductive error — is that the many-worlds interpretation, for all its intellectual appeal and its growing popularity among physicists who work on quantum foundations, is an *interpretation*, one of several, and every interpretation makes precisely the same experimental predictions. Copenhagen, many-worlds, pilot-wave theories, and various others all agree, to the last decimal, on what any experiment will show, including any quantum computation. Nothing a quantum computer does can distinguish between them. So while it is delicious to imagine your future laptop's quantum successor borrowing computational muscle from a trillion sibling universes, that image is a *choice of words*, not a fact established by the machine. The computer works identically whether or not the other worlds are "really there." Deutsch may be right that they are; he may be wrong; the mathematics of quantum computing is serenely neutral on the question, and one can be a superb quantum programmer while believing any interpretation at all, or none.

This neutrality matters enormously for the chapters ahead, because the parallel-universes story, romantic as it is, turns out to be a genuinely misleading guide to how quantum algorithms actually achieve their speedups — misleading enough to have spawned a whole literature of wrong intuitions and inflated promises. A quantum computer does *not* simply try every possible answer in a separate world and then read off the right one; if it did, it would be vastly more powerful than it actually is, and designing quantum algorithms would be trivial, which it emphatically is not. The truth is subtler, more constrained, and in its own way more beautiful: the machine prepares a superposition over many possibilities and then must carefully choreograph interference so that the wrong answers cancel and the right one survives to be measured. Whether that choreography happens "in many worlds" or "in the mathematics of one" is a matter of interpretation; that it is hard, delicate, and possible only for specially structured problems is a matter of fact. To build and wield the machine, we must therefore set the seductive but false picture firmly aside and replace it with an accurate one — beginning with a clear look at the fundamental unit the whole enterprise is built upon: not a bit, but a qubit.

Part IV — The Qubit

Chapter 15: Two States and a Maybe

Everything so far has been prologue. We have watched the strangeness of the quantum world assemble itself over a century of physics, and we have watched the idea of computation grow from Turing's tape to Deutsch's universal quantum machine. Now it is time to open the machine and look at its fundamental part — the thing that is to a quantum computer what the bit is to an ordinary one. It is called the qubit, a contraction of "quantum bit," a term coined by the physicist Benjamin Schumacher in 1995, half in jest, in conversation with his colleague William Wootters. The joke stuck, and the qubit is now the central object of a global industry. To understand quantum computing at all, you must understand the qubit, and to understand the qubit you must first understand exactly how it differs from, and resembles, the ordinary bit it is named after.

An ordinary bit is the simplest possible carrier of information: a system that can be in one of two states, which we label 0 and 1. It might be a switch that is off or on, a capacitor that is discharged or charged, a patch of magnetic material pointing one way or the other, a hole punched or not punched in a card. The physical substrate does not matter — that was Turing's lesson — only that there are two distinguishable states and that we can read and set them reliably. Every photograph, every song, every document, every program on every computer you have ever used is, at bottom, a long string of these two-state bits, and the entire edifice of digital technology is built on the discipline of keeping them sharply 0 or sharply 1, never smearing in between. The whole art of classical digital engineering is the art of *avoiding* ambiguity, forcing every signal firmly to one side or the other so that noise cannot accumulate.

A qubit also has two special states, and we write them, in the notation physicists use, as $|0\rangle$ and $|1\rangle$ — the vertical bar and angle bracket are called a "ket," a piece of shorthand invented by Paul Dirac, and you may simply read $|0\rangle$ as "the state zero" and $|1\rangle$ as "the state one." A qubit might be physically embodied as the spin of an electron pointing down or up, the polarization of a photon horizontal or vertical, an atom in its lowest energy level or an excited one, or a tiny superconducting circuit with current flowing one way or the other. If that were the whole story, a qubit would be just a bit dressed in fancy notation. But it is not the whole story, and the difference is the entire point. A qubit, being a quantum system, can exist not only in the state $|0\rangle$ or the state $|1\rangle$ but in any **superposition** of the two — a blend written as $\alpha|0\rangle + \beta|1\rangle$, where α and β are numbers (in general complex numbers, involving the square root of minus one) that measure "how much" of each state is present. The qubit is not secretly 0 or secretly 1 with the blend expressing our ignorance. It is genuinely, physically both at once, in the way that the electron of Part I was genuinely spread across many positions until measured.

Now comes the feature that trips up almost everyone, and that the introduction to this book promised to set straight. You might think that a qubit in superposition, being "both 0 and 1 at once," lets you read out two answers at the price of one — that a register of many such qubits, in a superposition of all possible bit strings, lets you read out exponentially many answers for free. This is the "tries all answers at once" fallacy, and it is wrong, because of what happens when you *look.* When you measure a qubit — when you actually interrogate it to find out its value — you do not get α and β, the amplitudes of the blend. You get a single classical bit: either 0 or 1, and nothing else. The superposition collapses, exactly as in Part I, and the two amplitudes α and β determine only the *probabilities* of the two outcomes: you get 0 with probability equal to the size of α squared, and 1 with probability equal to the size of β squared. After the measurement the superposition is gone, collapsed to whichever value you happened to get, and the delicate blend that held both possibilities is irrecoverably destroyed. Measurement is a one-way door. The rich quantum state goes in; a single humble classical bit comes out; the richness is lost.

This is the central frustration and the central discipline of quantum computing, and it deserves to be stated as sharply as possible. A qubit can hold a superposition of two possibilities, and a register of n qubits can hold a superposition of all 2^n possible bit strings at once — a genuinely enormous amount of "information" in some sense, exponentially more configurations than there are particles in the universe once n passes a few hundred. But you cannot *get that information out.* Any measurement yields only n ordinary bits, one collapsed value per qubit, chosen at random according to the amplitudes. The exponential wealth of the superposition is real while it lasts, but it is locked behind the wall of measurement, and a naive attempt to read it simply returns one random sample, no more useful than a guess. If quantum computing were only superposition, it would be a magnificent dead end.

The escape from this dead end — the thing that turns the locked wealth of superposition into usable answers — is **interference**, and it is the subject that the rest of this part builds toward. The amplitudes α and β are not mere probabilities; they are more like the heights of a wave, and, crucially, they can be *negative,* or complex, which ordinary probabilities never can. Because they can be negative, they can *cancel.* When two different computational paths lead to the same wrong answer, and one arrives with a positive amplitude and the other with an equal negative amplitude, the two contributions annihilate, and that wrong answer becomes impossible — its probability drops to zero. A quantum algorithm is a carefully engineered pattern of such cancellations, arranged so that the amplitudes for all the wrong answers destructively interfere and cancel out, while the amplitudes for the right answer reinforce, piling up until, when you finally measure, the right answer is overwhelmingly the one you get. The superposition explores; the interference selects. That is the real engine, and it is nothing like "trying everything at once and reading off the list."

Everything in a classical computer, from a blockbuster film to a bank balance, is built from a single humble atom of information: the bit, a thing that is either 0 or 1, off or on, low voltage or high. The entire digital world is bits by the trillion, herded and flipped. The quantum computer is built from a different atom, the quantum bit or qubit, and understanding exactly how it differs from an ordinary bit — no more, and crucially no less — is the key that unlocks the rest of this book. Get this right and the machine makes sense; get it wrong, as most popular accounts do, and you will be left with a caricature that promises magic and explains nothing.

A qubit, like a bit, has two basic states, which we label 0 and 1. But unlike a bit, which must sit squarely in one of them, a qubit can exist in a superposition — a specific, weighted blend of 0 and 1 at the same time. The temptation is to say "it is both 0 and 1 at once," and while that phrase is not exactly wrong, it invites the fatal misunderstanding. A qubit in superposition is not secretly 0 and also secretly 1, like a coin that is somehow heads and tails together. Nor is it merely a coin spinning in the air, definitely one or the other but not yet seen — that would be ordinary ignorance, a classical probability, and it would give you nothing new. The superposition is a genuinely different, genuinely quantum state of affairs, described not by a probability but by a pair of amplitudes: numbers that can be positive or negative (indeed complex), and that can therefore add up or cancel out. This capacity to cancel — interference — is the whole difference between a qubit and a random coin, and it is the source of every quantum advantage there is.

Consider what happens when you look. Measure a qubit and you never see a blend; you get a plain 0 or a plain 1, just as with an ordinary bit, and the superposition vanishes — collapses — to whichever answer you got. The amplitudes only tell you the PROBABILITY of each outcome (their squares, to be exact). So a single qubit, measured, yields exactly one bit of ordinary information, no more. This is the great disappointment that trips up every newcomer: you cannot read out the rich, continuous superposition; you can only sample it once and destroy it. A qubit is like a question you may ask nature exactly once, receiving a single yes or no, after which the question is gone. If a qubit merely stored a superposition that you could read at leisure, quantum computers would be trivial and infinitely capacious. They are not, because measurement is a brutal bottleneck, and the entire art of quantum algorithm design is the art of arranging the amplitudes, before you look, so that the answer you want is overwhelmingly likely to be the one you get.

So where does the fabled power come from, if a single qubit gives just one bit? It comes from combining qubits, and here the growth is explosive in exactly the way the simulation problem foretold. Two qubits together have four basic states — 00, 01, 10, 11 — and a general two-qubit superposition is a weighted blend of all four at once, described by four amplitudes. Three qubits: eight amplitudes. Ten qubits: 1,024. Three hundred qubits: more amplitudes than there are atoms in the observable universe. A modest register of qubits holds, in its superposition, an astronomically vast collection of amplitudes, all evolving together when you operate on the qubits. THIS is the exponential resource, the same one that defeats classical simulation, now turned to our advantage. The catch, always the catch, is that when you finally measure those three hundred qubits you still get only three hundred ordinary bits — one sample from that vast space. The universe hands you a single ticket from a lottery of unimaginable size.

Two features make qubits able to turn this cruel arithmetic into real computational power, and they are the two strange phenomena from the opening of this book. The first is interference: because amplitudes can be negative and cancel, a cleverly designed quantum computation can make the amplitudes of all the wrong answers destructively interfere and shrink toward zero, while the amplitude of the right answer constructively reinforces and grows large — so that when you finally sample, the ticket you draw is very probably the winning one. The second is entanglement: qubits can be linked so that their fates are correlated in ways no separate description can capture, letting the register behave as a single indivisible whole rather than a bag of independent coins. Superposition provides the vast space, entanglement stitches it into a connected structure, and interference sculpts it toward the answer. Miss any one of the three and the magic evaporates.

It is worth restating, as plainly as possible, the single correct sentence that should anchor everything the reader takes away about how a quantum computer works, because it is so often replaced by a wrong one. A quantum computer does *not* try all answers in parallel and read off the best; it prepares a superposition over all possibilities and then choreographs interference so that the useful answer emerges, with high probability, from a single measurement. Every word of that sentence matters. "Prepares a superposition" — the easy part, done with a few gates, and worthless by itself. "Choreographs interference" — the hard part, the whole art, arranging for wrong answers to cancel and the right one to reinforce. "Single measurement" — the brutal bottleneck, one sample, one bit per qubit, the rest destroyed. "High probability" — never certainty, only odds engineered to favor the answer. Hold that sentence firmly and the rest of quantum computing becomes comprehensible; replace it with "tries everything at once" and you will misunderstand every headline you ever read. The difference between the two sentences is the difference between magic and physics, between hype and understanding.

The cruelty of the measurement bottleneck deserves one more concrete illustration, because it is what makes quantum computing hard rather than trivially omnipotent. Imagine a register of three hundred qubits holding, in superposition, more amplitudes than there are atoms in the observable universe — an almost incomprehensible wealth of information, all present at once in the physical state of the machine. Now measure it. The universe hands you back three hundred ordinary bits: a single string of zeros and ones, one ticket drawn from a lottery of unimaginable size, and the entire rest of that astronomical superposition vanishes without a trace, forever inaccessible. This is why a quantum computer is not a magic parallel machine with infinite memory you can read at leisure. The vast superposition is real and does real work — it lets the amplitudes interfere and reinforce — but you are permitted only the crudest possible glimpse of the result, a single sample, after which the state is gone. The entire discipline of quantum algorithm design exists to make that one permitted glimpse count: to arrange, through interference, that the single ticket you are allowed to draw is overwhelmingly likely to be the winning one. Superposition provides the vast space; entanglement stitches it into a connected whole; interference sculpts it toward the answer; and measurement, the tyrant at the end, grants you one look. Miss any of the first three and the last look shows you only noise.

Chapter 16: The Bloch Sphere

A classical bit has two states, and there is nothing between them: 0 or 1, a coin lying flat on the table heads or tails. A qubit has a whole continuum of states, and to see them we need a richer picture than a coin. The picture that physicists use, and that has proved indispensable for building intuition, is a sphere — the **Bloch sphere**, named after the physicist Felix Bloch — on whose surface every possible state of a single qubit corresponds to a point. Learning to read this sphere is the closest a non-mathematical reader can come to seeing a qubit directly, and it repays the small effort it takes.

Picture a globe. The north pole represents the state |0>; the south pole represents the state |1>. These are the two "classical" states, the qubit's versions of a definite 0 and a definite 1, and they sit at opposite ends of the world, as far apart as can be. So far this is just a coin stood on its edge, with heads at the top and tails at the bottom. The novelty is everything *else* on the sphere. Every other point on the surface — the entire globe between the poles — represents a genuine superposition state, a legitimate condition a qubit can be in. Points near the north pole are superpositions weighted heavily toward |0>, likely to give 0 when measured; points near the south pole lean toward |1>. And the **equator** is the realm of perfect balance: any point on the equator represents a superposition that is exactly fifty-fifty, equally likely to yield 0 or 1 when measured. The equator is where a qubit is most fully "in between," most fully a maybe.

Now here is the subtle and beautiful part, the part that captures why a qubit holds more than a mere probability. All the points on the equator give the same measurement statistics — fifty-fifty — and yet they are *different states*, physically distinct, distinguishable by the right operations. What distinguishes them is their position *around* the equator, their longitude, and this longitude represents something with no classical counterpart at all: the qubit's **phase**. Two qubits can be equally balanced between 0 and 1, so that a direct measurement cannot tell them apart, and yet differ in phase — differ in *where* they sit around the equator — and this hidden difference has real, measurable consequences once you perform further operations. Phase is the qubit's secret dimension, invisible to a naive measurement but central to everything. It is the aspect of the quantum state that carries the "negative amplitudes" of the previous chapter, the ingredient that makes interference possible. When people say a qubit holds more than a bit, the "more" is, in large part, this phase — a continuous angle, a full circle of distinct states hidden behind identical measurement odds.

The two special equatorial states that sit at "east" and "west," directly opposite each other, deserve names, because they appear constantly in quantum computing. One, an equal blend of |0> and |1> with a plus sign, is called the **plus state**, written |+>. The other, an equal blend with a minus sign — the same fifty-fifty balance but with the opposite phase — is called the **minus state**, written |->. Measured directly, |+> and |-> are indistinguishable: both give 0 half the time and 1 half the time. But they are as different as north and south in the geometry of the sphere, sitting at opposite points of the equator, and a suitable operation can convert one into the other, or convert each into a *different* pole, revealing their distinctness. This pair — |+> and |-> — will be the workhorses of our explanation of quantum algorithms, because the difference between them is exactly a difference of phase, and phase is where the magic lives.

The Bloch sphere also gives us a clean way to picture what quantum **gates** do, which is the subject of the next chapter, so it is worth previewing here. Because every state of a qubit is a point on the sphere, every operation you can perform on a single qubit — every quantum gate — is simply a **rotation** of the sphere, turning one point into another. Flipping a qubit from |0> to |1>, the quantum version of a NOT gate, is a half-turn of the globe that swaps the north and south poles. Other gates are rotations by other angles about other axes, sliding a state from pole toward equator, or spinning it around the equator to change its phase. The whole repertoire of single-qubit operations is the whole repertoire of rotations of a sphere, an image of pleasing concreteness: to compute with a qubit is, geometrically, to turn a globe this way and that, steering a point across its surface. A quantum algorithm on a single qubit is a choreography of rotations, and the answer you read out at the end depends on where, after all the turning, the point has come to rest — how close to the north pole (likely to read 0) or the south (likely to read 1).

To manipulate qubits, physicists needed a way to picture them, and the picture they settled on is one of the most useful images in the whole field: the Bloch sphere, named for the physicist Felix Bloch. It turns the abstract algebra of a single qubit into a shape you can hold in your mind's eye and rotate with your hands, and once you have it, a great deal of the mysterious machinery of quantum computing becomes almost intuitive. Imagine an ordinary globe, a ball with a north pole and a south pole. Assign the north pole the state 0 and the south pole the state 1. A classical bit could only ever sit at one pole or the other. A qubit, by contrast, can point to ANY location on the entire surface of the sphere — any latitude, any longitude — and every one of those infinitely many points represents a distinct, valid quantum state.

This immediately reveals something profound and something humbling. The profound part: a single qubit contains, in a sense, an infinite amount of information, because specifying a point on a sphere requires two continuous angles, real numbers with endless decimal places. A point on the equator, halfway between the poles, is an equal superposition of 0 and 1; slide it around the equator and you change the phase, the relative sign between the two amplitudes that makes interference possible; tilt it toward a pole and you weight the superposition toward that outcome. Every subtlety of a qubit's state is a direction in this three-dimensional ball. The humbling part, which we have met before and must never forget: when you measure the qubit, you cannot read off its latitude and longitude. You get only a single bit — is it 0 or 1 — with a probability set by how close the point sits to the north or south pole. All that continuous richness collapses, at the moment of looking, to one of two poles. The sphere is real, but you are permitted only the crudest glimpse of it.

Here is where the Bloch sphere earns its keep, because it makes the operations of quantum computing visual and concrete. Every quantum gate that acts on a single qubit — every fundamental step of a quantum program — is simply a ROTATION of the sphere. Flip the state from 0 to 1? That is a half-turn about a horizontal axis, carrying the north pole to the south. Create an equal superposition, the workhorse operation that puts a qubit on the equator? That is a particular quarter-ish rotation, the famous Hadamard gate, which takes a definite pole and tips it out onto the equator into a balanced blend. Adjust the phase, sliding the point around the equator to set up an interference pattern? Another rotation, about the vertical axis. The entire repertoire of single-qubit manipulation, which sounds so forbidding in the language of complex amplitudes, becomes the familiar business of spinning a globe to point in a chosen direction. Physicists steering real qubits with precisely timed pulses of microwave or laser light are, quite literally, rotating points on Bloch spheres, and they think and speak in exactly these terms.

An everyday analogy helps cement it. Think of the qubit's state as the orientation of a spinning top or a compass needle free to point anywhere in three dimensions. Operations tilt and turn it; measurement, brutally, asks only "is it pointing more up or more down?" and knocks it to whichever pole it was leaning toward. You can aim the top with exquisite precision, but you may interrogate it only with the bluntest possible question, and the asking destroys the aim. Computation, on this picture, is a choreography of delicate, precise rotations performed in the dark, followed by a single crude glance — and the choreographer's whole skill lies in arranging the dance so that, by the time the lights come on, the top is leaning hard toward the answer.

The Bloch sphere also quietly dispels a common confusion. People often ask how a qubit can "hold more than a bit" if measuring it yields only a bit. The sphere answers: the qubit's state is a rich, continuous thing, a direction in space, but our access to it is impoverished. The information is genuinely there — it governs how the qubit will behave under further operations and how it will interfere with others — but it is not information you can simply download. It is more like the aim of an arrow than the number written on a card: it shapes what happens next without ever being directly legible. This is why quantum information cannot be copied (a fact, the no-cloning theorem, that will matter enormously for both error correction and cryptography) and why it is so devilishly hard to preserve.

The Bloch sphere repays a second visit because of a subtlety it makes vivid: the profound mismatch between how much information a qubit's state *contains* and how little you are permitted to *read out*. A point on the sphere is specified by two continuous angles — real numbers with endless decimal places — so in one sense a single qubit holds an infinite amount of information, a precise direction in three-dimensional space. Yet when you measure, all that continuous richness collapses to a single bit: is it nearer the north pole (0) or the south (1), with a probability set by how the point leans. This is the humbling truth at the heart of quantum computing: the information is genuinely there, governing how the qubit will behave under further operations and how it will interfere with others, but it is not information you can simply download. It is more like the aim of an arrow than a number written on a card — it shapes what happens next without ever being directly legible. This gap between the rich hidden state and the impoverished readout is exactly why quantum algorithms are so hard to design, and why measurement is the bottleneck the whole art must overcome.

The picture also quietly explains one of the most important facts in all of quantum information — the no-cloning theorem — and it is worth drawing out the connection. Because a qubit's state is a precise direction on the sphere that you cannot fully determine by measuring (one measurement gives one bit, not the two continuous angles), you also cannot *copy* it. To copy an unknown qubit you would first have to learn its state, and learning its state fully is impossible from a single copy; measuring it only tells you which pole it collapsed toward, destroying the very direction you wanted to duplicate. This impossibility of copying an unknown quantum state, which follows naturally from the geometry of the sphere, turns out to be foundational: it is why quantum error correction cannot simply store backup copies of a qubit, forcing the ingenious delocalization schemes of later chapters, and it is why quantum cryptography can detect eavesdroppers, since a spy cannot copy the quantum states passing by without disturbing them. A humble picture of a globe thus encodes some of the deepest constraints on what can and cannot be done with quantum information.

The limitation of the Bloch sphere is as instructive as its power, and it points directly toward the source of quantum computing's might. The sphere pictures a *single* qubit perfectly — but it cannot draw two. Two entangled qubits share a state that lives in a higher-dimensional space with no such tidy geometric portrait; the very entanglement that gives a quantum computer its power resists being pictured as two little globes side by side, because the qubits have no separate individual states to picture. This is not a failure of imagination but a reflection of something real: the information in an entangled system lives in the *relationships* between qubits, in connections that no picture of the qubits separately can capture. To harness that power we must move from the single sphere to operations that reach across qubits and bind them together — the quantum equivalent of the logic gates from which all computers are built, but gates that operate without wires, on states that can be blended and linked, obeying the strict reversibility that quantum law demands. The single sphere has taught us all it can; the real computing begins where the picture breaks down.

Chapter 17: Gates Without Wires

A classical computer computes by pushing bits through logic gates — small circuits that take one or two bits in and put a bit out according to a fixed rule. The AND gate outputs 1 only if both its inputs are 1; the OR gate outputs 1 if either input is 1; the NOT gate flips its single input. From these humble elements, combined in their billions on a silicon chip, every computation you have ever run is assembled. A quantum computer computes in an analogous way, by passing qubits through **quantum gates** — but the gates obey the rules of quantum mechanics, and those rules change everything. Understanding quantum gates is understanding how a quantum program actually works, and it is less forbidding than it sounds, especially with the Bloch sphere in hand.

Recall the two constraints quantum mechanics imposes, established in earlier chapters. First, operations on quantum states must be **unitary**, which above all means they must be **reversible**: no quantum gate may destroy information, because between measurements the quantum world conserves it. This immediately rules out the ordinary AND and OR gates, which are irreversible — knowing an AND gate output 0 does not tell you its inputs. Quantum computers must be built from reversible gates, exactly the kind that Bennett and Toffoli studied in the classical setting, plus new gates with no classical analog at all. Second, because a gate acting on a single qubit is, geometrically, a **rotation** of the Bloch sphere, the single-qubit quantum gates form a rich continuous family, far richer than the classical repertoire, which offered essentially only the NOT.

Consider the most important single-qubit gates. The **X gate** is the quantum NOT: it swaps |0> and |1>, a half-turn of the Bloch sphere exchanging the poles. There are also **Y** and **Z** gates, its cousins, rotations about the other axes; the Z gate is especially important because it does something with no classical meaning — it leaves |0> alone but flips the *phase* of

|1>, sliding a state around the equator without changing its measurement odds, a pure manipulation of that hidden longitude. But the single most important gate in all of quantum computing is the **Hadamard gate**, written H, and it deserves close attention because it is the gate that creates superposition and, later, that harvests interference. The Hadamard gate takes a definite pole and moves it to the equator: apply H to |0> and you get the plus state |+>, the perfect fifty-fifty superposition; apply H to |1> and you get the minus state |->, the same balance with the opposite phase. Run a Hadamard on a qubit sitting definitely at |0> and you have manufactured a qubit that is equally 0 and 1 — you have put it "into superposition," the first move of nearly every quantum algorithm. And here is the property that makes H the secret of interference: it is its own reverse. Apply H a second time and it undoes itself, carrying |+> back to |0> and |-> back to |1>. This means Hadamard gates can be used both to *spread* a qubit out into superposition at the start of a computation and to *gather* it back in at the end — and in the gathering, the phases that were accumulated in between cause the amplitudes to interfere, constructively for the right answer and destructively for the wrong ones. Superposition out, interference back: the Hadamard gate is the breath of a quantum computer.

Single-qubit gates alone, however, cannot make a powerful computer, because they never let qubits *talk* to one another, and without interaction there is no entanglement, and without entanglement no exponential advantage. The essential multi-qubit gate is the **controlled-NOT**, or **CNOT**, which we met in the classical reversible setting and which reappears here transformed. The CNOT acts on two qubits, a "control" and a "target." Its rule is simple: if the control qubit is |1>, flip the target; if the control is |0>, do nothing. Applied to definite classical inputs, this is a modest, reversible, almost boring operation. But apply it when the control qubit is in *superposition*, and something remarkable happens. Suppose the control is in the plus state, an equal blend of |0> and |1>, and the target starts at |0>. The CNOT must "do nothing" for the |0> part of the control and "flip the target" for the |1> part — and since the control is both at once, the result is a superposition of "control 0, target 0" and "control 1, target 1," blended together. The two qubits are now **entangled**: neither has a state of its own, but they are perfectly correlated, so that measuring either one instantly determines the other. This particular entangled pair is so fundamental it has a name, a **Bell state**, after John Bell, and the little circuit that makes it — a Hadamard on one qubit followed by a CNOT onto a second — is the "hello, world" of quantum computing, the first thing anyone runs on a real quantum machine. From a Hadamard and a CNOT, superposition and entanglement, all the strangeness of Part I is summoned to order on demand.

With these pieces we can finally say precisely what a quantum computation *is*. You begin with a register of qubits, all set to a definite starting value, usually all |0>. You apply a sequence of quantum gates — Hadamards to create superposition, phase gates and controlled operations to sculpt the amplitudes and entangle the qubits, all chosen according to the algorithm you are running. This sequence of gates is the **quantum circuit**, the quantum program, and it is often drawn as a diagram of horizontal lines (the qubits) crossed by symbols (the gates), read left to right like a musical score. Throughout this sequence the register holds an ever-shifting superposition of many possibilities, the amplitudes flowing and interfering under the gates' influence, exactly as a wave sloshes and reshapes under the forces acting on it. Then, at the end, you **measure** the qubits, collapsing the superposition and reading out a string of ordinary classical bits. If the circuit was well designed, the interference will have arranged for the desired answer to have almost all the amplitude, so that the string you read out is, with high probability, the solution you sought. If the answer is only probably right, you run the whole circuit again and take a vote. That is the entire scheme: prepare, apply gates, measure. Everything else — Shor's algorithm, Grover's search, quantum chemistry — is a matter of *which* gates, in *what* order, to make the interference come out right.

A classical computer, for all its sophistication, is at bottom an enormous array of logic gates — tiny switches that take one or two input bits and produce an output bit according to a fixed rule. The AND gate outputs 1 only if both its inputs are 1; the OR gate outputs 1 if either is; the NOT gate simply flips its input. From these humble operations, wired together by the billions on a silicon chip, everything digital is constructed. A quantum computer is built the same way in spirit — from quantum gates wired into circuits — but the gates themselves are creatures of a stranger world, and the differences are exactly the features that give the machine its power.

The first difference we have already glimpsed: a quantum gate must be reversible. Because the underlying quantum evolution never destroys information, every quantum gate must be an operation you can undo — from the outputs you must always be able to reconstruct the inputs. This immediately rules out the ordinary AND and OR gates, which throw information away (knowing an AND gate output 0 does not tell you which of three input combinations produced it). Quantum computing therefore uses reversible gates, the very kind that Bennett and Toffoli had worked out for classical reversible computing, now joined by wholly new gates with no classical counterpart. A single-qubit gate, as the Bloch sphere taught us, is just a rotation of that sphere; the most important is the Hadamard gate, which takes a definite 0 or 1 and tips it onto the equator into an equal superposition — the operation that first summons the quantum-ness into being, converting a plain bit into a genuine blend.

But single-qubit gates alone, rotating each qubit in isolation, can never produce the entanglement that is the machine's secret fuel. For that you need gates that act on two qubits at once, letting one qubit's state influence another's. The canonical example is the controlled-NOT, or CNOT: it flips a "target" qubit if and only if a "control" qubit is 1, and leaves the target alone if the control is 0. On plain 0s and

1s this is unremarkable, a simple conditional flip. But feed the control qubit a superposition — put it on the equator of its Bloch sphere — and something remarkable happens: the CNOT entangles the two qubits, weaving them into a single shared state in which neither has a definite value of its own but their values are perfectly correlated. This is the gate that manufactures entanglement on demand, and it is the two-qubit workhorse of essentially every quantum algorithm and every error-correcting code. With Hadamard gates to create superposition and CNOTs to create entanglement, you have the essential ingredients.

Here lies one of the most reassuring and important results in the whole theory: universality. Just as any classical circuit, however complex, can be built from a small handful of gate types (NAND gates alone suffice for everything), any quantum computation whatsoever can be built from a small, fixed set of quantum gates — for instance, a few single-qubit rotations plus the two-qubit CNOT. This is the quantum version of Turing's and Deutsch's universality, made concrete at the level of hardware. An engineer does not need to invent a bespoke device for every algorithm; build a machine that can perform this modest universal set of gates reliably, string them together in the right sequence — the "circuit" — and you can, in principle, run any quantum program at all. The staggering diversity of quantum algorithms all reduce, in the end, to choreographed sequences of the same few elementary rotations and controlled flips.

It is worth making the picture concrete. A quantum program is drawn as a circuit diagram: horizontal lines, one per qubit, read left to right like a musical score, with gate symbols placed along them marking which operation happens when. The qubits begin, conventionally, all in the state 0. A layer of Hadamard gates lifts them into superposition, spreading the register across a vast space of possibilities. Then a carefully designed sequence of gates — rotations and CNOTs and their kin — sculpts the amplitudes, entangling qubits and shifting phases so that, through interference, the amplitudes of the desired answers grow and the rest shrink. Finally, the qubits are measured, collapsing the whole elaborate superposition to a string of ordinary 0s and 1s that, if the choreography was right, encodes the answer with high probability. Design the dance, run it, and look: that is a quantum computation, from the outside.

The concept of universality deserves a final, emphatic word, because it is what makes quantum computing an engineering possibility rather than an impossible dream of building a bespoke device for every problem. Just as any classical circuit, however baroque, can be constructed from a small handful of gate types — indeed, the humble NAND gate alone suffices to build every classical computation there is — any quantum computation whatsoever can be built from a small, fixed set of quantum gates, for instance a few single-qubit rotations plus the two-qubit CNOT. This is a deep and reassuring result. It means an engineer does not need to invent new hardware for each algorithm; build a machine that can perform reliably this modest universal set of operations, string them together in the right sequence, and you can in principle run any quantum program at all, from Shor's algorithm to a molecular simulation to a program not yet imagined. The staggering diversity of quantum algorithms all reduce, in the end, to choreographed sequences of the same few elementary rotations and controlled flips — which is why the entire industrial effort can focus on one concrete goal: making a small set of gates work with high enough fidelity, on enough qubits, for long enough. Universality turns the open-ended problem of "build a quantum computer" into the tractable problem of "perfect these few operations."

It is worth making the physical reality of a quantum gate vivid, because the abstract language of "gates" and "circuits" can mislead into imagining something like a classical chip. There are no wires and no switches. A quantum "gate" is a precisely controlled physical manipulation of the qubits themselves — a carefully shaped pulse of microwave radiation delivered to a superconducting circuit, or a flash of laser light striking a trapped ion, timed and tuned to enact exactly the desired rotation of the Bloch sphere or the desired entangling operation. The quantum "circuit" is not a wiring diagram but a *schedule* — a score, read left to right in time, specifying which pulse strikes which qubit at which moment. When you picture a quantum computer running, do not picture current flowing through wired gates; picture a fantastically precise sequence of microwave or laser pulses, choreographed to the nanosecond, tilting and entangling the qubits in the dark, followed by a final measurement that reads out the result. The "gates" are events in time, not objects in space, and the whole computation is a dance performed on the qubits by precisely timed light.

A final caution corrects a natural but wrong intuition: that more gates always mean more computing power. In classical computing, adding more operations generally helps. In quantum computing, every single gate is an opportunity for error and for the fragile quantum state to decohere and leak away, so real quantum programming is a relentless fight to accomplish the most computation with the *fewest*, shortest operations. A quantum algorithm that needs a billion gates is far harder to run than one that needs a thousand, not merely because it takes longer but because each gate erodes the delicate superposition, and the errors accumulate until the answer dissolves into noise. This is why the "depth" of a quantum circuit — the length of its longest sequence of operations — is such a critical measure, and why so much ingenuity goes into finding shorter, shallower ways to compute the same thing. The fragility that shadows every gate is the reason a useful quantum computer has been so agonizingly hard to build, and it is the enemy the second half of this book is largely about. But before confronting that fragility, we must understand more deeply the resource these gates exist to create — entanglement, the strangest and most powerful ingredient of all, which a young generation learned to treat not as a philosophical puzzle but as a fuel to be spent.

Chapter 18: Entanglement as a Resource

We have now met entanglement twice: first in Part I, as the "spooky action at a distance" that appalled Einstein and was vindicated by Bell and Aspect, and again in the previous chapter, as the correlation a CNOT gate produces when its control qubit is in superposition. It is time to put entanglement at the center of the stage, because it is, more than superposition alone, the true source of a quantum computer's power. This is a subtle point, often muddled in popular accounts, and getting it right is worth a chapter.

Superposition by itself is not enough to beat a classical computer. This is easy to overlook, since superposition is the flashy, counterintuitive feature, but it is important. A single qubit in superposition, or even many qubits each independently in superposition but not entangled with one another, can be efficiently tracked and simulated by a classical computer; there is a well-known result to this effect, that certain quantum computations using only limited kinds of gates, without the full power of entanglement, can be mimicked classically without exponential cost. The exponential blow-up that defeats classical simulation — the 2^n numbers needed to describe *n* qubits — arises only when the qubits become *entangled*, woven into joint states that cannot be factored into independent pieces. It is entanglement that fills the exponentially large state space, entanglement that creates correlations with no classical shortcut, entanglement that Feynman identified (through the Bell inequalities) as the thing a classical computer cannot cheaply reproduce. A quantum computer with superposition but no entanglement would be a curiosity; a quantum computer that entangles its qubits is a genuinely new kind of machine.

What does it *mean* for the state space to be filled by entanglement, in concrete terms? Consider three qubits. If they are unentangled, each in its own definite or superposed state, then describing the trio requires only describing each qubit separately — a modest amount of information, three little

Bloch-sphere points. But if the three qubits are fully entangled, their joint state is a superposition spread across all eight possible three-bit combinations — 000, 001, 010, 011, 100, 101, 110, 111 — each with its own amplitude, and these eight amplitudes cannot be reconstructed from any description of the qubits individually. The information lives in the *pattern of correlations among all three*, not in any one of them. For three qubits this means eight numbers; for three hundred qubits it means 2^300 numbers, more than the atoms in the universe, all bound together in a single entangled state that a quantum computer can hold and manipulate as one object. Entanglement is the glue that binds the qubits into a whole whose description is exponentially vast — and the manipulation of that exponentially vast description, in a controlled way, is quantum computation.

Entanglement is not only the engine of computational speed; it is a *resource* in a more concrete, almost economic sense, one that quantum information scientists learned to quantify and spend. In the 1990s a series of discoveries revealed that a shared entangled pair — two qubits, one held by each of two distant parties, conventionally named Alice and Bob — is a kind of currency that can be exchanged for communication tasks otherwise impossible. The most famous of these is **quantum teleportation**, discovered in 1993 by a team including Charles Bennett. Despite its science-fiction name, teleportation does not transport matter or send signals faster than light. What it does is transfer the exact, unknown quantum *state* of one qubit to a distant qubit, using up one shared entangled pair and two ordinary classical bits of communication. Alice has a qubit in some mysterious quantum state she wishes to send to Bob; she cannot simply measure it (measurement would collapse and destroy the state) and she cannot copy it (the **no-cloning theorem**, a fundamental result, forbids making a copy of an unknown quantum state). Instead she performs a joint measurement on her mystery qubit and her half of the entangled pair, obtaining two classical bits, which she sends to Bob by ordinary means — telephone, email, anything. Bob, applying a simple operation dictated by those two bits to his half of the entangled pair, finds that his qubit has become an exact replica of Alice's original, which has meanwhile been destroyed at her end. The state has been "teleported," disassembled at one place and reconstructed at another, its journey carried partly by the pre-shared entanglement and partly by two humble classical bits. It works, it has been demonstrated countless times in laboratories and even between a ground station and a satellite, and it is a fundamental building block of quantum networks and of some quantum- computer architectures, which use teleportation internally to move quantum information around a chip.

The **no-cloning theorem**, mentioned in passing, deserves emphasis, because it is one of the sharpest ways the quantum world differs from the classical and it shapes everything about how quantum computers and quantum communication must work. In a classical computer, copying is trivial and ubiquitous: every time you save a file, forward an email, or read a bit of memory, you make copies, and the ease of copying is so basic we never think about it. In the quantum world, copying an unknown state is *impossible* — not difficult, but forbidden by the mathematics, a theorem proved in 1982. You cannot build a device that takes an arbitrary unknown qubit and produces two identical copies of it. This has profound consequences, some limiting and some liberating. It is limiting because it means you cannot back up quantum information the way you back up a file, cannot make copies to guard against errors in the naive way, which is part of why quantum error correction (a later chapter) had to be so cleverly indirect. It is liberating because it is the foundation of quantum cryptography: an eavesdropper cannot copy the quantum signals passing between Alice and Bob without disturbing them, so any attempt to spy is detectable, giving a form of security guaranteed by physics rather than by the mere difficulty of a computation. No-cloning is the quantum world's refusal to be duplicated, and it runs through the whole subject.

The great shift in thinking that made quantum computing possible was to stop treating entanglement as a philosophical embarrassment and start treating it as a resource — something you possess in measurable quantities, produce with gates, consume in protocols, and spend to accomplish tasks impossible without it. To Einstein, entanglement was "spooky action" and a sign that the theory was incomplete. To the builders of quantum computers, entanglement is fuel. This reframing — from paradox to power source — is as important to our story as any piece of hardware, and it happened gradually through the 1980s and 1990s as theorists discovered concrete things you could DO with entangled pairs.

To see why entanglement is powerful, recall what makes it different from ordinary correlation. Two ordinary correlated objects — a pair of gloves in separate boxes — share a single fixed fact: one is left, the other right. Entangled qubits share something far richer: they are correlated no matter which of many incompatible questions you ask them, along any axis you choose, even one selected at the last instant. This "correlation in every direction at once" cannot be reproduced by any pre-arranged classical instructions, as Bell proved and Aspect confirmed. It means that a collection of entangled qubits holds information in the RELATIONSHIPS between qubits that simply is not present in the qubits individually. You can know everything measurable about each qubit alone and still not know the shared state, because the information lives in the correlations, in the connective tissue between them. This is why the state of many entangled qubits cannot be written down as a list of separate qubit states — and why it holds exponentially more than the parts.

The first triumphant demonstration that entanglement is a usable resource was quantum teleportation, described in 1993 by a team including Charles Bennett — the same Bennett of reversible computing. The name is unfortunate, because it summons Star Trek, but the achievement is real and it is precise. Suppose you hold a qubit in some unknown quantum state and you wish to transfer that exact state to a distant partner, without physically shipping the qubit. The no-cloning theorem forbids you from simply copying it. But if you and your partner share an entangled pair beforehand, you can perform a joint measurement on your qubit and your half of the pair, obtain two ordinary bits of result, phone those two bits to your partner over an ordinary channel, and have them apply a corresponding operation to their half — whereupon their qubit springs into precisely the state yours was in, while your original is destroyed in the process. The quantum state has been transmitted using shared entanglement plus two classical bits, and no physical particle carrying the state ever traveled between you. Entanglement was spent — the shared pair is used up — to accomplish something otherwise impossible. It behaved exactly like a fuel: consumed to do work.

This resource framing pays off across the whole field, and it is worth cataloguing why, because the theme recurs in every chapter to come. In a quantum computer, entanglement generated by CNOT gates is what allows the register of qubits to act as a single coherent system exploring an exponentially large space, rather than a bag of independent coins; without entanglement, a "quantum" computer could be efficiently simulated by a classical one and would offer no advantage at all. In quantum error correction, entanglement is used to spread the information of one fragile logical qubit across many physical qubits, so that errors can be detected and undone without ever looking at — and destroying — the protected information. In quantum cryptography, shared entanglement lets two parties generate a secret key whose security is guaranteed by the very Bell-inequality violations that refuted Einstein. In quantum sensing, entangled probes measure fields and times more precisely than any independent probes could. The same phenomenon that Einstein flagged as a defect turns out to be the master resource of an entire technology.

There is a beautiful and humbling subtlety, though, that guards against overexcitement. Entanglement is necessary for quantum advantage but not, by itself, sufficient — you cannot simply pile up entangled qubits and expect miracles. The entanglement must be of the right structure, generated and consumed by a well-designed algorithm, and orchestrated together with interference to steer the computation toward its answer. Merely entangled states can be useless or even, in certain highly structured cases, efficiently simulable classically. Entanglement is the fuel, but fuel alone is not a journey; you also need an engine and a map. This is why, despite decades of work, useful quantum algorithms remain rare and precious: designing the choreography that turns raw entanglement into a solved problem is genuinely hard.

The reframing of entanglement from philosophical embarrassment to spendable resource is worth marking as one of the great conceptual shifts in the history of physics, comparable in its way to the moment heat was recognized as a form of energy that could be measured, converted, and put to work. To Einstein, entanglement was "spooky action at a distance," a sign that quantum mechanics was incomplete, a defect to be explained away. To the builders of quantum computers, entanglement is *fuel* — something you possess in measurable quantities, manufacture with gates, consume in protocols, and spend to accomplish tasks impossible without it. This shift, from paradox to power source, happened gradually through the 1980s and 1990s as theorists discovered concrete things you could *do* with entangled pairs, and it is as essential to our story as any piece of hardware. Once entanglement was seen as a resource rather than a riddle, the whole enterprise of quantum computing came into focus: a quantum computer is, in large part, a machine for generating, manipulating, and spending entanglement to compute what classical machines cannot.

The demonstration that most vividly established entanglement's status as a usable resource was quantum teleportation, and it repays a closer look because the name so badly misleads. Described in 1993 by a team including Charles Bennett, quantum teleportation has nothing to do with Star Trek; it is a precise protocol for transferring an unknown quantum state from one place to another without physically shipping the qubit that carries it. The no-cloning theorem forbids simply copying the state. But if the two parties share an entangled pair beforehand, the sender can perform a joint measurement on her qubit and her half of the pair, obtain two ordinary bits of result, transmit those two classical bits over an ordinary channel, and have the receiver apply a corresponding operation to his half of the pair — whereupon his qubit springs into precisely the state the sender's qubit was in, while her original is destroyed. The quantum state has been transmitted using shared entanglement plus two classical bits, and no particle carrying the state ever traveled between them. The shared entanglement is *consumed* in the process — used up, spent — behaving exactly like a fuel burned to do work. This was the moment entanglement's resource character became undeniable.

The catalogue of what entanglement, as a resource, makes possible is worth holding in view, because the same phenomenon recurs as the enabling ingredient across every corner of quantum technology. In a quantum computer, entanglement generated by CNOT gates lets a register of qubits act as a single coherent system exploring an exponentially large space, rather than a bag of independent coins — and without it, a "quantum" computer could be efficiently simulated classically and would offer no advantage at all. In quantum error correction, entanglement spreads one fragile logical qubit's information across many physical qubits, so errors can be detected and undone without ever reading the protected data. In quantum cryptography, shared entanglement lets two parties generate a secret key whose security rests on the very Bell-inequality violations that refuted Einstein. In quantum sensing, entangled probes measure with a precision beyond any independent probes. The one phenomenon Einstein flagged as a defect turns out to be the master resource of an entire technology. Yet a humbling subtlety guards against overexcitement: entanglement is *necessary* for quantum advantage but not, by itself, *sufficient* — it must be of the right structure, generated and consumed by a well-designed algorithm, orchestrated together with interference toward an answer. Fuel alone is not a journey; you also need an engine and a map. With superposition, gates, interference, and now entanglement all in hand, we possess the complete conceptual toolkit of quantum computation — and it is time to put the seductive but false "parallel universe" story firmly to rest and see clearly what a quantum computer really does when it computes.

Chapter 19: Why "Parallel Universes" Is the Wrong Story

Here is the sentence you have heard, in one form or another, every time quantum computing appears in the newspaper: "Whereas an ordinary computer tries each possibility one at a time, a quantum computer tries all the possibilities simultaneously, and that is why it is so fast." It is a wonderfully vivid image. It is also, as a guide to how quantum computers actually work, essentially wrong, and the wrongness matters, because it leads people to expect the impossible from these machines and to misunderstand both their real power and their real limits. This chapter is devoted to replacing the false image with a true one. If you remember nothing else from this book, remember this chapter.

Start with why the naive picture fails. Suppose you have a hard problem with a million possible answers, only one of them correct — say, a password, or the solution to a puzzle. The "tries all at once" story says: put a quantum register into a superposition of all one million candidates, run them all through the check simultaneously, and read off the winner. The first half of this is genuinely possible. With a modest number of qubits and a few Hadamard gates you really can create a superposition of all one million (or all a billion, or all 2^300) candidate answers, and you really can, with a single pass of a suitable circuit, compute the "correct or not" verdict for every candidate at once, the results all held together in one giant superposition. So far the story seems to hold. The catastrophe is in the last step — the reading off. When you measure that superposition, you do not get the winner. You get *one random candidate*, drawn according to the amplitudes, and since the correct answer started with the same tiny amplitude as every other, your chance of happening to measure it is one in a million — exactly the same as if you had simply guessed at random without any quantum computer at all. The superposition did contain the answer, in some sense, but measurement hands you a single random sample from it, and the answer drowns in the crowd. Quantum parallelism, by itself, buys you nothing. This is a theorem, not a temporary engineering limitation, and it is why the naive picture is not merely incomplete but actively misleading.

So what is the true picture? The true picture is **interference**, and the right mental model is not a committee of parallel workers but a wave. Recall that the amplitudes attached to each possibility in a superposition are not probabilities but wave-like quantities that can be positive, negative, or complex, and that when contributions to the same outcome are added up, positive and negative amplitudes can *cancel*. A quantum algorithm is a device for arranging cancellation. The art is to design the sequence of gates so that, for every *wrong* answer, the various computational paths leading to it arrive with amplitudes that cancel out — some positive, some negative, summing to nothing — while for the *right* answer, the paths arrive in phase, amplitudes reinforcing, piling up. By the end of the computation, the wrong answers have had their amplitudes destroyed by destructive interference, and the right answer has had its amplitude amplified by constructive interference. Now when you measure, the right answer is not one candidate among a million with equal odds; it is the one candidate that has survived the great cancellation, holding almost all the amplitude, overwhelmingly likely to be what you read out. The quantum computer did not succeed by trying everything and looking at the list. It succeeded by making the wrong answers annihilate one another, leaving the right answer standing.

The analogy that captures this best is not a parallel committee but a well-known trick of waves: **noise-cancelling headphones**, or the interference of light in a soap bubble or a hologram. Noise-cancelling headphones do not identify and delete the sound of an airplane engine; they generate a second sound wave precisely out of phase with the engine's, so that crest meets trough, and the two waves cancel to silence. Interference — the meeting of waves that reinforce or annihilate depending on their relative phase — is an everyday phenomenon, visible in the ripples of a pond, the colors of an oil slick, the dark bands where light passes two slits. Quantum computing is interference applied to the amplitudes of computational possibilities. The qubits' phases (the hidden longitude of the Bloch sphere) are the relative phases of the waves; the gates choreograph those phases; and the final measurement reads out where the amplitude has been made to pile up. This is why phase, that secret dimension of the qubit invisible to a naive measurement, is so central: it is the steering wheel of interference, and interference is the whole game.

Seen this way, the difficulty of quantum algorithm design comes into focus, and so does the reason quantum computers are not magic all-purpose speed machines. Making the wrong answers cancel is *hard*. It requires that the problem have some special mathematical structure that a clever designer can exploit to arrange the cancellation — a hidden periodicity, a symmetry, an oracle with the right properties. For most problems, no one knows how to set up the interference so that the right answer emerges, and for many problems it is believed to be impossible; a quantum computer offers no useful speedup at all for the great majority of computational tasks. The famous quantum algorithms

— Shor's factoring, Grover's search, quantum simulation — are famous precisely because they are rare cases where someone found a way to make the interference work. This is the honest reason quantum computers will not simply replace classical ones or "solve everything faster." They are specialized instruments that achieve dramatic speedups on the narrow class of problems whose structure permits the necessary interference, and little or nothing on the rest. A quantum computer is less a faster computer than a *different* computer, good at a peculiar and important set of tasks and unremarkable at the others.

There is a final subtlety worth naming, because it connects back to Deutsch and the many-worlds interpretation of the previous part. When you describe interference in terms of amplitudes cancelling, you are giving the mathematics, which everyone agrees on. When you describe it in terms of parallel universes doing the computation and then interfering, you are giving one *interpretation* of that mathematics, Deutsch's, which not everyone accepts. Both descriptions predict exactly the same experimental results. The "parallel universes" story is not wrong as a piece of interpretation — it may even be true — but it is wrong, or at least dangerously misleading, as an account of *why quantum computers are fast*, because it suggests that the parallelism alone does the work, when in fact the parallelism is useless without the interference that selects the answer. Even in the many-worlds picture, the point is not that many universes compute in parallel; it is that they are made to *interfere*, cooperating so that the wrong-answer universes cancel and the right-answer universes reinforce. Whatever story you prefer about what is "really" happening, the operative concept is the same, and it is not parallelism. It is interference. Hold onto that, and you understand quantum computing more deeply than most of what you will read about it.

It is time to confront, head-on and without mercy, the single most misleading sentence ever written about quantum computing: "A quantum computer tries all possible answers at the same time." You have read it in newspapers, heard it from television scientists, perhaps repeated it yourself. It is seductive, it is memorable, and it is wrong — wrong in a way that matters, because it makes quantum computers sound like magic and leaves you unable to understand why they are hard to build, why they help with only certain problems, and why designing a quantum algorithm is one of the most difficult tasks in computer science. If quantum computers truly tried all answers at once and let you read off the best, they would solve almost everything instantly, and the field would have finished decades ago. They do not, and it did not, and the reason why is the deepest and most beautiful idea in this book.

Here is the trap. It is true that you can put a register of qubits into a superposition of all possible inputs at once — with a few Hadamard gates, three hundred qubits hold a superposition spanning more possibilities than there are atoms in the universe. And it is true that if you then apply some computation, in a sense every one of those possibilities is processed. So far the "parallel universes" story seems vindicated: the machine really has, in some sense, computed the answer for every input simultaneously. But now comes the measurement, and the measurement is where the fantasy dies. When you look at the register, you do not receive all those answers. You receive exactly one, chosen at random, with a probability set by its amplitude — and all the astronomically many other results, all that supposed parallel labor, vanish without a trace, collapsed away, forever inaccessible. The vast parallel computation happened, if you like, but you are permitted to keep only a single random grain of its output. That is worse than useless; it is a lottery.

So the true challenge of quantum computing is not achieving the parallelism — that part is easy and, by itself, worthless. The true challenge is EXTRACTING a useful answer from a state you can only sample once and destroy. And the only tool for doing so is interference. Recall that quantum amplitudes, unlike classical probabilities, can be negative (indeed complex), and so they can cancel. The entire art of quantum algorithm design is to arrange the computation so that the amplitudes of the wrong answers interfere destructively — cancel out, shrink toward zero — while the amplitudes of the right answer or answers interfere constructively, reinforcing and growing large. Then, when you finally measure, the single random sample you are permitted to draw is overwhelmingly likely to be the answer you wanted, because you have quietly arranged for it to hog nearly all the probability. The parallelism sets up an enormous superposition; interference sculpts that superposition, before you look, so that the useful outcome dominates. Miss the interference and you have nothing but noise.

An analogy helps, and it is worth savoring because it captures why quantum computing is both powerful and narrow. Think of the amplitudes as ripples on the surface of a vast pond, one ripple for every possible answer. A classical random guess is like dropping a single pebble and seeing where it lands — one shot, one answer. A quantum computation is like generating a precisely engineered pattern of ripples across the entire pond at once, and then arranging, through the geometry of the problem, for all the ripples to cancel each other everywhere EXCEPT at the location of the answer, where they pile up into a towering wave. Look at the pond and you will almost certainly see the wave — the answer — because everywhere else the water has gone flat. But — and this is the crux — you can only pull off this feat of engineered cancellation for problems that have a special hidden structure, a mathematical regularity that the interference pattern can lock onto. For a problem with no such structure — a truly random haystack — there is nothing for the ripples to reinforce, no pattern to exploit, and quantum computers offer little or no advantage.

This is why quantum computers are not magic all-purpose accelerators, and why the honest experts are so careful to temper the hype. They help dramatically with a small number of problems that possess exactly the right kind of hidden mathematical structure — factoring large numbers, simulating quantum physics, certain search and optimization tasks — and they offer little or nothing for the vast majority of computing you do every day. Your email, your spreadsheets, your video calls will never run faster on a quantum machine; there is no interference pattern to exploit in showing a web page. The parallel-universe story, by contrast, implies a universal speedup for everything, and that implication is simply false. Understanding this one point separates those who grasp quantum computing from those who merely marvel at it.

David Deutsch, as we saw, would object that the parallelism IS real and IS performed across the branches of the multiverse, and that interference is what brings the branches back together. He may even be right about the metaphysics; nothing here refutes him. But notice that even on his view, the hard and essential part is the interference — the recombination — not the mere spreading out. Whether you believe the parallel computation happens in other worlds or simply in the mathematics of one, the lesson for the would-be quantum programmer is identical: raw superposition is free and useless, and the whole game is arranging the interference. That is the correct mental model, and armed with it we can finally do what this entire first act of the book has been preparing us for. We can watch a real quantum algorithm work — a small, clean, almost playful one, invented by Deutsch himself, in which we can see with our own eyes the amplitudes of the wrong answers cancel and the right answer emerge from a single measurement. It is the "Hello, World" of quantum computing, and it changes everything.

Part V — The Algorithms

Chapter 20: Deutsch's Coin

The first quantum algorithm ever devised solves a problem so artificial that no one, before or since, has ever needed its answer for any practical purpose. This is not a criticism. Deutsch's algorithm, published by David Deutsch in 1985 and refined with Richard Jozsa in 1992, was never meant to be useful. It was meant to be a *proof* — a clean, indisputable demonstration that a quantum computer can accomplish some definite task with fewer operations than any classical computer possibly could. It is the quantum equivalent of the Wright brothers' first hop at Kitty Hawk: twelve seconds, a hundred and twenty feet, of no transport value whatsoever, and the beginning of everything. To understand how quantum algorithms work, there is no better place to start than this deliberately tiny puzzle, because it contains, in miniature, the entire logic of every quantum algorithm that came after.

Here is the problem. Imagine a mysterious machine — a "black box," or in the jargon an **oracle** — that takes a single bit as input (0 or 1) and returns a single bit as output. There are only four possible such machines. Two of them are "constant": one always outputs 0 no matter the input, the other always outputs 1. The other two are "balanced": one copies the input (0->0, 1->1), the other flips it (0->1, 1->0). Now suppose you are handed one of these four machines, sealed, and asked a single question about it: is it *constant* (same output for both inputs) or *balanced* (different outputs for the two inputs)? You are not asked *which* of the four it is, only whether it belongs to the constant pair or the balanced pair. How many times must you run the machine to be sure?

Classically, the answer is obvious: twice. You must feed it a 0 and see what comes out, then feed it a 1 and see what comes out, and compare. One query is not enough — if you feed it a 0 and get a 0, the machine could still be the constant-0 machine (balanced verdict: no) or the copying machine (balanced verdict: yes), and you cannot tell which without also testing the input 1. Two queries are necessary and sufficient. There is no classical shortcut; to compare the outputs for two inputs, you must look at both.

Deutsch's astonishing result is that a quantum computer can answer the question with a *single* query. Not by being faster at each query, but by asking a fundamentally different kind of question — a question that interrogates *both* inputs at once and returns not their individual values but the *relationship* between them, which is exactly what the problem asks for. Here is the shape of it, and watch for the now-familiar rhythm: superposition, then a phase-sensitive operation, then interference, then measurement. You prepare a qubit and, with a Hadamard gate, put it into a superposition of 0 and 1 — so that it can, in a sense, carry both inputs into the oracle at once. You then run the oracle, but arranged cleverly (using an extra qubit and a trick that turns the oracle's output into a *phase* on the input qubit rather than a separate answer). The effect is that the oracle stamps the two branches of the superposition with phases that depend on its outputs for the two inputs. If the machine is constant, both branches get the same phase; if it is balanced, they get opposite phases. Now you apply a second Hadamard gate, which — recall from the chapter on gates — gathers the superposition back together and lets the two branches *interfere*. If the phases were the same (constant machine), the branches interfere constructively toward one measurement outcome; if the phases were opposite (balanced machine), they interfere toward the other. You measure the single qubit, and its value — 0 or 1 — tells you directly, with certainty, whether the machine was constant or balanced. One query. One measurement. Done.

Look closely at what happened, because it is the template for everything. The quantum computer did not "try both inputs and compare the answers" in the classical sense; it never learned the individual outputs at all. Instead it encoded the two outputs as phases on the two branches of a superposition, and then used interference to extract a single global property — the relationship between the outputs, same-or-different — that no single classical query could reveal. The superposition let both inputs into the machine together; the oracle wrote the relevant information into the phases; the final Hadamard turned that phase information into an interference pattern; the measurement read the answer out of the pattern. Every quantum algorithm in this book, from Shor's to Grover's, follows this same four-beat structure: spread into superposition, compute a function so that the answer is encoded in phases, interfere to convert phases into measurable amplitudes, measure. Deutsch's coin is the purest, smallest example of the pattern, which is why it comes first.

The generalization Deutsch worked out with Jozsa in 1992 made the advantage dramatic rather than merely factor-of-two. In the **Deutsch–Jozsa problem**, the oracle takes not one input bit but n of them, and is promised to be either constant (the same output for all 2^n inputs) or balanced (output 0 for exactly half the inputs and 1 for the other half). Classically, in the worst case, you might have to query the oracle a little more than half of 2^n times — an exponentially large number — to be *certain* of the answer, because you could keep getting the same output and not know whether the next one will differ. The quantum algorithm settles the question with a *single* query, regardless of n, by exactly the superposition-phase-interference trick scaled up. An exponential gap between quantum and classical, proven rigorously, for a definite problem. It was the first time anyone had shown such a gap, and though the problem itself was artificial, the demonstration was electrifying to the handful of people paying attention. It proved that Deutsch's universal quantum computer was not just a philosophical construct but a genuinely more powerful computational device, at least for *some* problems.

Every field needs its first clean example, the toy problem simple enough to hold in your head yet rich enough to reveal the essential trick, and for quantum computing that example is Deutsch's algorithm — often illustrated with a coin, or more precisely with a mysterious black box that computes a one-bit function. It is deliberately artificial; no one needs the answer it gives. Its glory is that in it, and nowhere simpler, you can watch quantum interference solve a problem with fewer steps than any classical method, and understand exactly how. It is the "Hello, World" of the subject, and generations of students have felt the ideas click into place upon seeing it.

Set up the puzzle. Imagine a black box — a function — that takes a single bit as input, 0 or 1, and returns a single bit as output. There are only four such functions possible. Two of them are "constant": they ignore the input and always return the same value (always 0, or always 1). The other two are "balanced": they return different values for the two inputs (one returns the input unchanged, the other returns its opposite). Now the question, and it is the whole game: you are handed one of these four boxes but not told which, and you must determine whether it is CONSTANT or BALANCED. You do not need to know exactly which function it is — only whether it belongs to the constant family or the balanced family. How many times must you use the box?

Classically, the answer is obvious and unavoidable: twice.

You must feed in 0 and see what comes out, then feed in 1 and see what comes out, and compare. One query tells you nothing — if you put in 0 and get 0, the box could be the constant-0 function or the balanced function that returns its input; you cannot tell without checking the other input. So a classical computer must open the box, run it, open it again, run it again: two uses, minimum. There is no cleverer classical strategy; the information simply is not there after a single query.

Deutsch's astonishing result is that a quantum computer can answer the question — constant or balanced — with just ONE use of the box. Here is how the interference works, and it is worth following slowly because every quantum algorithm that follows, up to and including Shor's world-shaking one, is this same trick writ large. You prepare your input qubit not as 0 or 1 but, using a Hadamard gate, as an equal superposition of both at once, so that when the box acts, it acts on 0 and 1 simultaneously. Crucially, you also arrange (through a second qubit and a phase trick) that the box's output is written not as a value but as a PHASE — a sign, plus or minus — attached to each branch of the superposition. Now the two branches, one carrying the box's answer for input 0 and the other for input 1, are brought back together with a final Hadamard gate, and they interfere. And here is the magic: if the function is constant, the two branches carry the same phase and interfere CONSTRUCTIVELY, steering the qubit deterministically to one definite outcome; if the function is balanced, the two branches carry opposite phases and interfere DESTRUCTIVELY on one outcome and constructively on the other, steering the qubit to the opposite result. Measure the qubit, and its single value tells you, with certainty, which family the box belongs to — from one query.

Look closely at what happened, because it dispels the parallel-universe fantasy in the most concrete possible way. The quantum computer did NOT learn the full function — it never tells you exactly which of the four boxes you held.

It learned only a single global PROPERTY of the function, the constant-versus-balanced distinction, a property that depends on BOTH of the function's values at once. The interference computed a relationship between the two outputs — do they agree or differ? — without ever revealing either output on its own. This is the deep pattern of all quantum speedups: they compute global properties, relationships that a classical machine could only discover by examining many values one at a time, by orchestrating an interference among all those values simultaneously and reading off the single bit of relationship that survives the cancellation. You trade away the ability to see the details in exchange for the ability to see, in one glance, a pattern that spans them all.

A final reflection on Deutsch's little algorithm explains why so modest a result deserves so prominent a place in this history. The problem it solves is admittedly artificial — determining whether a black-box function is constant or balanced, a question no practical task hands you — and in its original 1985 form the algorithm was even slightly flawed, sometimes returning an inconclusive "don't know," a blemish that Deutsch and Richard Jozsa cleaned up in 1992 when they generalized it to functions of many bits and made the quantum advantage exponential. Skeptics were right to note that Deutsch-Jozsa solved a puzzle nobody actually had. But the point was never the puzzle; it was the *proof of principle*, blazingly clear and rigorous, that quantum interference could deliver a genuine, provable, unbounded speedup over any classical method. Before Deutsch's algorithm, "quantum speedup" was a suggestive intuition; after it, it was a demonstrated mathematical fact. That is why the little algorithm, useless in itself, is the seed from which Shor's and Grover's world-shaking algorithms grew — it was the first firm foothold on the mountain, the existence proof that the climb was possible at all.

The deep pattern that Deutsch's algorithm first revealed — that quantum computers compute *global properties* rather than individual values — deserves to be carried forward as the single most useful idea for understanding every quantum algorithm that follows. Deutsch's machine never tells you exactly which of the four functions you held; it tells you only a single global property, the constant-versus-balanced distinction, that depends on *both* of the function's values at once. The interference computed a relationship spanning the whole function — do its outputs agree or differ? — without ever revealing either output on its own. This is the essence of every quantum speedup there is. Shor's algorithm does not learn the whole factoring function; it extracts one global property, the period. Grover's algorithm does not examine each item; it amplifies one global feature, the location of the target. Quantum computers trade away the ability to see the details in exchange for the ability to see, in a single glance, a pattern that spans them all — a pattern a classical machine could discover only by examining many values one at a time. Grasp this, and you grasp both the power of quantum computing and its narrowness: it excels precisely when the answer you want is a global property that interference can surface, and it offers nothing when no such exploitable pattern exists. The door Deutsch opened led, step by step, toward a problem the world would care about intensely — and the next figures to walk through it carried the idea steadily closer to the beating heart of modern cryptography.

Chapter 21: Bernstein, Vazirani, and the Complexity of the Quantum

Deutsch and Jozsa had shown that a quantum computer could beat a classical one on a specific artificial problem. But their result left a large theoretical question unanswered, and answering it was necessary before quantum computing could be taken seriously as a branch of computer science rather than of physics. The question was foundational: what is the *right theory* of quantum computational power? Computer scientists do not just want a few examples of quantum speedup; they want a rigorous framework — a "complexity theory" — that classifies problems by how hard they are for quantum computers, that relates quantum power to the established classical categories, and that lets one prove general theorems about what quantum machines can and cannot do. Building that framework was the achievement of two theorists at Berkeley, the graduate student Ethan Bernstein and his advisor Umesh Vazirani, in the early 1990s, and it is the reason quantum computing became a respectable and rigorous discipline rather than a collection of clever tricks.

Classical complexity theory sorts problems into classes with cryptic names. **P** is the class of problems a classical computer can solve efficiently — in "polynomial time," meaning the effort grows only as some fixed power of the problem size, not exponentially. **NP** is the class of problems whose *answers*, once guessed, can be efficiently *checked*, even if finding them seems hard — the class that contains thousands of important practical puzzles and whose relationship to P (the famous "P versus NP" question) is the deepest open problem in computer science, carrying a million-dollar prize. **BPP** is the class of problems a classical computer can solve efficiently if it is allowed to flip coins and accept a tiny chance of error — "bounded-error probabilistic polynomial time," widely regarded as the true class of "efficiently solvable" problems in practice. These classes are the map by which theorists navigate the landscape of computational difficulty.

Bernstein and Vazirani's contribution, in a paper first presented in 1993, was to define the quantum analog rigorously and to place it on this map. They gave a careful, robust definition of the quantum Turing machine and of the class of problems it can solve efficiently with bounded error — a class now universally called **BQP**, "bounded-error quantum polynomial time." BQP is the formal answer to the question "what can a quantum computer do efficiently?" Everything this book celebrates a quantum computer for — factoring, searching, simulating — is a claim that some problem lies in BQP, in the quantum efficient class, whether or not it lies in the classical one. Bernstein and Vazirani proved the basic properties one needs for such a class to be meaningful: that quantum computers can be made robust against small imperfections, that the definition does not depend on fiddly details of the model, that BQP contains the classical efficient classes (a quantum computer can do anything a classical one can, at least as fast) and sits inside certain larger classes (a quantum computer is not *all-powerful* — it cannot, for instance, efficiently solve every problem in NP, and it is bounded by classical resources of a higher but still limited kind). They mapped quantum power into the existing landscape and showed it was a coherent, well-behaved new territory.

To demonstrate that this new territory contained something classical computers could not efficiently reach, they offered a sharpened oracle problem, now called the **Bernstein–Vazirani problem**. The setup: a hidden binary string s of n bits is locked inside an oracle, and the oracle, when queried with any input string, returns a single bit — the result of a certain simple combination of the input with the hidden s. Classically, to discover all n bits of the secret string, you must query the oracle at least n times, extracting essentially one bit of the secret per query — there is no way to learn more than a bit at a time. The quantum algorithm recovers the *entire* hidden string with a *single* query. The mechanism is by now familiar in outline: put all the input qubits into superposition with Hadamard gates, run the oracle once so that the hidden string is stamped into the phases of the superposition, apply Hadamard gates again to convert that phase pattern into a definite output by interference, and measure — reading off, in one shot, all n bits of the secret that a classical computer would need n queries to extract. It is Deutsch's four-beat structure again, tuned to pull an entire hidden string out of the interference pattern at once.

After Deutsch and Jozsa showed that a quantum computer

could beat a classical one at a clean, well-posed problem, a new question sharpened in the minds of theorists: was this a fluke of one artificial puzzle, or the first sighting of a whole landscape of quantum advantages? Two computer scientists at Berkeley, Ethan Bernstein and Umesh Vazirani, set out in the early 1990s to put quantum computing on a rigorous footing as a branch of theoretical computer science — not physics, but complexity theory, the mathematical study of what can be computed efficiently. Their work, and the problem that carries their names, marked the moment quantum computing stopped being a physicists' curiosity and became a subject that hard-nosed computer scientists had to take seriously.

The Bernstein-Vazirani problem is another black-box puzzle, and like Deutsch's it is artificial, but it isolates the quantum trick with beautiful clarity. Imagine a hidden secret string of bits — say a particular sequence of 0s and 1s, unknown to you, of some length. You are given a black box that computes a specific function: hand it any input string, and it returns a single bit formed by comparing your input against the secret string in a certain precise way (technically, the bitwise dot product modulo two — but the details matter less than the shape of the problem). Your task is to discover the hidden secret string. How many queries to the box do you need?

Classically, the answer is exactly the length of the string. If the secret is a hundred bits long, you must query the box a hundred times, each query cleverly designed to reveal one bit of the secret — probe the first position, then the second, and so on. There is no shortcut; each query, at best, extracts one bit of the hidden string, and there are as many bits to find as the string is long. This is intuitive and, classically, unavoidable: to learn an n-bit secret you need n bits of information, and each query yields at most one.

The quantum result is once again startling: a quantum computer recovers the ENTIRE secret string with a single query. Just one. Whether the secret is ten bits or ten thousand, one query suffices. The mechanism is the by-now-familiar choreography — put the input register into an equal superposition of all possible strings with a layer of Hadamard gates, let the box stamp its comparison onto the phases of the superposition, then apply a second layer of Hadamard gates that makes all the branches interfere. When the dust settles, the interference has arranged matters so that the amplitude piles up entirely on one particular outcome: the secret string itself. Measure the register and it reads out the hidden secret, in full, in one shot. The interference pattern was, in effect, tuned by the box to point directly at the answer, and the final Hadamard layer focused it there.

What makes Bernstein-Vazirani important beyond its own artificiality is what it demonstrated about the SOURCE of quantum power. The problem contains no entanglement in the usual sense at its heart — its speedup comes almost purely from superposition and phase interference, the ability to query all inputs at once and let their phases combine. It showed that even this simplest quantum resource, cleanly deployed, buys a real advantage: an n-fold reduction in queries, one instead of n. And it did something subtler and more consequential for the theory. Bernstein and Vazirani used it, along with deeper work, to lay the formal foundations of quantum complexity theory — to define rigorously the class of problems a quantum computer can solve efficiently (later christened BQP, "bounded-error quantum polynomial time") and to begin mapping how that class relates to the familiar classical classes P and NP. They gave the field the vocabulary and the tools with which every subsequent claim of quantum advantage would be stated and judged.

The transition that Bernstein and Vazirani effected — moving quantum computing from the realm of physics into the rigorous discipline of theoretical computer science — is worth dwelling on, because it changed who took the field seriously and how its claims would thereafter be judged. Before their work, quantum computing was largely the province of physicists intrigued by the strange information-processing possibilities latent in quantum mechanics. Bernstein and Vazirani, working as computer scientists, insisted on treating the quantum computer as a formal object of complexity theory — the mathematical study of what can be computed efficiently — subject to the same rigorous definitions and proofs as any classical model. They defined precisely the class of problems a quantum computer can solve efficiently, later christened BQP (bounded-error quantum polynomial time), and began the still-unfinished work of mapping how that class relates to the familiar classical classes P and NP. In doing so they gave the field the vocabulary and the standards of proof by which every subsequent claim of quantum advantage would be stated and tested. Quantum computing acquired a rigorous spine, and the hard-nosed theoretical computer scientists could no longer dismiss it as physicists' hand-waving.

The Bernstein-Vazirani problem itself, artificial as it is, isolates a crucial point about the *source* of quantum power that is easy to miss amid the more famous algorithms. Its dramatic speedup — recovering an n-bit hidden string in a single query where a classical computer needs n queries — comes almost purely from superposition and phase interference, with none of the entanglement-heavy machinery of Shor's algorithm. This demonstrated that even the simplest quantum resource, cleanly deployed, buys a real advantage, and it helped theorists begin to dissect exactly which quantum ingredients — superposition, interference, entanglement, the ability to query all inputs at once — contribute what to a computation's power. Understanding the anatomy of quantum speedup, taking it apart to see which piece does the work, was essential groundwork for the deeper question of why quantum computers help with some problems and not others, and the Bernstein-Vazirani problem was one of the cleanest specimens on the dissecting table.

The pattern of discovery visible across these early algorithms — Deutsch, Deutsch-Jozsa, Bernstein-Vazirani, and soon Simon and Shor — is worth naming explicitly, because it corrects a common myth about how breakthroughs happen. The public imagines scientific advance as a series of bolts from the blue, lone geniuses struck by inspiration. The reality, laid bare in this sequence, is a relay race: Deutsch proved a quantum speedup possible on a contrived problem; Deutsch and Jozsa made it exponential; Bernstein and Vazirani sharpened the underlying theory and showed superposition alone was potent; Simon would find the periodicity trick; and Shor would point that trick at the heart of cryptography. Each runner carried the baton a little further, building precisely on the last, and each artificial black-box puzzle — useless in itself, solving no problem anyone actually had — was a rung on a ladder the community was steadily climbing toward a result the world would care about. None of these problems mattered in themselves; together they mapped the terrain and honed the tools that made the great applications possible. The next rung, and the decisive one, was Simon's.

Chapter 22: Simon's Problem — the Exponential Hint

Every so often in science a modest paper turns out to contain, in embryo, an idea of world-shaking consequence, and its author is remembered less for the paper itself than for what it inspired in someone else. Daniel Simon's 1994 result is such a paper. Simon, then a young researcher, posed and solved another oracle problem — abstract, artificial, of no direct practical use — and in doing so he demonstrated the first *exponential* separation between quantum and classical computing for a problem where the classical difficulty was robust, surviving even randomized, error-tolerant classical methods. Deutsch–Jozsa's exponential gap had evaporated if you allowed the classical computer to guess and accept a small error; Simon's gap did not. A classical computer, even one permitted to flip coins and tolerate mistakes, needed exponentially many queries to solve Simon's problem, while a quantum computer needed only a modest, polynomial number. This was the strongest evidence yet that quantum computers were genuinely, dramatically more powerful than classical ones, and it was noticed by exactly the right person at exactly the right moment.

Simon's problem concerns a hidden *periodicity*, and since periodicity is the golden thread running toward Shor, it is worth describing its flavor. You are given an oracle — a function — with a secret property: it is guaranteed to be "two-to-one" with a hidden pattern. That is, the function maps its inputs to outputs in such a way that inputs come in secret *pairs* that produce the same output, and the two members of each pair are related by a fixed hidden string s: if you know one member of a pair, you get the other by combining it with s. The task is to discover the secret string s — the hidden period that governs which inputs collide. It is, once again, a puzzle about uncovering a concealed regularity buried inside a function you can only probe from the outside, one query at a time. Classically, finding s is hopeless without brute force: you must keep querying different inputs and watching for a "collision," two inputs that yield the same output, which reveals a pair and hence s — but collisions are exponentially rare, so you must make exponentially many queries before you stumble on one. The hidden period stays hidden.

Simon's quantum algorithm finds the period with dramatically fewer queries, and it does so by a method that is the direct ancestor of Shor's. The now-familiar four-beat rhythm applies, but with a new and crucial move at the end. You put the input register into a superposition of all possible inputs and run the oracle once, so that the register now holds a superposition entangled with the function's outputs. Then you measure the output part — and here is the trick — which, because the function is two-to-one, collapses the input register not to a single input but to a superposition of *ex-*

actly the two inputs that share that output, the two members of a hidden pair. You have manufactured a small superposition whose structure secretly encodes the period s. Applying Hadamard gates and measuring now yields, through interference, not s itself but a *clue* about s — a linear constraint that s must satisfy. Repeat the whole process a modest number of times, gathering a fresh clue each time, and after roughly n repetitions you have enough constraints to solve for the hidden string s by ordinary arithmetic. Polynomially many quantum queries suffice where exponentially many classical queries would fail. The quantum computer found the hidden period by preparing superpositions whose interference patterns leak information about the period's structure — and it is precisely this capacity, detecting a hidden period through interference, that Shor would weaponize.

The significance of Simon's result within the small community of quantum-computing theorists was immediate and large, but its true importance is what it did to one reader in particular. **Peter Shor**, a mathematician at AT&T Bell Laboratories in New Jersey, encountered Simon's work — indeed, Shor has recounted that he saw Simon's result before it was even published, learning of it through the informal channels of the research community — and it set him thinking. Shor grasped the essential lesson of Simon's algorithm with a clarity that would change history: that a quantum computer is a superb machine for finding the hidden *period* of a function, the secret length after which its behavior repeats. Simon's period was a hidden binary string under an exotic kind of addition; but periods, Shor knew, are everywhere in mathematics, and some of them are anything but academic. In particular, Shor knew a fact from number theory that most quantum physicists did not: that the problem of *factoring a large number into its prime components* — the problem whose presumed difficulty is the foundation of the RSA cryptosystem that secures the internet — can be *transformed* into the problem of finding the hidden period of a certain simple, ordinary numerical function. Factoring, dressed up correctly, is a period-finding problem. And period-finding is exactly what Simon had just shown quantum computers do exponentially well.

Daniel Simon's contribution in 1994 is the hinge on which this entire history turns, though his name is far less famous than the one that immediately follows it. Simon, then a young researcher, devised a black-box problem — artificial, like its predecessors — that delivered an exponential quantum speedup of a new and more powerful kind, and, crucially, did so using a mathematical structure that would prove to be the master key. Peter Shor has said plainly that it was Simon's problem that showed him the way to factoring. Without Simon's insight, the algorithm that made quantum computing a matter of national security might have been years or decades later in coming. This is a chapter about the last stepping-stone before the flood.

Simon's problem is a hidden-pattern puzzle, and the pattern it hides is periodicity — repetition. Imagine a black box that computes some function, and you are promised that the function has a special hidden regularity: there is a secret string such that the function gives the same output for any two inputs that differ by that secret string. In other words, the function repeats itself with a hidden period, and pairs of inputs related by the secret collide onto the same output. Your task is to find the secret period. It is the kind of structure that sounds abstract until you realize that periodicity is everywhere in mathematics — and, fatefully, that the security of the world's dominant encryption rests on a periodicity that is fiendishly hard to find classically.

How hard is Simon's problem classically? Extraordinarily hard: it requires, on average, a number of queries that grows exponentially with the length of the secret string. The reason is intuitive — to find the hidden period you must, in effect, hunt for a "collision," two different inputs that give the same output, and among an exponentially large space of inputs such matching pairs are astronomically rare. You would have to query enormous numbers of random inputs and wait for a coincidence, and the waiting grows exponentially. Classically, the hidden periodicity is buried too deep to dig out in any reasonable time. This exponential classical hardness is what makes the quantum result so dramatic.

The quantum algorithm finds the hidden period in a number of steps that grows only GENTLY — polynomially — with the length of the secret, an exponential speedup over any classical method. And the mechanism is a decisive advance over everything before it. Simon's algorithm uses superposition to evaluate the function on all inputs at once, then — and here is the new ingredient — it exploits the function's hidden periodicity to make the interference pattern depend on the secret period, so that measuring the register yields, each time, a random clue constraining the secret. Run the quantum routine a handful of times, gather a handful of such clues, and then solve a small system of ordinary linear equations classically to pin down the period exactly. The quantum computer does the exponentially hard part — surfacing information about the global periodic structure — while a trivial classical calculation finishes the job. This hybrid pattern, a quantum core wrapped in classical pre- and post-processing, is exactly how Shor's algorithm, and most practical quantum algorithms since, are built.

The truly historic feature of Simon's algorithm is the technique at its heart: using quantum interference to extract the PERIOD of a function. This is the essential move, and Peter Shor, hearing of Simon's work, recognized that the security of RSA encryption reduces to a periodicity problem — that cracking RSA means finding the period of a particular function related to the number you wish to factor. Simon had found period-finding for an abstract black box; Shor would find period-finding for the specific function that guards the world's secrets, replacing Simon's artificial oracle with a concrete, computable function and swapping in a piece of mathematics called the quantum Fourier transform, exquisitely suited to detecting periodicity, in place of Simon's simpler interference. The lineage is direct and acknowledged: Deutsch proved a speedup possible, Deutsch-

Jozsa made it exponential, Bernstein-Vazirani built the theory, and Simon found the periodicity trick — and then Shor pointed that trick at the beating heart of modern cryptography.

The direct line from Simon's problem to Shor's algorithm is one of the best-documented and most instructive stories of influence in the history of computing, and it repays a closer look because it shows exactly how abstract mathematics becomes world-altering technology. Peter Shor has said plainly that it was Daniel Simon's work that showed him the way to factoring. Simon had devised an artificial black-box problem whose solution required finding a hidden *periodicity* — a secret repetition in a function — and had shown that a quantum computer could find it exponentially faster than any classical method, by using superposition to evaluate the function everywhere at once and then exploiting interference to surface the period. Shor, learning of this, recognized that the security of RSA encryption *also* reduces to a periodicity problem — that cracking RSA means finding the period of a particular function related to the number one wishes to factor. Simon had found period-finding for an abstract oracle; Shor would find period-finding for the specific function guarding the world's secrets, swapping in a piece of mathematics called the quantum Fourier transform, exquisitely suited to detecting periodicity, in place of Simon's simpler interference. Within months, an artificial puzzle had become a threat to global cryptography.

The hybrid structure of Simon's algorithm — a quantum core doing the exponentially hard part, wrapped in trivial classical processing — deserves emphasis, because it is the template for nearly every practical quantum algorithm that followed, and it corrects the fantasy of the all-powerful quantum machine. Simon's algorithm does not solve the whole problem quantum-mechanically. It uses the quantum computer for exactly one thing that classical machines cannot do: surfacing, through interference, a clue about the function's hidden periodic structure. Each run of the quantum routine yields one random constraint on the secret period; run it a handful of times, gather a handful of clues, and then a trivial classical calculation — solving a small system of ordinary linear equations — pins the period down exactly. The quantum machine does the one impossible thing; ordinary computation does everything else. Shor's algorithm has precisely this shape, and so do most quantum algorithms of practical interest: a classical scaffold with a single decisive quantum step bolted into its center. Understanding this structure is understanding why quantum computers help only with problems that can be recast in terms of some hidden structure that interference can detect.

There is a poignant asymmetry of fame worth registering here. Daniel Simon's name is known to specialists but to almost no one outside the field, while Shor's algorithm is famous enough to have entered the general vocabulary of technology and national security. Yet Simon's problem is arguably the true conceptual origin of the most consequential quantum algorithm ever devised — the moment the essential technique of period-finding by quantum interference was first isolated and understood. This is how science usually works, and how credit usually flows: the famous final sprinter wins the glory partly because of where the earlier runners had already carried the baton, and the crucial penultimate insight, without which the celebrated breakthrough could not have happened, is remembered mainly by the historians. When Shor put the final pieces together and pointed Simon's periodicity trick at the beating heart of modern cryptography, the abstract game of black-box puzzles suddenly acquired terrifying real-world stakes, and quantum computing changed overnight from an elegant amusement into something governments and banks could not afford to ignore.

Chapter 23: Shor's Bombshell

In the spring of 1994, Peter Shor was a thirty-five-year-old mathematician at Bell Labs, the fabled research arm of AT&T in New Jersey where the transistor, the laser, and information theory itself had been born. Shor was quiet, undramatic, formidably talented, the kind of mathematician who chews on a problem in silence for weeks and then produces something that reorganizes a field. He had been mulling Simon's period-finding result and the number-theoretic fact that factoring could be recast as period-finding. Over a period of weeks, in his office and on his walks, he worked out how to fit them together — how to build a quantum algorithm that would find the period that unlocks a factorization, and thereby crack the problem that the security of the digital world was built on. When he had it, and had convinced himself and then a few skeptical colleagues that it truly worked, the news traveled through the small research community with the speed of a rumor of buried treasure. Within months, quantum computing was transformed from an obscure theoretical amusement into one of the hottest topics in science, and the transformation was almost entirely the doing of this one result.

To feel the force of the bombshell, you must understand what factoring has to do with secrets. Nearly all secure communication on the internet — every time you see a padlock in your browser, every online purchase, every encrypted message, every software update, every exchange of diplomatic or military or financial secrets — relies, at some layer, on **public-key cryptography**, and the most widespread public-key system is **RSA**, named for its inventors Rivest, Shamir, and Adleman in 1977. RSA's security rests on a beautifully simple asymmetry. It is easy, even for a pocket calculator, to multiply two large prime numbers together to get a bigger number. But it is fiendishly hard, given only that bigger number, to work backward and recover the two primes that were multiplied — to *factor* it. Multiply two 300-digit primes and you get a 600-digit number; hand that 600-

digit number to all the classical supercomputers on earth and set them factoring, and they will still be grinding away when the sun dies, because the best known classical factoring methods take a time that grows almost exponentially with the number's size. RSA turns this asymmetry into a lock: your public key is a big number that is secretly the product of two primes; anyone can use it to *encrypt* a message to you, but only someone who knows the secret primes — you — can *decrypt* it. The whole edifice stands or falls on one assumption: that factoring large numbers is, and will remain, practically impossible.

Shor's algorithm demolishes that assumption, for anyone who can build a large enough quantum computer. His algorithm factors a number not by any cleverer version of trial division, but by the indirect, quantum-mechanical route through period-finding. The connection, known to number theorists, is this: to factor a number N, it suffices to pick a random number and study the sequence of its powers modulo N — the remainders you get when you divide its successive powers by N. That sequence of remainders always eventually *repeats*; it is periodic, and the length of its period is a hidden number that, once known, hands you the factors of N by a short piece of ordinary arithmetic. So factoring reduces to finding the period of this "modular exponentiation" sequence. Finding that period classically is as hard as factoring itself — no shortcut. But finding the period is *exactly* the kind of task Simon had shown quantum computers excel at. Shor's algorithm prepares a superposition over a vast range of exponents, computes the modular remainders in superposition so that the periodic structure is imprinted across the amplitudes, and then applies the mathematical heart of the method — the **quantum Fourier transform**, a quantum circuit that acts as an exquisitely sensitive detector of periodicity, causing the amplitudes to interfere so that measurement yields the hidden period with high probability. From the period, classical arithmetic finishes the job. What would take a classical computer longer than the age of the universe, Shor's algorithm accomplishes in a time that grows only as a modest power of the number's size — polynomial, not exponential. RSA, in the presence of a large fault-tolerant quantum computer, is broken.

The **quantum Fourier transform** at the algorithm's core deserves a word, because it is the single most important subroutine in quantum computing and it embodies the interference principle in its most refined form. The ordinary Fourier transform is a mathematical tool, used everywhere in science and engineering, that takes a signal and reveals its hidden frequencies — the pure tones hidden in a chord, the periodicities hidden in a wiggling curve. It is how your phone recognizes a song and how an engineer finds the resonances in a bridge. The quantum Fourier transform does the same thing to the amplitudes of a quantum superposition, and — this is the miracle — it does so exponentially faster than the classical version, because it operates on all the amplitudes at once through interference rather than crunching them one by one. When Shor's algorithm has loaded the periodic sequence of remainders into a superposition, the quantum Fourier transform sifts that superposition for its hidden frequency, making the amplitudes at the true period reinforce and all others cancel, so that a measurement reads out the period. It is the four-beat structure — superpose, compute-into- phase, interfere, measure — executed with a subroutine of extraordinary power, and it recurs throughout the field: the quantum Fourier transform is also the engine behind quantum algorithms for the discrete logarithm (which breaks the other main public-key systems, Diffie–Hellman and elliptic-curve cryptography) and for the broad "hidden subgroup problem" that unifies most of the known exponential quantum speedups.

In the spring of 1994, a thirty-five-year-old mathematician at Bell Labs named Peter Shor produced the result that transformed quantum computing from an intellectual curiosity into a matter of urgent global consequence. He showed that a quantum computer, if one could be built, would be able to factor large numbers efficiently — to break them apart into their prime constituents in a reasonable time — and in doing so would shatter the cryptographic system that protects nearly all digital communication on Earth. Overnight, a field that had been the playground of a few dozen theorists became the concern of intelligence agencies, banks, and standards bodies. Shor's algorithm is the single most important result in the history of quantum computing, and the reason the machine is being pursued with billions of dollars and the full attention of nation-states.

To feel the force of it you must understand the peculiar mathematical fact on which modern security rests. It is easy to multiply two large prime numbers together to get a bigger number; a schoolchild, or a pocket calculator, can multiply two 300-digit primes in an instant. But going backwards — taking that 600-digit product and recovering the two primes that were multiplied to make it — is, as far as anyone knows, staggeringly hard for a classical computer. There is no known efficient method; the best classical algorithms must essentially search, and the effort explodes as the number grows. Factoring a number a few hundred digits long would take the fastest classical supercomputers longer than the age of the universe. This grotesque asymmetry — trivial one way, hopeless the other — is a "one-way function," and it is the foundation of the RSA cryptosystem, named for Rivest, Shamir, and Adleman, who devised it in 1977. When your browser shows a padlock, when you send a payment or a private message, RSA or a close relative is very likely standing guard, its security resting entirely on the belief that factoring is hard.

Shor's bombshell was to prove that belief false in a quantum world. His algorithm factors numbers not by searching but by a completely different route: it converts the factoring problem into a period-finding problem — the very periodicity trick Daniel Simon had pioneered — and then uses quantum interference, via a tool called the quantum Fourier transform, to find that period exponentially faster than any classical machine. The running time of Shor's algorithm grows only gently with the size of the number, so that a task that would take a classical computer longer than the lifetime of the cosmos would take a sufficiently large quantum computer mere hours or days. The one-way function is one-way no longer; the trapdoor swings open. Every secret ever protected by RSA — every encrypted message, every digital signature, every secured transaction — would, before a working large-scale quantum computer, lie exposed.

The reaction was electric and has never fully subsided. Almost immediately, funding for quantum computing research surged; agencies that had ignored the field now poured money into it, both to build the machine and to understand the threat. For here was the crucial realization: Shor's algorithm is a threat that exists the moment the mathematics is known, even before the hardware exists. An adversary can record your encrypted communications TODAY, store them, and simply wait until a quantum computer capable of running Shor's algorithm is built, then decrypt everything retroactively. This "harvest now, decrypt later" logic means that any secret which must stay secret for a decade or two is arguably already at risk, because the harvest is happening now and the decryption is only a matter of time. Shor did not merely threaten future security; he retroactively endangered the present.

It is worth appreciating the human moment. Shor has recounted that when he first worked out the algorithm he could scarcely believe it, and checked and rechecked, half-expecting to find an error that would deflate the whole thing. He gave an early informal talk, word spread with astonishing speed through the small community, and within days researchers around the world were poring over the result and confirming it. There is a poignant irony in the fact that RSA-129 — a specific factoring challenge posed in 1977, thought at the time to be safe for essentially forever — was cracked by classical means in that very same year, 1994, through a massive eight-month effort by six hundred volunteers pooling computers across the internet, revealing a hidden joke message: "The Magic Words Are Squeamish Ossifrage." That heroic classical effort factored a 129-digit number and strained the limits of the possible. Shor had just shown that a quantum computer would swallow such numbers, and far larger ones, without effort. The contrast could not have been starker: the summit of classical achievement was the quantum machine's warm-up.

It is worth appreciating the precise reason Shor's result landed with such force, transforming an obscure academic field overnight into a concern of intelligence agencies and boardrooms. It was not merely that a clever new algorithm had been found; it was that the algorithm threatened something the entire modern world unknowingly depends upon. The security of digital civilization — every encrypted message, every online payment, every software update your devices trust, every secured government and military communication — rests on a small number of mathematical problems believed to be hard, and factoring large numbers is chief among them. Shor did not chip away at this foundation; he showed that a sufficiently large quantum computer would demolish it entirely, turning the "impossible" problem of factoring into a routine calculation. When the most respected cryptographers understood what he had done, the reaction was not academic interest but something closer to alarm, and funding, attention, and talent poured into quantum computing almost overnight. A single mathematical result had made a speculative machine into a matter of national security.

The retroactive character of the threat is worth restating, because it is what makes Shor's bombshell urgent even though the machine to detonate it does not yet exist. The danger is not confined to the future moment when a large quantum computer is finally built. Because encrypted data can be intercepted and stored today, an adversary can harvest your secrets now — recording the scrambled ciphertext and filing it away — and simply wait until a quantum computer capable of running Shor's algorithm exists, then decrypt the entire archive at leisure. Any secret that must remain secret for a decade or two is therefore arguably already at risk, because the harvesting can happen today and the decryption is only a matter of time. Shor did not merely threaten the security of future communications; he retroactively endangered the security of present ones, casting the shadow of a future machine backward over everything encrypted today. This is why the world has begun, urgently, to replace its cryptographic foundations long before the threatening machine has been built — a defensive scramble we will follow in later chapters.

A crucial caveat must be stated clearly, lest the alarm curdle into panic, because the honest picture is one of a real but not imminent threat. Shor's algorithm requires a large, reliable, error-corrected quantum computer — one with thousands or even millions of high-quality qubits working in concert through billions of precise operations — and no such machine exists today, nor is one expected imminently. The largest number a real quantum computer has factored with Shor's algorithm remains laughably small, a testament to the staggering gap between the elegant algorithm on paper and the machine capable of running it at scale. The threat is a storm on the horizon, not overhead: real, certain to grow, reshaping how the world approaches security, but not yet upon us. Estimates of when a cryptographically relevant machine might arrive range from a decade to several decades to, in the view of some, never. Yet the mere certainty that the storm is coming has already changed the world, and before we follow the engineers into their thirty-year struggle to build the machine that could make Shor's threat real, it is worth looking more closely at the algorithm itself — at how, exactly, factoring becomes the detection of a hidden rhythm, and why, three decades on, your secrets are, for now, still safe.

Chapter 24: Inside Shor's Algorithm

Having met the bombshell, it is worth slowing down to examine it more carefully, because Shor's algorithm is so central to this history — it is the reason the money and the urgency exist — that a reader deserves a clearer view of its machinery, its reach, and, just as importantly, its limits. Much of the public conversation about quantum computing and cryptography is confused, veering between the panic that "quantum computers will break all encryption any day now" and the complacency that "it's all decades away and hypothetical." The truth is more precise and more interesting than either, and it lives in the details.

Begin with what the algorithm actually needs. Shor's algorithm has two parts, a classical part and a quantum part, and only the quantum part requires a quantum computer. The classical part is the number theory: reducing factoring to period-finding, choosing a random base, and, after the quantum computer reports the period, finishing with a short arithmetic calculation to extract the factors. This part runs on an ordinary laptop. The quantum part is the period-finding: preparing a large superposition, computing modular exponentiation across it, and applying the quantum Fourier transform. This is where the quantum computer earns its keep, and it is demanding. To factor a number of the size used in real RSA keys — say, 2048 bits, the current standard — the quantum computer must manipulate a superposition over an enormous range, perform an intricate sequence of reversible arithmetic operations on thousands of qubits, and hold the whole delicate quantum state coherent, free of the decoherence that destroys quantum information, throughout a computation involving *billions* of individual gate operations. This is the crux of why RSA is still safe today, thirty years after Shor: not because the algorithm doesn't work — it does, provably — but because no one has built, or is close to building, a quantum computer with enough high-quality, error-corrected qubits to run it on a real key. The gap is entirely one of engineering, and it is a vast gap, as later parts of this book will detail.

How vast? Consider the numbers. The largest number ever factored by a real quantum computer running Shor's algorithm, as of the mid-2020s, is a humbling reminder of the distance still to travel. In 2001, IBM researchers used a seven-qubit machine to factor 15 into 3 × 5 — a result any schoolchild knows, achieved to demonstrate the principle. Later experiments factored 21, and there have been various claims, some using mathematical shortcuts that arguably don't constitute a genuine run of the full algorithm, of factoring somewhat larger numbers. But no quantum computer has come remotely close to factoring a number of cryptographic size. Careful estimates by researchers of how many physical qubits it would take to factor a 2048-bit RSA key with a realistic, error-corrected machine have ranged into the *millions* of physical qubits, running for hours or days — because each of the few thousand "logical" qubits the algorithm needs must, given current error rates, be built from a thousand or more noisy physical qubits through quantum error correction (a story told in Part VI). The best real machines of the mid-2020s have on the order of a few hundred to about a thousand *physical* qubits, noisy and uncorrected. The gap between what exists and what Shor's algorithm demands is not a gap of a little more effort; it is a gap of many orders of magnitude, and closing it is the central engineering quest of the field. Encouragingly for the machine-builders and worryingly for the cryptographers, those qubit-count estimates have been *falling* over the years as algorithms and error- correction schemes improve, but the target remains far beyond present reach.

Now consider precisely what Shor's algorithm threatens, because it is more specific than "all encryption." Shor's algorithm breaks the public-key systems whose security rests on factoring or on the closely related discrete-logarithm problem: RSA, Diffie–Hellman key exchange, and elliptic-curve cryptography. These are the systems used to *establish* secure connections and to make *digital signatures* — the handshake at the start of every secure web session, the signatures that verify software and certificates. Their fall would be catastrophic, because they are the foundation on which the rest of internet security is bootstrapped. But Shor's algorithm does *not* directly break the other main pillar of cryptography, **symmetric** encryption — systems like AES, which use a single shared secret key to scramble bulk data, and which secure the actual content of communications once a connection is established. Symmetric ciphers are threatened only by Grover's algorithm (the subject of two chapters hence), and Grover offers merely a *quadratic* speedup, which weakens symmetric encryption but does not shatter it: doubling the key length restores security, so that AES-256 remains safe against quantum attack. The precise picture, then, is this: quantum computers, when large enough, break today's public-key cryptography (the key exchanges and signatures) completely, and weaken symmetric cryptography manageably. It is the public-key layer that must be replaced, and replacing it is exactly what the post-quantum cryptography effort, standardized by NIST in 2024, is about — a story we will tell in Part X.

Finally, a word about the *timing* of the threat, because it is more urgent than the "millions of qubits, not yet built" framing suggests, and this is a point often missed. You might think that since no quantum computer can break RSA today, there is no need to worry until one exists. This is dangerously wrong, for a reason with a memorable name: **harvest now, decrypt later**. An adversary — a nation-state intelligence agency, say — does not need a quantum computer *today* to benefit from Shor's algorithm *tomorrow*. It need only record and store your encrypted communications now, patiently, in vast data warehouses, and wait. When a cryptographically relevant quantum computer eventually exists, the adversary can decrypt the entire hoard retroactively, reading today's secrets years or decades after they were sent. For information that must stay secret for a long time — state secrets, medical records, the design of weapons systems, anything with a shelf life measured in decades —

the quantum threat is therefore effectively *present tense*, even though the quantum computer is not. This is why governments and security-conscious organizations are already, urgently, migrating away from the vulnerable systems, not waiting for Q-Day but racing to beat it. The clock started ticking the day Shor published, and it has been ticking ever since.

Having felt the force of Shor's bombshell, we can now look inside it, gently, to see the machinery turn — because the way Shor's algorithm works is one of the most beautiful pieces of reasoning in all of computer science, and understanding even its outline demystifies both its power and its limits. You do not need advanced mathematics to grasp the shape of the idea; you need only follow how a problem about factoring is transformed, by a chain of clever substitutions, into a problem about detecting a repeating rhythm, which quantum interference is uniquely suited to solve.

Begin with the surprising first move, which belongs to old-fashioned number theory and has nothing quantum about it at all. Shor's algorithm does not attack the number to be factored directly. Instead it picks some other number at random and looks at its powers, taken with respect to the number you wish to factor — that is, you repeatedly multiply the random number by itself and keep only the remainder after dividing by the target. Here is the remarkable classical fact, known to mathematicians for centuries: this sequence of remainders always eventually REPEATS, cycling back on itself with some period. And — this is the pivot on which everything turns — once you know that period, a little further classical arithmetic (using the greatest-common-divisor calculation that Euclid described two thousand years ago) hands you, with high probability, a genuine factor of the original number. Factoring, astonishingly, reduces to finding the period of a repeating sequence. The whole problem has been transformed: crack the rhythm, and the number falls apart.

So the entire difficulty is concentrated into a single question:

what is the period of that sequence of remainders? Classically, finding this period is as hopeless as factoring itself — the sequence is astronomically long, and you would have to compute vast numbers of terms and wait for the repeat, which takes exponential time. This is precisely where the quantum computer enters, and where Simon's periodicity trick, scaled up and sharpened, does its work. The quantum machine prepares a superposition over an enormous range of exponents, computes all the corresponding remainders at once into an entangled register, and then applies the quantum Fourier transform — a quantum operation exquisitely designed to detect periodicity, the mathematical cousin of how a prism separates light into its component frequencies or how the ear picks out the pitch of a note from a jumble of sound. The Fourier transform makes the interference pattern sharply concentrate on values related to the hidden period, so that when you measure the register, the result you draw is a clue that points at the period. Repeat a few times, do a little classical cleanup, and the period — and hence the factors — emerges.

Notice the division of labor, because it is characteristic of nearly all real quantum algorithms and it corrects the fantasy of the all-powerful quantum machine. The quantum computer is not doing the whole job. It is doing exactly one thing that classical machines cannot: finding the period of a function, by orchestrating a global interference that senses the function's repeating rhythm all at once. Everything else — choosing the random number, the modular arithmetic setup, the greatest-common-divisor finish — is ordinary classical computation, done on an ordinary computer. Shor's algorithm is a classical scaffold with a single, decisive quantum step bolted into its center, and that step is period-finding via the quantum Fourier transform. This is why quantum computers help only with problems that can be recast in terms of some hidden structure — here, periodicity — that interference can detect. Strip away the beautiful reduction to period-finding and there would be no quantum advantage at all.

Now we can understand precisely why your secrets are, for the moment, still safe, and it comes down to a brutal accounting of resources. To factor a number large enough to matter — a 2,048-bit RSA key, say, the size guarding real transactions — Shor's algorithm needs to perform its delicate period-finding on a quantum register of thousands of qubits, executing millions upon millions of gate operations in sequence, every one of them precise, with the fragile superposition held intact throughout the entire marathon. Real qubits, as the coming chapters will lay bare in painful detail, are appallingly fragile; they lose their quantum character — decohere — in fractions of a second, long before millions of operations could complete, and every gate introduces error. Bridging the gap between the few dozen noisy qubits of today and the thousands or millions of reliable ones Shor's algorithm demands is the central engineering challenge of the field, and estimates for when it will be met range from a decade to never. The algorithm is proven; the machine to run it at scale is not built.

The division of labor within Shor's algorithm — a classical scaffold with a single decisive quantum step bolted into its center — deserves a final, emphatic word, because it is the key to understanding both the algorithm's power and the limits of quantum computing generally. The quantum computer is not doing the whole job of factoring. It is doing exactly one thing that classical machines cannot: finding the period of a function, by orchestrating a global interference that senses the function's repeating rhythm all at once, using the quantum Fourier transform. Everything else — choosing the random number, setting up the modular arithmetic, running Euclid's ancient greatest-common-divisor calculation to turn the period into a factor — is ordinary classical computation, performed on an ordinary computer. This hybrid structure, a quantum core wrapped in classical pre- and post-processing, is not peculiar to Shor's algorithm; it is the template for nearly every practical quantum algorithm. The quantum machine is a specialized co-processor summoned to perform the one impossible sub-task — surfacing hidden structure through interference — while classical computers do all the surrounding work. Understanding this is understanding why quantum computers help only with problems that can be recast in terms of some hidden structure that interference can detect, and why they are co-processors for particular hard kernels rather than replacements for the general-purpose machines on our desks.

The precise reason your secrets remain safe, for now, comes down to a brutal accounting of resources that is worth laying out concretely. To factor a number large enough to matter — a 2,048-bit RSA key, the size guarding real transactions — Shor's algorithm must perform its delicate period-finding on a quantum register of thousands of qubits, executing millions upon millions of gate operations in sequence, every one precise, with the fragile superposition held intact throughout the entire marathon. Real qubits, as the coming chapters will lay bare, decohere in fractions of a second and err with every gate. Bridging the gulf between the few dozen noisy qubits of today and the thousands or millions of reliable, error-corrected ones Shor's algorithm demands is the central engineering challenge of the entire field, and estimates for when it will be met range from a decade to never. The algorithm is proven and exact; the machine to run it at scale is not built, and building it is the hardest technological problem the field faces. This is the reassuring gap between the terror of Shor's mathematics and the safety of your present-day encryption.

The vivid contrast between promise and present poverty is worth ending on, because it captures the state of the art with a single unforgettable image. In 2001, a team at IBM and Stanford ran Shor's algorithm on a tiny quantum computer and successfully factored the number 15 into 3 times 5 — a result a kindergartner could beat with mental arithmetic, but a genuine proof that the algorithm runs on real quantum hardware. In the more than two decades since, despite enormous investment and effort, the largest numbers reliably factored by Shor's algorithm remain almost comically small, a testament to how staggeringly hard scaling up has proven. The gap between factoring 15 and factoring a 617-digit RSA key is the gap this book's second half is about — a gap measured in error rates and coherence times and the heroic, still-unfinished quest to protect quantum information from a universe that seems bent on destroying it. But Shor was not the only one to find treasure in the mid-1990s. While his algorithm threatened one precious and specific target, another researcher at the same Bell Labs found a quantum speedup of a humbler but far more universal kind — one that helps not with a single structured problem but with the vast, generic task of searching a haystack.

Chapter 25: Grover's Search

Two years after Shor upended the world with factoring, another mathematician at Bell Labs found the second great quantum algorithm, and it could hardly be more different in character. Where Shor's algorithm attacks a narrow, special problem — factoring — with a spectacular exponential speedup and world-shaking consequences, Lov Grover's algorithm attacks the most general problem imaginable — searching for a needle in a haystack — with a modest but *universally applicable* speedup and no menace at all. If Shor's is the field's terrifying specialist, Grover's is its friendly generalist. Together they mark out the two poles of what quantum computers are good for, and understanding the difference between them is essential to a clear-eyed view of the technology's promise.

The problem Grover solved in 1996 is called **unstructured search**, and it is as basic as computing gets. Suppose you have a huge collection of items — a database, a list, a space of possibilities — with no organization, no sorting, no structure to exploit, and you are looking for the one item that satisfies some condition: the name that matches a phone number in an unsorted directory, the input that makes a function output "yes," the key that decrypts a message. Classically, with no structure to guide you, there is nothing cleverer to do than check the items one by one until you find the right one. If there are a million items, you will need, on average, half a million checks, and in the worst case a million. The effort grows in direct proportion to the size of the haystack: search a list N long, expect to do about N checks. There is no classical shortcut for a truly unstructured search; the linear cost is a hard floor.

Grover's algorithm searches the same unstructured haystack of N items in about the *square root* of N steps. For a million items, that is about a thousand steps instead of half a million — a five-hundred-fold improvement. For a trillion items, it is about a million steps instead of half a trillion. This is a **quadratic** speedup: not the exponential magic of Shor, but a genuine and broadly useful acceleration, and, remarkably, it applies to *any* search with no structure to exploit, which is an enormous class of problems. Many important computational tasks, including a great many of the hard problems in the class NP, can be cast as searches through a space of candidate solutions, and Grover's algorithm speeds all of them up quadratically. It will not make an intractable problem tractable — a quadratic speedup on an exponentially large search space still leaves you with an exponentially large (if square-rooted) amount of work — but it is a real and general gift, and it has been proven optimal: no quantum algorithm can search an unstructured haystack faster than Grover's. The square root is the quantum limit.

The mechanism of Grover's algorithm is the most geometrically beautiful in the field, and it is a perfect illustration of interference at work, so it is worth picturing. Begin, as always, by putting the register into an equal superposition of all N possible items, every candidate present with the same small amplitude. One of these candidates is the target you seek; the rest are not. Grover's algorithm then repeats a simple two-step maneuver, over and over, that gradually shifts amplitude away from the wrong answers and onto the right one. The first step uses an oracle — a check that recognizes the target when it sees it — to flip the *phase* of the target's amplitude, marking it with a minus sign while leaving all the others alone. This does not yet change any measurement probabilities (recall that phase is invisible to a direct measurement), but it has secretly singled out the target. The second step performs a clever operation, sometimes called "inversion about the average," that reflects all the amplitudes around their mean value — and because the target's amplitude was flipped negative, this reflection has the effect of *increasing* the target's amplitude while slightly decreasing all the others. Each repetition of these two steps nudges more amplitude onto the target, like a pendulum swinging steadily toward the answer. After about the square root of N repetitions, the target's amplitude has grown so large — and all the others shrunk so small, by the destructive interference of the reflection — that a measurement returns the target with high probability. The wrong answers have been made to cancel; the right answer has been made to dominate; the four-beat structure of superpose, mark-with-phase, interfere, measure is executed as an elegant geometric rotation of the amplitude vector, step by step, toward the solution.

One subtlety of Grover's algorithm is worth savoring because it is so counterintuitive: you must not over-search. The amplitude swings toward the target like a pendulum, and if you keep applying the steps past the optimal number of repetitions, the amplitude swings *past* the target and back down again, and a measurement becomes *less* likely to find the answer. There is a precise, correct number of iterations — about the square root of N — at which the target's amplitude peaks, and you must stop there and measure. Continue, and you spoil it. This is unlike anything in classical search, where more effort never hurts; in the quantum version, the interference that concentrates amplitude on the answer will, if pushed too far, un-concentrate it. It is a small, vivid reminder that a quantum algorithm is a wave phenomenon, governed by the rhythms of constructive and destructive interference, and not a simple accumulation of work.

Grover's algorithm has one further significance for the themes of this book: it is the reason symmetric cryptography must take modest account of quantum computers, though it need not fear them. Breaking a symmetric cipher like AES by brute force means searching the space of all possible keys for the right one — an unstructured search, exactly Grover's domain. Grover's quadratic speedup means a quantum computer could brute- force a symmetric key in about the square root of the classical time, which effectively *halves* the security level: a 128-bit key offers only 64 bits of protection against a quantum attacker, which is uncomfortably weak, while a 256-bit key still offers a robust 128 bits and remains secure. The fix is simply to use longer keys, which is cheap and easy, and it is why AES-256 is considered quantum-safe while AES-128 is being phased out for long-term security. Grover threatens to weaken symmetric encryption, but only by a factor that a longer key comfortably absorbs — in stark contrast to Shor, which does not merely weaken but utterly destroys the public-key systems it targets. The two algorithms together define the precise shape of the quantum threat to cryptography: catastrophic for public keys, manageable for symmetric ones.

Two years after Shor's bombshell, in 1996, another researcher at Bell Labs — Lov Grover — discovered the second of the two foundational quantum algorithms, and where Shor's was a rapier aimed at one precious target, Grover's is a broadsword useful against a whole class of everyday problems. Grover's algorithm speeds up SEARCH: the generic task of finding a needle in a haystack, of locating one item that satisfies some condition among a vast collection of possibilities. Because searching is so ubiquitous — it lurks inside countless computational problems — Grover's algorithm has a breadth of application that Shor's, for all its drama, does not.

Consider the purest version of the problem. You have an

unstructured list of N items — unstructured meaning there is no order, no index, no clever way to narrow the hunt, like a phone book whose entries have been scrambled into random order. Somewhere in it is the one entry you seek, and your only tool is the ability to check any single item and ask "is this the one?" How many checks do you need? Classically, the answer is grim and obvious: on average about N/2, and in the worst case all N. With no structure to exploit, you can do no better than to examine the items one by one until you stumble on the target; a list of a million scrambled entries demands, on average, half a million checks. There is no shortcut, because there is nothing to be clever about — the haystack is genuinely featureless.

Grover's algorithm finds the target in roughly the SQUARE ROOT of N steps. For a million items, that is about a thousand checks instead of half a million — a five-hundred-fold speedup; for a trillion items, about a million steps instead of half a trillion. This is a "quadratic" speedup, dramatically less than Shor's exponential leap but applicable to an enormously wider range of problems, since so many tasks reduce to searching an unstructured space of possibilities. The mechanism is a lovely piece of geometry and gives the clearest possible picture of quantum interference at work. The algorithm begins, as always, by placing the register in an equal superposition of all N possibilities at once, each with a small, equal amplitude. Then it repeats a two-step maneuver — often called the Grover iteration — that acts like a pump. First it invisibly "marks" the target by flipping the sign of its amplitude (a phase kick delivered by the checking function). Then it performs an operation called "inversion about the average" that, in effect, reflects all the amplitudes about their mean value. The combined effect of these two reflections is to nudge a little more amplitude onto the target and pull a little off the rest, with each iteration. Repeat the pump about the square root of N times and the target's amplitude has grown until it dominates — whereupon you measure, and out pops the needle with high probability.

The image of two reflections steadily rotating the state toward the answer captures something essential and often misunderstood about Grover: it is NOT the parallel-universe fantasy of checking all items at once and reading off the winner. If it were, search would take a single step, not the square root of N. Instead the algorithm slowly, iteratively amplifies the target's amplitude through repeated interference, and you must run it just the right number of times — and here is a delightful and counterintuitive fact — because if you run it too MANY times, you overshoot, the amplitude begins to swing back down, and your probability of success DECREASES. The optimal number of iterations is precisely tuned; more is not better. This is unlike any classical search, where continuing to look can only help. In Grover's algorithm the state rotates like a hand sweeping around a clock face toward the answer, and if you keep pumping past the target you sail right by it and out the other side.

Remarkably, Grover's quadratic speedup was later proven OPTIMAL: no quantum algorithm can search a truly unstructured space faster than the square root of N. This is a striking and clarifying result, because it draws a firm line around quantum power. For unstructured problems — problems with no hidden regularity for interference to lock onto — the square-root speedup is the absolute best a quantum computer can ever do, and it can do no better because there is simply no structure to exploit beyond that. This is the deep reason quantum computers are not magic: their dramatic exponential advantages, like Shor's, come only from special problems with rich hidden structure, while for generic search they offer a real but merely quadratic gain, and for many everyday tasks nothing at all. Grover marks the ceiling of the unstructured world.

The optimality of Grover's speedup deserves a final, emphatic reflection, because it draws one of the sharpest and most clarifying lines around what quantum computers can do. It was later proven — rigorously, as a theorem — that no quantum algorithm can search a truly unstructured space faster than the square root of N. Grover's algorithm is not merely a good approach to unstructured search; it is the *best possible* approach, and no cleverer quantum algorithm will ever beat it, because there is simply no structure in the haystack to exploit beyond what Grover already uses. This is a profoundly important result, and it should be held alongside Shor's exponential triumph as the two poles of quantum capability. Shor achieves an exponential speedup, but only because factoring has rich hidden structure — periodicity — that quantum interference can lock onto. Grover achieves merely a quadratic speedup, but it applies to the vast class of unstructured problems precisely because it demands no structure at all. And the quadratic speedup is a hard ceiling for such problems: where there is no pattern, there is no exponential magic, only the square-root gain and no more. This is the deepest reason quantum computers are not universal accelerators — their dramatic powers are confined to structured problems, and for the generic, patternless tasks that make up so much of computing, the best they offer is Grover's modest square root.

The practical reach of that quadratic speedup deserves a clear-eyed accounting, because it is both broad and, on sober inspection, limited. Because so many hard problems reduce to searching a large space of possibilities — database lookup, constraint satisfaction, breaking certain codes by brute force, solving puzzles by trial and error — Grover's algorithm offers a general-purpose acceleration across an enormous front, and in that breadth it is more widely applicable than Shor's narrow rapier. But a quadratic speedup, while genuine, is often not enough to matter in practice, and honesty requires saying so. The reason is overhead: a quantum computer's individual operations are far slower and vastly more expensive than a classical computer's — the slow gates, the error correction, the refrigeration, the sheer cost — so a mere square-root reduction in the number of operations can easily be swamped by the enormous per-operation penalty, leaving a well-tuned classical machine to finish first anyway. For Grover's speedup to be worth the trouble, the problem must be large enough that halving the exponent of an already-astronomical search still leaves a worthwhile gain after paying the quantum overhead. This is a real but demanding condition, and it means Grover's algorithm, for all its elegance and generality, will transform far fewer practical problems than its breadth might suggest.

Its most concrete and certain impact is on cryptography, and it is worth stating precisely because it contrasts so sharply with Shor's. Grover's algorithm halves the effective strength of symmetric encryption keys — the kind used by standards like AES to do the bulk of data encryption once a connection is established — so that a 128-bit key offers only about 64 bits of protection against a quantum adversary. But this is a gentle, easily remedied threat: cryptographers simply recommend doubling key lengths to 256 bits, restoring full security at trivial cost. Symmetric cryptography survives the quantum era essentially intact, needing only longer keys, whereas public-key cryptography faces total destruction from Shor. Together, Shor and Grover define the two archetypes of quantum advantage — the rare, spectacular, exponential speedup for specially structured problems, and the broad, moderate, quadratic speedup for generic search — and between them they mark out most of what quantum computers are genuinely good for. With the two great algorithms in hand, the field had proven, beyond doubt, that the machine would be worth building. Now came the vindication of Feynman's original dream, and the far harder question of whether such a machine could actually be made to exist against the relentless fragility of the quantum world.

Chapter 26: Lloyd and the Vindication of Feynman

The two algorithms that made quantum computing famous — Shor's and Grover's — both concern classical problems:

factoring numbers, searching lists. It is a curious irony that the application Feynman had originally proposed, the one that started the whole enterprise, received its rigorous foundation only after the code-breaking and searching results had stolen the spotlight. Feynman had *conjectured* in 1981 that a quantum computer could simulate quantum physics efficiently, fighting the exponential explosion of quantum states with a machine that had the same exponential resources on its side. But he had not proved it. He had not shown that a general, programmable quantum computer could be instructed to mimic *any* quantum system, or that it could do so with the modest, non-exponential overhead that would make the simulation worthwhile. That proof came in 1996 from Seth Lloyd, and it closed the theoretical circle, vindicating the founding vision and establishing quantum simulation as the application many experts still consider the most likely to deliver real value first.

Seth Lloyd is one of the more colorful figures in the field, a professor of mechanical engineering at MIT with a physicist's mind and a philosopher's taste for the largest questions — he has written, only half- provocatively, that the universe itself is a quantum computer, ceaselessly computing its own next state. In 1996 he provided the missing rigor behind Feynman's dream. Lloyd proved that a universal quantum computer — a programmable, general-purpose machine of the kind Deutsch had defined — can efficiently simulate a broad and important class of quantum systems: those whose parts interact through *local* forces, each piece influencing only its neighbors, which describes an enormous range of real physical systems, from molecules to magnetic materials to the exotic matter studied in physics laboratories. He showed how to break the continuous, tangled evolution of such a quantum system into a sequence of small, simple quantum gates that a quantum computer can execute, with the number of gates growing only modestly — polynomially — with the size of the system and the accuracy demanded. The exponential wall that defeats classical simulation simply does not arise, because the quantum computer represents the quantum system directly, qubit for particle, entanglement for entanglement, rather than trying to write out the exponentially long classical description. Feynman's fight-fire-with-fire intuition was not just a nice idea; it was a theorem.

The importance of this result for the practical future of quantum computing is hard to overstate, and it is worth being concrete about why quantum simulation matters so much. Recall the simulation problem from Part II: chemists and materials scientists cannot exactly compute the behavior of molecules and materials because the quantum behavior of their electrons explodes exponentially beyond classical reach. This is not an academic frustration; it is a bottleneck on real-world progress. The design of new drugs depends on predicting how candidate molecules bind to proteins, a quantum problem. The design of better batteries, more efficient solar cells, high-temperature superconductors, and novel catalysts depends on understanding the quantum behavior of electrons in materials. The industrial process that fixes nitrogen from the air into fertilizer — the Haber–Bosch process, which feeds a large fraction of humanity — consumes something like one to two percent of the world's total energy, in part because we cannot design a better catalyst, because we cannot accurately simulate the relevant quantum chemistry. Nature performs these computations effortlessly in every leaf and every enzyme; a quantum computer, Lloyd showed, could in principle perform them too, opening the door to rational design of molecules and materials that classical computation cannot reach. Of all the promised applications of quantum computing, this — the simulation of quantum chemistry and materials — is the one that most experts regard as both the most valuable and the most likely to arrive, because it plays directly to the machine's deepest strength and, unlike code-breaking, requires fewer and less perfect qubits to begin yielding useful results.

There is a satisfying intellectual symmetry in Lloyd's result closing the circle that Feynman opened. The entire field began with the recognition that the quantum world is too rich for classical computers to simulate — that its exponential vastness overwhelms any classical machine. Feynman turned that liability into an opportunity: build the computer out of the same quantum stuff, and the vastness becomes a resource rather than an obstacle. Lloyd proved the opportunity was real and general. And so the quantum computer's most natural application turns out to be the very thing that motivated its invention: understanding quantum nature herself. There is something fitting in the idea that the first truly useful thing these machines do may be to help us design medicines and materials by simulating the molecular quantum mechanics that governs them — that the strangeness of the quantum world, having been harnessed into a machine, is turned back upon the quantum world to comprehend it. The snake, elegantly, bites its tail.

Amid the excitement over Shor and Grover — algorithms that broke codes and searched haystacks — it was easy to forget that quantum computing had been born from a different dream entirely. Feynman's original vision at Endicott House was not about cracking encryption; it was about simulating nature. He wanted a machine that could imitate quantum systems too complex for any classical computer, and thereby unlock the secrets of molecules and materials. For over a decade this founding purpose remained more slogan than science: Feynman had gestured at a universal quantum simulator but never shown, rigorously and generally, that one could be built or how it would work. In 1996, Seth Lloyd supplied the missing proof, and in doing so vindicated the very idea that had launched the field.

Lloyd — a physicist of famously wide-ranging imagination, who would later write a book arguing the universe itself is a giant quantum computer — proved that a quantum computer could efficiently simulate any system governed by local quantum interactions, which is to say essentially any real physical system: molecules, materials, exotic states of matter. He showed explicitly how to take the laws governing such a system and break their continuous evolution into a sequence of small, discrete quantum gate operations that a programmable quantum computer could execute, step by step, approximating the true dynamics as closely as desired. The trick, drawing on a mathematical result called the Trotter decomposition, was to slice time into tiny intervals and, in each slice, apply the various interactions one after another in quick succession; do this finely enough and the piecewise imitation converges on the real thing. Feynman's handwaving universal simulator was now a concrete, general algorithm with provable guarantees. The founding promise was redeemed.

This matters enormously because quantum simulation, not code-breaking, is very likely to be the FIRST genuinely useful, world-changing application of quantum computers — the one that pays for all the rest. The reasons are practical. Shor's algorithm demands enormous, near-perfect machines to threaten real encryption, and may be decades away. But simulating a modestly sized quantum system — a tricky molecule, a novel magnetic material — is far more forgiving: it needs fewer qubits, tolerates more noise, and even imperfect, near-term machines may deliver scientifically valuable results. And the payoff is immense. Recall the exponential wall: classical computers choke on the quantum behavior of entangled electrons, which is exactly why chemistry and materials science lean on crude approximations. A quantum simulator speaks nature's own language and could, in principle, predict molecular behavior with an accuracy classical methods can never reach.

Consider the concrete prizes dangling just out of reach. There is nitrogen fixation: the natural enzyme nitrogenase pulls nitrogen from the air and turns it into fertilizer at ordinary temperature and pressure, while our industrial Haber-Bosch process, which feeds roughly half the human race, must crush and heat the air at vast energy cost, consuming a notable slice of the world's natural gas. Nobody fully understands how the enzyme does it, because its active site is a knot of entangled electrons beyond classical simulation — but perhaps within reach of a quantum one, and cracking it could revolutionize how the planet feeds itself. There is high-temperature superconductivity, a decades-old mystery whose solution could remake power grids and transport. There is drug design, where predicting how a candidate molecule binds a target protein could compress years of laboratory trial and error into hours of computation. There is the search for better batteries, catalysts, and materials of every kind. All of these are, at bottom, problems of simulating entangled quantum matter — the very task Feynman said demanded a quantum machine.

It is worth dispelling a misconception that inflates expectations. Quantum simulation does not mean the computer produces a little movie of atoms bouncing around that you watch and understand. It means the machine is prepared to embody the same quantum state as the system of interest, evolved under the same rules, so that measuring the machine yields the quantities you want — energies, reaction rates, magnetic properties — that are otherwise incalculable. And there are real caveats: reading useful numbers out of the simulation efficiently is its own challenge, and preparing the right starting state can be hard. Quantum simulation is not an oracle; it is a powerful but delicate instrument. Still, of all the applications, it is the one that most experts believe will justify the entire enterprise, and the one closest to practical value.

The distinction between what Feynman *envisioned* and what Lloyd *proved* is worth drawing sharply, because it illustrates how a beautiful intuition becomes solid science. Feynman, in 1981, had gestured at a universal quantum simulator — a controllable quantum system that could be tuned to imitate any other quantum system — but he had left the idea hazy, offering no general construction and even doubting whether a fully universal version was possible. For fifteen years the founding purpose of the field remained more slogan than theorem. Lloyd's 1996 achievement was to make it rigorous and general: he proved explicitly that a programmable quantum computer could efficiently simulate any system governed by local quantum interactions — essentially any real physical system — and he showed *how*, by slicing the continuous evolution into a sequence of small discrete quantum gates (using a mathematical result called the Trotter decomposition) that a gate-based machine could execute step by step, approximating the true dynamics as closely as desired. Feynman's evocative hand-waving became a concrete algorithm with provable guarantees. The dream had been given a spine.

The reason quantum simulation is so widely regarded as the *first* application likely to deliver genuine, world-changing value — ahead of code-breaking, ahead of optimization — deserves emphasis, because it inverts the popular ranking. Code-breaking with Shor's algorithm is the most dramatic application, but it demands enormous, near-perfect machines to threaten real encryption, and may be decades away. Quantum simulation is far more forgiving: simulating a modestly sized quantum system — a tricky molecule, a novel material — needs fewer qubits, tolerates more noise, and even imperfect, near-term machines may deliver scientifically valuable results long before a code-breaking machine exists. And the payoff is immense, because the exponential wall that defeats classical simulation of entangled electrons is exactly why chemistry and materials science lean on crude approximations that fail for the hardest, most interesting systems. A quantum simulator speaks nature's own language and could predict molecular behavior with an accuracy classical methods can never reach. Of all the promised applications, this is the one closest to practical value and the one most experts believe will justify the entire enterprise.

There is a satisfying symmetry in the field's arc that Lloyd's proof completes, and it is worth naming as we prepare to leave the theoretical foundations behind. Quantum computing was conceived, by Feynman, as a tool to simulate nature;

it detoured, thrillingly, through Shor's assault on cryptography, which drew the money and the headlines and made the field a matter of national security; and it is now circling back to its origin, with simulation regarded as the likeliest first triumph. The machine born to understand molecules may earn its keep by understanding molecules after all. But every one of these dreams — simulation, factoring, search — has so far rested on an assumption we were allowed to make freely and must now confront: that the qubits will do as they are told, holding their delicate superpositions steady through long computations. They will not. Real quantum states are almost unimaginably fragile, and this fragility very nearly killed quantum computing in its cradle. To understand why building the machine has been so agonizing, we must look squarely at the enemy the engineers have battled for thirty years.

Part VI — Taming the Noise

Chapter 27: The Fragility Problem

In the mid-1990s, just as Shor's algorithm was drawing brilliant minds and serious money into quantum computing, a chorus of doubt arose from some of the most distinguished physicists alive, and the doubt was grave enough that it might have strangled the field in its cradle. The objection was not about the algorithms, which were plainly correct, nor about the theory, which was sound. It was about the physics of building the thing, and it went to the heart of what makes quantum states powerful: their fragility. Quantum computing, the skeptics argued, was a beautiful fantasy that could never survive contact with the real, warm, noisy world, because the quantum states it depends on are so delicate that they cannot be maintained long enough, or precisely enough, to compute anything useful. This chapter lays out that objection in its full force, because only by feeling how devastating it seemed can one appreciate the miracle of its solution.

The enemy has a name: **decoherence**. We met it briefly in Part I, but now it moves to center stage. Decoherence is the process by which a quantum system, through its unavoidable interactions with its surroundings, loses the delicate superpositions and phase relationships that make it quantum, and degrades into something ordinary and classical. Recall that a qubit's power lies partly in its *phase* — the hidden longitude on the Bloch sphere, the relative alignment of the wave-like amplitudes that makes interference possible. This phase is exquisitely sensitive. Every time a stray photon bounces off the qubit, every time a whisper of heat jostles it, every time a faint magnetic field wavers nearby, the environment "measures" the qubit a little, and each such interaction blurs its phase and nudges its superposition toward collapse. A qubit in isolation might hold its quantum state indefinitely, but perfect isolation is impossible; the qubit must be built of real matter sitting in a real world, coupled however weakly to a vast environment of jostling particles and fields, and that coupling relentlessly leaks the qubit's quantum information out into the surroundings, where it is lost. The characteristic time over which a qubit retains its coherence — its **coherence time** — is, in real hardware, achingly short: microseconds to milliseconds, in the machines of the era, a flicker before the quantum state dissolves into classical mush.

Now weigh that against what the algorithms demand. Shor's algorithm, run on a cryptographically meaningful number, requires *billions* of individual gate operations performed in sequence, each one on qubits that must remain coherent throughout, their superpositions and entanglements intact from the first gate to the last. The mismatch is staggering. If a qubit loses its coherence after a few thousand operations — and early qubits were far worse than that — then a computation requiring billions of operations has no hope of completing before the quantum state has decohered into noise and the answer has turned to garbage. It is like trying to carry out a symphony on instruments that fall out of tune within the first few bars, or to build a cathedral of cards in a gale. Worse, quantum errors are subtler and more various than classical ones. A classical bit can only suffer one kind of error: a flip, 0 becoming 1 or vice versa. A qubit, living on the continuous surface of the Bloch sphere, can drift in infinitely many ways — its phase can slip by any amount, its superposition can tilt in any direction, errors can be small and continuous rather than discrete. The space of possible quantum errors is vastly richer than the space of classical ones, and it seemed impossible to catch and correct such a continuum of subtle deviations.

And here the objection sharpened into something that looked like a proof of impossibility, drawing on the very principles established earlier in this book. In classical computing, the basic defense against errors is *redundancy through copying*: store each bit in triplicate, and if one copy is corrupted, a majority vote of the three recovers the correct value. It is simple, robust, and universal. But this defense is *forbidden* in the quantum world, by the **no-cloning theorem** we met in Part IV. You cannot copy an unknown quantum state; you cannot make the backup copies that classical error correction relies on. Worse still, the standard way to check for errors is to *look* — to measure the system and see if it has gone wrong. But measurement, in the quantum world, is catastrophic: measuring a qubit collapses its superposition, destroying the very quantum information you were trying to protect. So the quantum error-corrector seemed trapped between two impossibilities. It could not copy the information to create redundancy, because of no-cloning. And it could not look at the information to check for errors, because measurement destroys superposition. Redundancy forbidden, inspection forbidden — the two pillars of all classical error correction knocked away. It seemed to follow, almost inescapably, that quantum information simply could not be protected, and that quantum computers of any useful scale were therefore physically impossible, however sound the theory. Serious physicists said so in print. The great experimentalist and theorist Rolf Landauer, whom we met earlier, was among the prominent skeptics, insisting that quantum computing enthusiasts were ignoring the brutal realities of noise. For a moment in the mid-1990s it looked as though the whole edifice might be a castle in the air.

For all the elegance of the algorithms, quantum computing came within a hair of being still-born, killed by a single brutal fact of physical reality: quantum states are almost inconceivably fragile. The very features that give a quantum computer its power — delicate superpositions, sprawling entanglement, precisely tuned phases — are exquisitely sensitive to the slightest disturbance from the outside world, and they fall apart, blur into meaninglessness, almost the instant they are created. This process is called decoherence, and in the mid-1990s many serious physicists believed it was an insurmountable barrier, that quantum computers were a beautiful theoretical fantasy that the laws of thermodynamics would never permit anyone to build. Understanding decoherence is understanding why the second half of this book — the story of the machine — is a thirty-year war against a relentless enemy.

The core problem is that a qubit, to compute, must be simultaneously two contradictory things: perfectly isolated from its environment, so that its fragile superposition is not disturbed, and yet perfectly accessible to the outside world, so that you can prepare it, manipulate it with gates, and measure it. These demands are in direct opposition. Any coupling to the environment strong enough to let you control the qubit is also a channel through which the environment can disturb the qubit, and the environment is vast, warm, and endlessly meddlesome. A stray photon, a whisper of heat, a faint magnetic field, a vibration, a cosmic ray — any of these can interact with the qubit, and when it does, information about the qubit's state leaks out into the environment. And here is the fatal subtlety: the moment information about a superposition escapes into the surroundings, the superposition is effectively measured by the environment and collapses, just as surely as if a physicist had looked. The environment is always watching, and its watching destroys the very quantum-ness the computer needs.

To feel the magnitudes, consider that a superposition is like a soap bubble of coherent phase relationships, and decoherence is the world reaching in to pop it. Real qubits maintain their coherence for astonishingly short times — microseconds to at best seconds, depending on the technology — and every operation you perform, every gate you apply, must be completed within that fleeting window, before the bubble bursts. Worse, each gate is itself imperfect, nudging the qubit slightly off its intended state, and these small errors accumulate. A useful computation like Shor's algorithm on a real encryption key would require millions or billions of gate operations executed in sequence with the superposition held intact throughout — a demand that, against raw decoherence times, looks simply impossible, like asking someone to carry an overflowing cup of water at a dead sprint across a stadium without spilling a drop, and to do it a million times in a row.

There is a second, even nastier feature that makes quantum errors seem hopeless, and it is worth stating because it makes the eventual solution all the more miraculous. Classical bits suffer errors too — a 0 occasionally flips to a 1 — but classical errors are simple and discrete, and easy to fix by brute redundancy: store each bit three times, and if one copy disagrees, take the majority vote. Quantum errors are far more insidious. A qubit's state is a continuous point on the Bloch sphere, so errors are continuous too — the state can drift by any small angle, in any direction, a little bit-flip here, a little phase-flip there, in infinitely many shades. You cannot simply take a majority vote of continuous quantities. And worse still, the obvious classical fix — copy the qubit three times — is flatly forbidden by the no-cloning theorem, which proves that an unknown quantum state cannot be duplicated. You cannot make backup copies of a qubit. The two pillars of classical error correction, redundancy by copying and majority voting, are both unavailable. It seemed you could not protect quantum information at all.

The mood in the mid-1990s was, accordingly, one of deep pessimism among many hard-headed physicists, and their skepticism was not foolish — it was grounded in the most reliable physics we have. Rolf Landauer, the same IBM thinker who proved information is physical, was a prominent and persistent skeptic, insisting that quantum computer proposals airily ignored the noise and imperfection that would swamp any real device. He half-jokingly suggested that every quantum computing paper should carry a warning label noting that it had not been shown to work in the presence of realistic errors. The objection had teeth. Superposition looked too delicate, decoherence too relentless, quantum errors too continuous and too un-copyable, for any machine to survive a computation of useful length. The algorithms of Shor and Grover, dazzling on paper, threatened to be forever unrealizable, cathedrals built on quicksand.

It is worth dwelling on the exquisite cruelty of the trade-off at the heart of the fragility problem, because it is not a mere engineering inconvenience but something close to a fundamental tension in the nature of quantum information. A qubit, to be useful, must be two contradictory things at once: perfectly isolated from its environment, so its fragile superposition survives; and perfectly accessible to the outside world, so you can prepare it, manipulate it, and read it out. These demands pull in opposite directions. Any coupling to the environment strong enough to let you control the qubit is also a channel through which the environment can disturb the qubit — and the environment is vast, warm, and endlessly meddlesome. You cannot have a qubit that listens to you but not to the universe; the same door that admits your commands admits the noise. Managing this tension — building qubits isolated enough to stay coherent yet connected enough to compute — is the central engineering struggle of the entire field, and it is why a quantum computer is so much harder to build than its classical cousin, whose bits happily ignore the world.

The insidious subtlety that the environment's mere *observation* destroys a superposition deserves careful statement, because it connects the fragility problem directly to the deepest features of quantum mechanics. The moment information about a superposition leaks into the surroundings — the instant a stray photon or a whisper of heat carries away a hint of which state the qubit is in — the superposition is effectively *measured* by the environment and collapses, just as surely as if a physicist had deliberately looked. The environment is always watching, and its watching is a kind of measurement, and measurement destroys the very quantumness the computer needs. This is why decoherence is not merely a matter of energy leaking away like a cooling cup of coffee; it is the continuous, unavoidable measurement of the qubit by an ever-present, ever-curious world. The superposition is a secret that the universe is constantly, inadvertently, discovering — and each discovery pops the bubble.

The difficulty of quantum errors compared with classical ones is worth restating, because it makes the eventual solution — quantum error correction — seem almost miraculous. Classical bits suffer errors too, but classical errors are simple, discrete, and easy to fix by brute redundancy: store each bit three times and take a majority vote. Quantum errors are far nastier on two counts. First, a qubit's state is a continuous point on the Bloch sphere, so errors are continuous too — the state can drift by any small angle in any direction, in infinitely many shades — and you cannot take a majority vote of continuous quantities. Second, the obvious classical fix, copying the qubit, is flatly forbidden by the no-cloning theorem. Both pillars of classical error correction — redundancy by copying, and majority voting — are unavailable, and for a time it seemed you could not protect quantum information at all. This double impossibility was the ground of the deep pessimism that gripped many hard-headed physicists in the mid-1990s, Landauer foremost among them, who suspected quantum computers were a beautiful fantasy that thermodynamics would never permit. That they were proven wrong — that quantum information could, after all, be shielded against a relentless, watching universe — is the great turnaround to which we now turn, and it began with the same man who had lit the fuse under cryptography.

Chapter 28: Shor's Second Miracle

Peter Shor had already given quantum computing its reason for existing with the factoring algorithm. In 1995, a year later, he gave it its reason for being *possible*, with a discovery that many physicists consider even more astonishing: the first **quantum error-correcting code**. If the factoring algorithm was a bombshell, the error-correcting code was a rescue, snatching the whole field back from the brink of the impossibility argument laid out in the previous chapter. That the same person did both, within twelve months, is one of the more remarkable one-two punches in the history of science. Shor showed that quantum information, though it cannot be copied and cannot be inspected, can nonetheless be protected — and in doing so he dissolved the objection that had threatened to kill quantum computing at birth.

The trick — and it is a genuinely beautiful trick, worth trying to grasp even in outline — is to sidestep both forbidden operations at once, by cleverly redefining what you protect and how you check it. Recall the two apparent impossibilities: you cannot copy the quantum state (no-cloning), and you cannot measure it to check for errors (measurement destroys superposition). Shor's insight was that you do not need to do either. Instead of copying a single qubit's state into identical backups, you *spread* the state of one "logical" qubit across several physical qubits in an *entangled* way — not making copies, which is forbidden, but distributing the information so that it lives in the correlations among many qubits rather than in any one of them. The precious quantum information is now delocalized, smeared across the whole cluster, present in the pattern of entanglement and nowhere in particular. This does not violate no-cloning, because no qubit holds a copy of the state; the state exists only in the joint condition of the group.

Then comes the second half of the trick, which evades the destructiveness of measurement. To check for errors, you do *not* measure the qubits to find out what state they are in — that would collapse the superposition and destroy the information. Instead you measure something cleverly indirect: the *relationships between* qubits, the parities, the answers to questions like "do these two qubits currently agree or disagree?" — questions whose answers reveal *whether an error has occurred and where,* without revealing anything about the actual value of the encoded data. These carefully chosen measurements are called **syndrome measurements**, and they are the heart of the method. A syndrome measurement is like a doctor detecting that a bone is broken, and even *which* bone, from an X-ray, without needing to know the patient's thoughts: it diagnoses the error without reading the protected information. Because the syndrome measurement asks only about relationships and parities, not about the data itself, it does not collapse the encoded superposition; the logical qubit sails on, its quantum state intact, while the error is diagnosed on the side. Learn from the syndrome that a particular physical qubit has flipped, and you can flip it back with a corrective operation, restoring the encoded state — all without ever having looked at the information you were guarding.

There remained the problem that quantum errors form a *continuum* — a qubit can drift by any amount in any direction, not merely flip discretely as a classical bit does — and it seemed hopeless to correct a continuous infinity of possible faults with a finite set of operations. Here Shor's scheme revealed a piece of quantum magic so elegant it feels like a conjuring trick. The very act of performing a syndrome measurement *discretizes* the error. A qubit that has drifted continuously into some slight superposition of "flipped" and "not flipped" is, by the syndrome measurement, forced to collapse into one definite alternative or the other — either "a flip happened here" or "no flip happened here" — because measurement always yields a discrete outcome. The continuous drift is projected, by the measurement itself, onto a discrete set of standard errors, each of which can then be corrected by a standard operation. You do not need to correct the infinite continuum of possible quantum errors; you need only correct a small finite set of standard error types, because the syndrome measurement collapses every continuous error onto one of them. This is the deep reason quantum error correction works at all, and it turns the continuity of quantum errors — which had seemed an insurmountable obstacle — from a liability into a non-problem. The measurement that diagnoses the error also tames it.

In 1995, one year after his factoring bombshell, Peter Shor performed a second feat that is, in the judgment of many physicists, even more important than the first — for while the factoring algorithm gave quantum computing its purpose, this second discovery gave it a future. Shor showed that quantum error correction is possible: that the fragile, un-copyable, continuously-erring quantum state can, in fact, be protected against the ravages of decoherence. It was a result so contrary to the prevailing pessimism that it changed the field's destiny overnight, converting quantum computing from a probable impossibility into a hard but achievable engineering goal.

The obstacles seemed insurmountable, and Shor had to defeat all of them at once. The no-cloning theorem forbids copying a qubit, so classical redundancy is out. Quantum errors are continuous, not discrete, so majority voting seems inapplicable. And — this is the subtlest trap — you cannot check a qubit for errors by looking at it, because looking is a measurement that collapses the very superposition you are trying to protect. To correct an error you must first detect it, but detection seems to require measurement, and measurement seems to destroy the information. It looked like a closed circle with no way in. Shor found the way in, and his solution is a masterpiece of lateral thinking.

His first insight was to spread the information of a single "logical" qubit across several physical qubits using entanglement, so that the quantum information is stored not in any one qubit but in the correlations among many. Because it is encoded in the collective, entangled state rather than in individual qubits, this respects no-cloning — you are not copying the state, you are delocalizing it, smearing one qubit's worth of information across nine physical qubits in Shor's original code — yet it creates the redundancy needed for protection. His second and more dazzling insight solved the measurement trap. You do not measure the qubits themselves; instead you make careful, cleverly designed measurements of RELATIONSHIPS between qubits — asking, for instance, "do these two qubits agree or disagree?" without asking what either one actually is. Such a measurement reveals whether an error has occurred and where, while revealing nothing about the protected information itself, so the superposition survives the check. These relationship-measurements are called error syndromes, and they are the heart of the trick: they extract information about the error without extracting information about the data.

The third insight disposed of the continuous-error nightmare with breathtaking economy. Quantum errors form a continuum — a qubit can drift by any small angle in any direction — and you cannot correct a continuum by discrete means. But Shor showed that the act of measuring the error syndrome DISCRETIZES the error: the measurement forces the continuous, blurry error to collapse into one of a few definite, discrete possibilities — essentially a bit-flip, a phase-flip, both, or neither — which can then be corrected by a simple, exact counter-operation. This is a genuinely astonishing turn. The measurement that detects the error also, in the same stroke, snaps the infinitely various error into one of a handful of clean, fixable cases. It is as if the mere act of asking "which way did it go wrong?" forced a hopelessly smeared mistake to choose a single definite answer, which you could then simply reverse. The continuous enemy was tamed by being made, at the moment of inspection, discrete.

Put together, Shor's nine-qubit code took one precious logical qubit, encoded it across nine physical qubits, and could detect and correct any single error on any of them — bit-flip, phase-flip, or any combination — all without ever disturbing the protected quantum information. The circle that had looked closed was open after all. Quantum information could be shielded. The impossibility was not an impossibility. One can hardly overstate the psychological effect on the field: the single most powerful argument of the skeptics, the fragility that Landauer and others had wielded as a likely death sentence, had been answered, and answered decisively, by the very man who had made the machine worth building.

It is worth dwelling on why Shor's 1995 error-correction result is regarded by many physicists as even more important than his famous factoring algorithm, because the judgment is counterintuitive and revealing. The factoring algorithm gave quantum computing its *purpose* — a dazzling application that drew money, attention, and national-security concern. But the factoring algorithm would have been worthless, a permanent curiosity, if quantum states could not be protected against decoherence, and in 1995 many serious physicists believed they could not — that the fragility of superposition was an insurmountable barrier fixed by the laws of thermodynamics. Shor's error-correction result gave quantum computing its *future*, proving that the fatal fragility could be overcome. An algorithm without a machine to run it is a mathematical amusement; Shor's second miracle showed that the machine was, in principle, buildable. The purpose was thrilling, but the future was essential, and the future was what the error-correction result secured.

The three obstacles Shor had to defeat simultaneously are worth restating, because their simultaneous defeat is what makes the achievement so remarkable. First, the no-cloning theorem forbids copying an unknown qubit, so the classical trick of storing three copies and taking a majority vote is unavailable. Second, quantum errors are *continuous* — a qubit's state can drift by any small angle in any direction — so there seems to be nothing discrete to correct. Third, and most insidiously, you cannot check a qubit for errors by looking at it, because looking is a measurement that collapses the very superposition you are trying to protect. Each obstacle alone looked fatal; together they seemed to prove that quantum information simply could not be protected. Shor defeated all three at once with a solution of astonishing ingenuity: spread the information across many entangled qubits (respecting no-cloning, since the state is delocalized rather than copied); measure cleverly-designed *relationships* between qubits — error syndromes — that reveal whether an error occurred without revealing the protected data; and let those syndrome measurements *discretize* the continuous error, snapping the infinitely various drift into one of a few definite, correctable cases.

The discretization of continuous errors is the subtlest and most beautiful part, and it deserves a second pass because it is so counterintuitive. A qubit's error lives on a continuum — any small rotation in any direction — and one cannot, it seems, correct a continuum by discrete means. But the act of measuring the error syndrome *forces* the continuous, blurry error to collapse into one of a handful of definite possibilities: essentially a bit-flip, a phase-flip, both, or neither. The measurement that detects the error also, in the same stroke, snaps the smeared mistake into a clean, fixable form, which a simple counter-operation can then reverse. It is as if merely asking "which way did it go wrong?" forced a hopelessly indefinite mistake to choose a single definite answer. This is why the psychological effect on the field was so profound: the skeptics' single most powerful argument — the fragility that Landauer and others had wielded as a likely death sentence — had been answered, decisively, by the very man who had made the machine worth building. The impossibility was not an impossibility, and other theorists immediately rushed in to improve on the first crude code and to answer the deeper question of whether correction could work even when the correction machinery itself is faulty.

Chapter 29: Steane, Stabilizers, and the Code Zoo

The months after Shor's nine-qubit code were a period of explosive creativity, one of those bursts in which a new idea, once shown to be possible, is rapidly generalized, systematized, and improved by a community suddenly aware that a door has opened. Within a year the isolated trick of a single error-correcting code had become a whole theory, with its own elegant mathematics, its own zoo of specimens, and its own systematic methods for breeding new ones. This chapter surveys that flowering, because the codes it produced are the raw material of every fault-tolerant quantum computer now being built, and the concepts it introduced — logical versus physical qubits, code distance, the stabilizer formalism — are the working vocabulary of the field.

Andrew Steane, an Oxford physicist, made the first great improvement almost immediately, in 1996. Where Shor's code used nine physical qubits to protect one logical qubit, Steane found a cleverer construction using only **seven**, and his code had a deeper elegance: it was built by adapting a classical error-correcting code (the "Hamming code," known to communications engineers for decades) into the quantum setting, showing that the vast existing body of classical coding theory could be imported, with modifications, to serve quantum purposes. This was a liberating realization. Classical engineers had spent half a century developing sophisticated codes to protect data on scratched CDs, noisy phone lines, and deep-space transmissions from distant probes; much of that hard-won knowledge could be carried across the bridge into the quantum world. The **Steane code** became a workhorse of the field, and the general recipe it exemplified — build quantum codes from pairs of compatible classical codes — produced a large family known as **CSS codes**, after Robert Calderbank, Shor, and Steane, who worked out the construction.

To navigate the proliferation of codes, physicists needed a

unifying language, and Daniel Gottesman supplied it in his 1997 doctoral thesis with the **stabilizer formalism**, one of those conceptual tools that make a hard subject tractable. Without the mathematics, the idea is that a large class of quantum codes can be completely described not by writing out their complicated entangled states directly — which would be hopeless — but by listing a set of simple "check" operations, called stabilizers, that the encoded states must satisfy, much as one might describe a complicated shape not by listing all its points but by giving the symmetries it obeys. The stabilizers are precisely the operations one measures in the syndrome measurements of the previous chapter; they are the questions whose answers diagnose errors without revealing data. The stabilizer formalism turned the design of quantum codes from an art into something closer to engineering, providing a systematic way to construct codes, to determine how many errors they can correct, and to work out the circuits that implement their correction. Nearly every quantum code of practical importance today, including the surface codes that dominate current hardware efforts, is a stabilizer code, and Gottesman's formalism is the language in which they are all described and analyzed.

Several concepts from this era have become the permanent vocabulary of quantum computing, and it is worth fixing them clearly, because they recur constantly in any discussion of the field's progress. The distinction between a **physical qubit** and a **logical qubit** is the most important. A physical qubit is an actual piece of hardware — a real superconducting circuit, a real trapped ion — noisy and error-prone. A logical qubit is the reliable, error-corrected qubit that many physical qubits are marshaled together to emulate, the qubit that actually stores the protected information and on which the computation is really performed. The entire engineering challenge of building a useful quantum computer can be summarized as: how many noisy physical qubits does it take to make one good logical qubit, and how do you keep the whole apparatus of correction running faster than errors accumulate? A second key concept is **code**

distance, a measure of how many simultaneous physical errors a code can withstand before the logical information is corrupted; higher-distance codes are more robust but require more physical qubits, and increasing the distance is the standard way to buy more protection. When you read that a machine has demonstrated a "distance-7" code, or that increasing the code distance suppressed the error rate, these are the concepts at work. The scoreboard of quantum computing has quietly shifted, over the years, from counting raw physical qubits to counting logical qubits and code distances — a sign that the field has matured from building bigger noisy machines to building reliable ones.

Shor's nine-qubit code proved that quantum error correction was possible, and the proof acted like a starting gun. Within months, theorists around the world threw themselves at the problem of building better codes — more efficient, more powerful, more elegant — and within a few years they had not only produced a menagerie of specific codes but discovered a beautiful unifying framework that organized them all and made designing new ones almost routine. This chapter is about that flowering: how quantum error correction went, in the space of two or three years, from a single ingenious trick to a mature mathematical theory.

The first great improvement came from Andrew Steane in Oxford, who in 1996 devised a code that protected one logical qubit using just seven physical qubits instead of Shor's nine — a leaner encoding with a more elegant structure, built cleverly by adapting a classical error-correcting code (the venerable Hamming code, used since the 1950s to protect ordinary data) into the quantum realm. The Steane code did something conceptually important: it showed how the rich, well-developed classical theory of error-correcting codes could be imported and transformed to protect quantum information, opening a highway between two previously separate disciplines. Around the same time, an even smaller code was found — a five-qubit code, proven to be the absolute minimum number of physical qubits that can correct any single arbitrary error on one logical qubit. You cannot do it with four; five is the floor. These specific constructions were valuable, but the deeper revolution was the discovery of the pattern behind them.

That pattern was the stabilizer formalism, developed principally by Daniel Gottesman in his 1997 doctoral thesis, and it is one of those unifying ideas that transforms a subject from a collection of clever tricks into a systematic science. The insight is to describe a quantum code not by writing out its complicated entangled states directly — which is unwieldy — but by specifying a set of measurement operators, called stabilizers, that the valid, error-free states all share in common. Think of the stabilizers as a set of yes-or-no questions with the property that every legitimate encoded state answers all of them "yes." To check for errors, you simply ask the questions — measure the stabilizers — and if any answer comes back "no," you know an error has occurred, and the precise pattern of yes and no answers, the error syndrome, tells you exactly which error and where, so you can reverse it. Crucially, these questions are designed to reveal the error without revealing the protected data, exactly as Shor's syndrome measurements did, but now within a clean, general, algebraic framework that lets you design and analyze codes almost mechanically.

The stabilizer formalism did for quantum error correction what the periodic table did for chemistry: it turned a bewildering variety into an ordered system. Suddenly, whole families of codes could be constructed to order, their properties predicted from their stabilizers, their error-correcting power computed by algebra rather than guessed. It also yielded a startling and practically vital side result, the Gottesman-Knill theorem, which showed that a certain large class of quantum operations — the so-called Clifford gates, which include many of the operations used in error correction itself — can actually be simulated EFFICIENTLY on a classical computer. At first this sounds deflating: if these quantum operations are classically simulable, they cannot be the source of quantum advantage. But the lesson is illuminating rather than discouraging. It tells us precisely where the quantum power does NOT lie, and by elimination, where it MUST lie: in the extra, "non-Clifford" gates that lift a computation beyond classical reach. Quantum advantage, it turns out, requires a specific magic ingredient beyond the Clifford operations, and identifying that ingredient — later formalized as "magic states" — became central to understanding what makes a quantum computer genuinely more powerful than a classical one.

The practical upshot of all this theoretical ferment was a growing zoo of codes with different trade-offs, and the trade-offs matter enormously for real machines. Some codes use fewer physical qubits per logical qubit but demand very low hardware error rates; others tolerate messier hardware but demand many more physical qubits. Some require qubits to interact across long distances, which is hard to build; others need only neighboring qubits to talk, which is far more practical. The art of choosing and tailoring codes to the strengths and weaknesses of a particular hardware platform became, and remains, a central discipline of quantum engineering. There is no single best code; there is only the best code for a given machine and a given goal.

The speed with which quantum error correction matured — from a single crude code to a rich, systematic theory in the space of two or three years — is worth marveling at, because it is a case study in how a proof of possibility can unleash a flood of progress. Shor's nine-qubit code of 1995 acted like a starting gun. Within months Andrew Steane had devised a leaner seven-qubit code by cleverly adapting a classical error-correcting code (the venerable Hamming code, in use since the 1950s), demonstrating that the rich, well-developed classical theory of error correction could be imported and transformed to protect quantum information. Soon after came the five-qubit code, proven to be the absolute minimum — you cannot correct an arbitrary single error on a logical qubit with fewer than five physical qubits. But these specific constructions, valuable as they were, mattered less than the unifying framework that emerged to organize them all: Daniel Gottesman's stabilizer formalism, developed in his 1997 doctoral thesis, which did for quantum error correction what the periodic table did for chemistry, turning a bewildering variety of clever tricks into an ordered, systematic science.

The stabilizer idea is elegant enough to restate, because it underlies nearly every code used in real machines today. Rather than describing a quantum code by writing out its complicated entangled states directly — which is unwieldy — you specify a set of measurement operators, the stabilizers, that every valid, error-free encoded state shares in common. Think of them as a set of yes-or-no questions with the property that every legitimate encoded state answers all of them "yes." To check for errors, you simply ask the questions; if any answer comes back "no," an error has occurred, and the precise pattern of yes and no answers — the error syndrome — tells you exactly which error and where, so you can reverse it. Crucially, the questions are designed to reveal the error without revealing the protected data. This clean, algebraic framework let theorists design and analyze whole families of codes almost mechanically, predicting their properties from their stabilizers rather than guessing, and it transformed error correction from an art into an engineering discipline.

The stabilizer formalism also yielded a startling and practically vital side result — the Gottesman-Knill theorem — whose lesson is worth spelling out because it illuminates the very source of quantum power. The theorem shows that a certain large class of quantum operations, the so-called Clifford gates (which include many of the operations used in error correction itself), can actually be simulated *efficiently* on a classical computer. At first this sounds deflating: if these quantum operations are classically simulable, they cannot be the source of quantum advantage. But the lesson is illuminating rather than discouraging, because it tells us precisely where quantum power does *not* lie, and by elimination where it *must* lie — in the extra, "non-Clifford" gates that lift a computation beyond classical reach. Quantum advantage, it turns out, requires a specific magic ingredient beyond the Clifford operations, later formalized as "magic states," and identifying that ingredient became central to understanding what makes a quantum computer genuinely more powerful than a classical one. Yet beneath all this theoretical elegance a fundamental anxiety still lurked: every code assumes you can measure the syndromes and apply corrections, but those operations are themselves performed by imperfect, error-prone hardware. What if the cure spread more disease than it healed? The entire future of the field hung on that single question.

Chapter 30: The Threshold Theorem

The most important theoretical result in the practical history of quantum computing is also one of the least known to the public, because it is abstract, its name is dull, and its content takes a moment to grasp. Yet without it, no one would be spending billions of dollars trying to build quantum computers, because without it there would be no reason to believe the effort could ever succeed. It is called the **threshold theorem**, or sometimes the quantum fault-tolerance theorem, and it was proved, in various forms, around 1996 to 1998 by several groups working independently — including Dorit Aharonov and Michael Ben-Or; Emanuel Knill, Raymond Laflamme, and Wojciech Zurek; and the Russian physicist Alexei Kitaev, whom we will meet again. The threshold theorem is the guarantee that the fight against noise can be won, and it specifies the condition for victory. It is the reason quantum computing is an engineering problem rather than a fool's errand.

The previous chapter ended on the nightmare that had to be dispelled: perhaps quantum error correction cannot work in practice, because the correction machinery is itself made of imperfect, noisy components, so that trying to fix errors introduces new ones, and the cure is as bad as the disease. If every attempt to correct an error had, say, a one-percent chance of introducing an error, and every gate you added to run the correction was itself faulty, then it was entirely possible that piling on more qubits and more correction circuitry would make matters *worse*, not better — that the noise of the repair shop would exceed the noise it was meant to repair, and the whole scheme would spiral into useless chaos. If that were the case, quantum computing would be impossible in practice regardless of how elegant the codes looked on paper. Everything depended on which way the balance tipped.

The threshold theorem settles the question with a precise and hopeful answer. It states that there exists a critical error rate — a **threshold** — with the following decisive property. *If* the error rate of the individual physical components (the gates, the measurements, the qubits) can be pushed *below* this threshold value, then quantum error correction wins, and wins arbitrarily well. Specifically, once you are below threshold, you can make the logical error rate — the rate at which the protected, encoded information gets corrupted — as small as you like, by using larger codes with more physical qubits per logical qubit. Each increase in the size of the code suppresses the logical error rate further, and the suppression is rapid: modest increases in overhead buy dramatic increases in reliability. Below threshold, adding resources makes the machine better, and you can drive the error rate down far enough to complete computations of any length — Shor's billions of gates and beyond. The dream is achievable. Conversely — and this is the sting — if the physical error rate is *above* the threshold, error correction *fails*: adding more qubits and more correction makes things worse, the noise of the repair machinery dominates, and no amount of resources can save you. The threshold is a sharp dividing line between a regime where quantum computing is possible and a regime where it is not. Everything comes down to getting the physical components good enough to cross the line.

What is the threshold, numerically? This is where the news was, at first, sobering, and later, encouraging. The early estimates of the threshold, in the late 1990s, were daunting: they suggested that physical error rates would need to be pushed down to around one error in ten thousand operations, or better — a level of precision far beyond what any hardware of the time could achieve, where qubits failed after a handful of operations rather than ten thousand. It looked as though the threshold, while it existed, might be practically unreachable. But over the following two decades, two things happened in parallel. On the theoretical side, better codes and cleverer fault-tolerant constructions raised the threshold, making the target more forgiving; the **surface code**, in particular, which we meet in the next chapter, turned out to have a threshold of around one percent — meaning physical components need only be right about ninety-nine times out of a hundred, a far more achievable standard. On the experimental side, the quality of physical qubits improved relentlessly, year after year, as engineers learned to isolate them better, control them more precisely, and reduce their errors, until the best qubits crossed from error rates of a few percent down toward and past that one-percent surface-code threshold. The two trends — a rising, more forgiving threshold and a falling physical error rate — were racing toward each other, and the great question of the 2010s and early 2020s was whether, and when, they would meet: whether real hardware would cross below the threshold and demonstrate that error correction actually *wins* in practice, on a real machine, and not just in a theorem.

The threshold theorem is the single most important result in the practical theory of quantum computing, the theorem that turned the machine from a probable impossibility into an engineering project with a clear, if daunting, target. Proven independently by several groups in the years around 1996 to 1998 — with key contributions from John

Preskill, Dorit Aharonov and Michael Ben-Or, Alexei Kitaev, Emanuel Knill, Raymond Laflamme, and others — it answered the terrifying question left hanging by the discovery of quantum codes: can error correction work even when the correction machinery itself is imperfect? The answer, astonishingly, is yes, and the theorem states the condition precisely.

Here is the crux of the worry it resolved. Every quantum code needs gates to encode the qubits, gates to measure the error syndromes, and gates to apply the corrections. But those gates are made of the same flawed physical hardware that is causing the errors in the first place. So each round of error correction introduces its own fresh crop of errors. It is a race: does a round of correction remove more errors than it creates, or fewer? If fewer — if the machinery adds errors faster than it heals them — then error correction is worse than useless, a cure that spreads the disease, and quantum computing is doomed no matter how clever the codes. If more — if each round is a net win — then something almost magical becomes possible. The threshold theorem proves that which side of the race you are on depends on a single number: the error rate of your physical hardware.

The theorem's statement is beautifully sharp. There exists a critical error rate — a threshold — such that IF the error rate of your individual physical qubits and gates is below that threshold, THEN quantum error correction removes errors faster than it creates them, and by layering the protection you can suppress the logical error rate as low as you please, enabling arbitrarily long and reliable quantum computations. But if your hardware's error rate is ABOVE the threshold, no amount of error correction can save you; adding more only makes things worse, and reliable computation is impossible. It is a stark, binary, all-or-nothing dividing line. Get your physical error rate below the magic number and the universe grants you, in principle, a perfect quantum computer of any size. Stay above it and you have nothing.

The mechanism by which sub-threshold hardware yields arbitrary reliability is worth savoring, because it is where the deepest magic lives, and it works by a hierarchy of nested protection called concatenation. Encode each logical qubit using a code — say, several physical qubits protecting one logical qubit, cutting the error rate somewhat. Now take those already-protected logical qubits and encode THEM inside another layer of the same code, and then encode those inside yet another layer, and so on, building a tower of error correction nested within error correction. If you are below threshold, each additional layer suppresses the error rate not merely a little but exponentially — a small increase in the number of physical qubits buys a dramatic drop in the logical error rate. So to make the logical error rate ten times smaller you need only a modest, manageable increase in overhead, and to make it a trillion times smaller you need only a few more layers. This exponential return on investment is what makes arbitrarily long computation possible with merely finite, if large, resources. The tower can be built as tall as the computation is long.

This is the theorem that saved quantum computing, and it did so by transforming a question of principle into a question of engineering. Before the threshold theorem, the fragility of quantum states looked like a fundamental barrier, a law of nature forbidding the machine. After it, fragility became a mere engineering specification: get your error rate below the threshold, pay the overhead in extra qubits, and the machine will work. The pessimists like Landauer had been answered not with a promise but with a proof. The entire subsequent history of quantum hardware — every experiment, every new qubit design, every incremental improvement in fidelity — can be read as a thirty-year assault on a single number, a grinding campaign to push physical error rates down below the threshold and then further and further beneath it, where the overhead becomes affordable.

The psychological transformation the threshold theorem

wrought upon the field is worth making explicit, because it converted despair into a roadmap. Before the theorem, the fragility of quantum states looked like a fundamental barrier — a law of nature that might forbid the machine outright, as the skeptics like Landauer insisted. The worry was concrete and terrifying: every round of error correction is performed by the same flawed hardware causing the errors, so each round introduces its own fresh crop of mistakes, and it was entirely possible that the correction machinery would add errors faster than it removed them, making error correction worse than useless — a cure that spread the disease. If that were so, no amount of cleverness could save quantum computing, and the beautiful algorithms of Shor and Grover would remain forever unrealizable, cathedrals built on quicksand. The threshold theorem answered this deepest fear with a proof, and in doing so transformed the fragility problem from a question of *principle* into a question of *engineering*.

The theorem's statement is worth stating once more in its stark, binary form, because everything about the field's subsequent trajectory follows from it. There exists a critical error rate — a threshold — such that *if* the error rate of your physical qubits and gates is below it, *then* error correction removes errors faster than it creates them, and by layering the protection you can suppress the logical error rate as low as you please, enabling arbitrarily long and reliable computation. But if your hardware sits *above* the threshold, no amount of error correction can save you; adding more only makes things worse. It is all-or-nothing, a sharp dividing line. Get below the magic number and the universe grants you, in principle, a perfect quantum computer of any size; stay above it and you have nothing. This is why the entire subsequent history of quantum hardware can be read as a thirty-year assault on a single number — a grinding campaign to push physical error rates down below the threshold and then further and further beneath it, where the overhead of correction becomes affordable.

The mechanism that makes sub-threshold hardware yield arbitrary reliability — concatenation, the nesting of codes within codes — is worth one more look, because it is where the deepest magic lives. Encode each logical qubit using a code, cutting the error rate somewhat; then encode those already-protected logical qubits inside another layer of the same code, and encode *those* inside yet another layer, building a tower of error correction nested within error correction. If you are below threshold, each additional layer suppresses the error rate not merely a little but *exponentially* — a small, manageable increase in the number of physical qubits buys a dramatic drop in the logical error rate. To make the logical error rate ten times smaller you add a layer; to make it a trillion times smaller you add only a few more. This exponential return on investment is what makes arbitrarily long computation possible with merely finite, if large, resources: the protective tower can be built as tall as the computation is long. Two honest caveats keep the triumph in proportion — the early proven thresholds were dishearteningly tiny, and the theorem guarantees only that it *can* be done, not that it can be done cheaply — but both would be eased dramatically by a new kind of code, invented by a Russian physicist thinking about the exotic physics of two-dimensional quantum matter, that raised the threshold to a forgiving one percent and needed only neighboring qubits to interact.

Chapter 31: Kitaev's Surface

Some ideas in science are valuable because they are useful; a few are valuable because they are both useful and profound, revealing an unexpected connection between distant regions of knowledge. Alexei Kitaev's invention of the surface code, and of the broader notion of topological protection, belongs to the second class. It gave the field the error-correcting code that, more than any other, has driven the hardware efforts of Google, IBM, and others, and it did so by importing into quantum computing an idea from the abstract mathematics of topology — the study of those properties of shapes that survive stretching and bending, the properties that make a coffee cup and a doughnut "the same" because each has exactly one hole. It is one of the most elegant chapters in the whole history, and its protagonist is among the most brilliant and reticent figures in the field.

Alexei Kitaev is a Russian-born physicist, now at Caltech, of legendary depth and characteristic modesty, the kind of theorist whose papers are studied for years for the riches packed into their terse pages. In the late 1990s he introduced two related ideas that would shape everything after. The first was **topological quantum computation**, a radically different vision of how to build a quantum computer, which we will meet properly in Part VIII when we survey the hardware modalities; its promise is to protect quantum information not by actively correcting errors but by encoding it in global, topological properties of an exotic physical system, properties so robust that local noise cannot disturb them, giving error resistance "for free" at the hardware level. The second idea, more immediately practical and the subject of this chapter, was the **surface code** (and its close relative the toric code), a quantum error-correcting code with a special and enormously advantageous structure: it requires each physical qubit to interact only with its immediate neighbors on a two-dimensional grid.

To see why that neighbor-only property is such a gift, consider the practical predicament of the hardware engineer. Building a quantum computer means fabricating many physical qubits and wiring them so they can interact to perform gates and syndrome measurements. The earlier error-correcting codes, elegant as they were, often demanded that arbitrary, distant qubits interact directly with one another — that qubit number three, say, perform a joint operation with qubit number ninety, across the chip. In real hardware this is a nightmare: you cannot easily run controlled interactions between arbitrary far-apart qubits without a tangle of connections that introduces noise and defies fabrication. The surface code sidesteps this entirely. Its qubits sit on a flat two-dimensional grid, like a checkerboard, and every operation the code requires — every stabilizer check, every syndrome measurement — involves only a qubit and its four nearest neighbors. Nothing reaches across the chip. This "locality" matches beautifully the reality of solid-state hardware, where qubits laid out on a chip naturally couple to their neighbors, and it is the single biggest reason the surface code became the favorite of the superconducting-qubit efforts at Google and IBM. It is a code designed with the grain of the hardware, not against it.

The surface code works, in essence, by tiling a two-dimensional surface with physical qubits and repeatedly measuring local checks across the grid — asking, everywhere, whether neighboring qubits agree in the ways the code requires. When an error strikes a qubit, it disturbs the checks in its vicinity, and the pattern of disturbed checks — the syndrome — pinpoints where the error occurred, allowing correction, all without ever measuring the encoded information itself. The logical information is stored not in any local cluster of qubits but in *global*, topological features of the whole grid — loosely, in patterns that stretch across the entire surface and cannot be created or destroyed by any local disturbance, only by an error that manages to string itself all the way across the array. This is the topological protection: to corrupt the encoded information, noise would have to conspire to produce a whole connected chain of errors spanning the grid, and if the physical error rate is below threshold, such spanning chains are exponentially unlikely, so the logical information is exponentially safe. Make the grid bigger — increase the code distance — and the protection grows exponentially stronger, exactly as the threshold theorem promised. The code distance is, quite literally, the length an error-chain would have to reach to do damage, and enlarging the grid lengthens that reach.

The surface code came with a price, and it is a price that dominates the economics of quantum computing to this day:

overhead. Protecting a single logical qubit well enough to run a long computation requires a large patch of the grid — hundreds, or with realistic error rates a thousand or more, physical qubits per logical qubit. This is the brutal arithmetic behind the earlier claim that factoring a 2048-bit RSA key might require *millions* of physical qubits: the algorithm needs a few thousand logical qubits, and each logical qubit, built with the surface code at realistic error rates, swallows a thousand or more physical qubits, and the product is millions. The surface code made fault-tolerant quantum computing *possible* and *practical in principle*, matched to real hardware and blessed with a forgiving threshold — but it made it *expensive*, setting a daunting target of physical-qubit counts that the machines of the 2020s, with hundreds to about a thousand qubits, are still far from meeting. Much of the most active current research, which we will reach in Part XII, aims precisely at reducing this overhead — finding new codes, the low-density parity-check or "qLDPC" codes chief among them, that protect logical qubits with far fewer physical ones, and so bring the era of useful fault-tolerant computing closer. But the surface code remains the benchmark, the code against which all others are measured, and the one on which the most celebrated recent milestones have been achieved.

Alexei Kitaev is one of the most original and influential physicists of the quantum computing era, a theorist of legendary depth who has repeatedly reshaped the field with ideas that seem to come from a slightly different angle than everyone else's. His invention of the surface code — and the broader idea of topological protection that underlies it — is arguably the single most important practical contribution to building real quantum computers, because the surface code is the error-correcting scheme that nearly every leading hardware effort, from Google to IBM, actually uses. If the threshold theorem promised that a machine could be built, the surface code is the blueprint that made the promise buildable.

Kitaev's genius was to bring a completely different branch of physics — topology, the mathematics of shape and connectedness, of properties that survive continuous deformation — to bear on the problem of protecting quantum information. Topology studies the features of an object that do not change when you stretch or bend it: a coffee cup and a doughnut are "the same" to a topologist because each has exactly one hole, and no amount of gentle reshaping can add or remove that hole. Kitaev's insight, first developed in the late 1990s, was that quantum information could be stored in the GLOBAL, topological properties of a system of many qubits rather than in the fragile local state of any individual qubit. And global topological properties are inherently robust: just as you cannot remove the hole from a doughnut by poking or squeezing it locally, you cannot corrupt topologically-encoded information by any local disturbance. Since decoherence acts locally — a stray photon here, a bit of heat there — information hidden in the global topology is naturally shielded from exactly the kind of small, local errors that destroy ordinary qubits. The protection is baked into the geometry.

The surface code is the practical embodiment of this idea, and its virtues for real hardware are almost too good to be true. Picture a two-dimensional grid of physical qubits laid out on a chip, like a checkerboard. The logical quantum information is encoded in the collective, topological state of the whole grid, and errors are detected by continuously measuring simple relationships — the stabilizers — between each qubit and its immediate neighbors. Two features make this a gift to engineers. First, the surface code needs only NEAREST-NEIGHBOR interactions: each qubit only has to talk to the handful of qubits right next to it, never to distant ones. This is precisely what solid-state hardware can naturally provide, since qubits fabricated on a chip are wired to their neighbors; codes requiring long-range interactions, by contrast, are nightmares to build. Second, and decisively, the surface code has a remarkably HIGH error threshold — around one percent, meaning physical qubits need only be about ninety-nine percent reliable for error correction to start winning. Compared with the punishing one-in-ten-thousand thresholds of the early concatenated codes, one percent is a rate that real, imperfect hardware could actually hope to achieve. Kitaev had lowered the bar from the impossible to the merely very hard.

There is a cost, and it is steep, and honesty requires naming it. The surface code's robustness and practicality are paid for in sheer numbers of qubits: protecting a single high-quality logical qubit may require hundreds or even thousands of physical qubits arranged in the grid, depending on how good the hardware is and how low you need the logical error rate. This is the daunting "overhead" that shapes every roadmap in the industry, and it is why building a machine capable of running Shor's algorithm on a real encryption key — which needs thousands of reliable logical qubits — may demand millions of physical qubits, a scale far beyond today's devices of a few hundred. When you hear that a useful, error-corrected quantum computer might need a million qubits, the surface code's overhead is the reason. The trade is brutal but, for now, worth it: the surface code asks for many mediocre qubits rather than a few perfect ones, and mediocre qubits are what we can build.

Kitaev's imagination did not stop at the surface code. He also proposed an even more radical vision, topological quantum computing, in which the qubits themselves would be built from exotic quasiparticles — hypothetical entities called anyons, and in particular Majorana particles — whose quantum information would be stored in the topological braiding of their world-lines through space and time, making them intrinsically, hardware-level protected against decoherence, error correction built into the physics rather than bolted on top. This audacious idea, which we will return to when we survey the hardware platforms, has inspired a major and difficult experimental program, most prominently at Microsoft. Whether it can be realized remains, decades on, one of the great open questions of the field. But even setting aside that far-horizon dream, Kitaev's surface code alone would secure his place among the founders, for it is the scheme on which the concrete hopes of the entire industry now rest.

It is worth restating, as we close the theoretical half of this book, just how completely Kitaev's surface code reshaped the practical prospects of quantum computing — for it converted a beautiful but forbidding theorem into a concrete, buildable blueprint. The threshold theorem had promised that a machine *could* be built if hardware crossed a certain quality line, but the early proven thresholds were dishearteningly tiny (error rates of one in ten thousand or better) and the codes required qubits to interact across long distances, both of which put a real machine seemingly out of reach. Kitaev's surface code eased both obstacles at a stroke: it raised the threshold to a far more forgiving one percent, meaning physical qubits need only be about ninety-nine percent reliable — a rate real hardware could plausibly attain — and it required only nearest-neighbor interactions, exactly what solid-state chips naturally provide. It lowered the bar from the impossible to the merely very hard, and in doing so it became the scheme on which essentially every serious hardware effort — Google's, IBM's, and others' — now stakes its concrete hopes. When Google's Willow chip demonstrated error correction below threshold in 2024, it was the surface code it was running; Kitaev's idea had become the working blueprint of the entire industry.

The price of the surface code's robustness — its brutal appetite for physical qubits — deserves a final, honest emphasis, because it shapes every roadmap and every timeline in the field. Protecting a single high-quality logical qubit with the surface code may require hundreds or even thousands of physical qubits arranged in a two-dimensional grid, depending on the hardware quality and the logical error rate required. This is the daunting "overhead" that stands between today's machines and useful computation, and it is why building a machine capable of running Shor's algorithm against a real encryption key — which needs thousands of reliable logical qubits — may demand a million or more physical qubits, a scale far beyond today's devices of a few hundred. The surface code makes a specific, pragmatic bargain: it asks for many mediocre qubits rather than a few perfect ones, trading qubit *quantity* for tolerance of qubit *imperfection*, because many imperfect qubits are exactly what we can currently build. Whether that bargain proves affordable at scale, or whether more efficient codes are needed to reduce the overhead, is one of the central open questions the final chapters of this book address.

With the surface code, the theoretical foundations of quantum computing were essentially complete, and it is worth taking stock of how much had been assembled before a single large machine existed. The field had algorithms worth running — Shor's factoring, Grover's search, Lloyd's universal simulation. It had a proof that errors could be corrected — Shor's first codes, refined into the elegant stabilizer formalism. It had a guarantee that correction works even when the correction machinery itself is faulty — the threshold theorem. And now it had a practical, buildable code with a forgiving threshold and only nearest-neighbor connectivity — Kitaev's surface code. Everything needed, in principle, was known; the remaining problem was no longer whether a quantum computer was possible or how to protect it, but the sheer, grinding engineering of actually building the physical qubits — real, tangible devices that could hold and manipulate quantum information in the laboratory, coherent enough and accurate enough to cross the threshold the theory demanded. The story now leaves the realm of pure thought and enters the workshop, beginning with the first technology to hold a qubit steady enough to compute: single atoms, stripped of an electron and suspended, motionless, in a cage of electric fields.

Part VII — Building the Impossible

Chapter 32: Ions in a Trap

The first physical system in which anyone genuinely computed with qubits — in which the abstractions of Part IV became real, controllable, laboratory objects — was a row of individual atoms, stripped of an electron and suspended in the empty space of a vacuum chamber by invisible fingers of electromagnetic force, chilled with laser light to within a whisker of absolute stillness, and addressed one by one with precisely tuned beams. The **trapped-ion** quantum computer was, and in many respects remains, the aristocrat of the field: the cleanest, the most precise, the highest in fidelity, the closest to the textbook ideal of a qubit. That it was also, for a long time, among the hardest to scale up to large numbers is part of the story too. But it was here, with trapped ions, that quantum computing first drew breath as an experimental science, and the people who made it possible had spent decades perfecting the art of holding a single atom still.

The foundational technology came from atomic physics, from a line of work on trapping and cooling individual charged atoms that long predated quantum computing and was pursued for its own sake, as a way to make the most precise measurements and the best atomic clocks in the world. To trap an ion — an atom with a net electric charge — you exploit that charge: carefully shaped, oscillating electric fields can confine the ion in a tiny region of space, holding it suspended in vacuum, isolated from every surface, floating in nothing. The device that does this, the **Paul trap**, earned its inventor Wolfgang Paul a share of the 1989 Nobel Prize. Then, to quiet the ion's residual jiggling — for even a trapped ion vibrates with thermal motion that would blur any quantum operation — physicists developed **laser cooling**, a beautiful technique in which precisely tuned laser light, striking the ion, saps its motion photon by photon until it is cooled to a fraction of a degree above absolute zero and sits almost perfectly still. The mastery of laser cooling and ion trapping, developed over the 1970s and 1980s by physicists including David Wineland at the U.S. National Institute of Standards and Technology, was recognized with Wineland's own Nobel Prize in 2012. This deep reservoir of atomic-physics expertise, built up for reasons having nothing to do with computing, turned out to be exactly what quantum computing needed, and the trapped-ion computer was its heir.

The qubit, in a trapped-ion machine, is encoded in the internal energy states of the ion — two particular electronic levels of the atom, one designated |0> and the other |1>, exquisitely stable and long-lived because they are properties of the atom's own structure, shielded within the ion and coupled only weakly to the outside world. This is the source of the trapped ion's superb quality: the qubit lives inside a single atom, one of nature's own perfect, identical, unchanging objects, rather than in a fabricated circuit subject to the imperfections of manufacture. Trapped-ion qubits boast among the longest coherence times and the highest gate fidelities of any technology — the best two-qubit operations reaching accuracies of 99.9 percent and beyond, comfortably past the surface-code threshold — precisely because the atom is such a clean and well-isolated home for quantum information. Single-qubit operations are performed by shining laser pulses (or microwaves) on an individual ion, nudging its internal state, rotating its Bloch sphere with great precision.

The crucial and beautiful trick — the thing that makes a *computer* rather than a collection of isolated atoms — is how you get two ions to interact, to perform the entangling two-qubit gates without which there is no quantum advantage. The ions, all sharing the same charge, repel one another, and so a string of them in a trap arranges itself into a neat line, evenly spaced, like beads on an invisible wire. Because they are coupled by their mutual electrical repulsion, the whole string can vibrate together in collective modes of motion, like the shared oscillations of a row of masses linked by springs. In 1995 — the same fertile year as Shor's error-correcting code — Ignacio Cirac and Peter Zoller proposed a method, now foundational, for using this shared vibration as a **quantum data bus**: a laser addressing one ion can couple its internal qubit state to the collective motion of the whole string, and another laser addressing a second ion can couple that motion back to the second ion's qubit, so that quantum information passes between the two ions *through their shared vibration*. In this way any two ions in the string can be made to interact and entangle, even without being neighbors, because they all share the same motional bus. The Cirac–Zoller gate turned a trapped string of ions into a programmable quantum computer, and it was demonstrated in the laboratory within months by Wineland's group, which produced the first entangling gate between trapped ions — one of the founding experiments of the field.

When theorists had finished proving what a quantum computer could do, the burden passed to experimentalists, who faced a question of staggering difficulty: what physical object, in the real, warm, jostling world, could actually serve as a qubit — held still enough, isolated enough, and controllable enough to compute? The first technology to answer convincingly, and still in many ways the gold standard for quality, was the trapped ion: a single atom, stripped of one electron so that it carries a net electric charge, suspended in empty space by carefully shaped electric and magnetic fields, and manipulated with precisely tuned laser light. In 1995, the same year Shor showed error correction was possible, Ignacio Cirac and Peter Zoller published a landmark proposal showing exactly how a string of trapped ions could be made into a working quantum computer, and within months experimentalists at the U.S. National Institute of Standards and Technology, led by David Wineland, demonstrated the first quantum logic gate between trapped ions. The machine had left the page and entered the laboratory.

The appeal of trapped ions begins with a profound advantage: the qubits are perfect. Every ion of a given species — a calcium ion, say, or an ytterbium ion — is exactly, fundamentally identical to every other, because they are the same kind of atom, manufactured not in a fab but by the universe to identical specification. There is no variation, no defect, no fabrication scatter; the qubits do not need to be built, only caught. The quantum information is stored in the internal energy states of the ion — for instance, whether an outer electron sits in one stable configuration (call it 0) or another (call it 1) — and because these atomic states are set by the immutable laws of atomic physics, they are extraordinarily stable and reproducible. A trapped ion, well isolated in ultra-high vacuum and shielded from stray fields, can maintain its quantum coherence for seconds, even minutes in the best cases — an eternity by the standards of decoherence, far longer than most rival technologies, giving ample time to perform many operations before the fragile state decays.

The manipulation is done with lasers, and the elegance of the scheme repays a close look. A finely tuned laser beam aimed at an ion can flip its internal state — a single-qubit gate, a rotation of the Bloch sphere enacted by light. The harder trick is making two ions interact to produce entanglement, and here Cirac and Zoller's insight was beautiful:

the ions in a trap, all bearing the same charge, repel one another and so line up like beads on a string, and because they are electrically coupled they can vibrate together in shared, collective modes of motion, rocking back and forth in unison. This shared vibration acts as a kind of data bus, a common channel through which the internal state of one ion can be conditionally linked to another. By playing laser light of exactly the right frequency, an experimentalist can couple an ion's internal qubit state to this collective motion and thereby transmit quantum information from one ion to another, entangling them. The ions never touch; they converse through their shared dance.

The results have been spectacular, and trapped ions still hold most of the records that matter for quality. They boast the highest gate fidelities of any technology — the accuracy of individual quantum operations has been pushed above 99.9 percent, meaning fewer than one error in a thousand operations, and in some demonstrations beyond 99.99 percent, comfortably past the surface-code threshold. They have the longest coherence times, the cleanest measurements, and, because every ion in a trap can in principle be made to interact with every other through the shared motion, they enjoy "all-to-all connectivity," a flexibility that many chip-based rivals lack. Companies built on this technology — IonQ, spun out of academic work at the University of Maryland, and Quantinuum, the trapped-ion venture backed by Honeywell — have become leaders precisely on the strength of these superb per-qubit numbers, and Quantinuum in particular has produced some of the most advanced demonstrations of error correction and logical qubits anywhere.

But trapped ions carry a serious handicap, and it is the mirror image of their virtue: they are SLOW and HARD TO SCALE. Because the entangling operations rely on the ions' collective, sluggish motion and on precisely aimed laser beams, gates take far longer than in solid-state rivals — microseconds to tens of microseconds, versus tens of nanoseconds for superconducting circuits, a thousandfold difference in raw speed. And the very feature that gives them all-to-all connectivity — a shared string of ions rocking together — becomes a liability as numbers grow: a single trap can hold only so many ions, perhaps a few dozen, before the collective vibrations become too complex and crowded to control cleanly. Scaling to the thousands or millions of qubits a useful machine requires means somehow linking many separate traps together, shuttling ions between zones or connecting distant traps with photons — a formidable architectural challenge that the field is actively, and only partially, solving. The trapped-ion bargain, then, is quality over quantity: the best qubits in the world, but agonizingly difficult to assemble into large numbers.

The historical significance of the first trapped-ion gate is worth underlining, because it marks the exact moment quantum computing crossed from theory into laboratory reality. In 1995 — the same remarkable year that Shor proved error correction possible and Cirac and Zoller published their blueprint — David Wineland's group at the U.S. National Institute of Standards and Technology demonstrated the first quantum logic gate between the internal and motional states of a trapped ion. Up to that point, the quantum computer had been a creature of chalkboards and papers: proven possible by Benioff, made desirable by Feynman, defined by Deutsch, made threatening by Shor, but never actually *built*, not even in miniature. Wineland's gate changed that. It was a single operation on a single ion, doing nothing a pocket calculator could not, but it proved that the abstract logic of quantum computation could be enacted in real matter, with real lasers, on a real atom held in a real trap. The machine had left the page. Wineland's decades of exquisite work controlling single ions would earn him a share of the 2012 Nobel Prize.

The elegance of how trapped ions are entangled deserves a second, appreciative look, because it is one of the prettiest ideas in experimental physics. The ions in a trap, all carrying the same electric charge, repel one another and settle into an evenly spaced line, like beads on a string. Because they are electrically coupled, they cannot move independently; they vibrate together in shared, collective modes, rocking back and forth in unison like a row of masses joined by springs. Cirac and Zoller's insight was to use this shared vibration as a *data bus* — a common channel through which the internal quantum state of one ion could be conditionally linked to another. By playing laser light of exactly the right frequency, an experimentalist couples an ion's internal qubit to this collective motion, transmits the information through the shared dance to another ion, and thereby entangles the two. The ions never touch; they converse through their common vibration, whispering quantum information to one another across the trap by nudging the shared rhythm of their motion. It is a mechanism of remarkable subtlety, and it gives trapped ions their prized ability to entangle any ion with any other.

The enduring lesson of trapped ions, restated for emphasis, is that they set the quality benchmark against which every rival technology is measured — a standing proof that in quantum computing the best is not always the biggest. Their gate fidelities are the highest of any platform, above 99.9 and even 99.99 percent; their coherence times are the longest, seconds against the microseconds of superconducting circuits; their measurements are the cleanest; and their all-to-all connectivity is the most flexible. On every metric that reflects genuine quality rather than mere size, trapped ions lead or contend. Their handicap — slow gates and the difficulty of scaling a single trap beyond a few dozen ions — is real and serious, and it is why the largest machines by qubit count are built from other technologies. But their excellence is a permanent rebuke to the qubit-counting hype, a reminder that a smaller machine of higher quality can outperform a larger, noisier one on the tasks that truly test a quantum computer. Even as trapped ions were demonstrating the first gates, though, another approach was briefly seizing the headlines with a spectacular stunt — computing with the nuclear spins of countless molecules in an ordinary test tube, read out by the same kind of machine that scans a patient's knee.

Chapter 33: NMR and the Factoring of Fifteen

The first quantum computations — the first times anyone ran a real quantum algorithm on real qubits and got the right answer — were performed not with trapped ions or superconducting circuits but with an older, more familiar technology repurposed for a strange new task: **nuclear magnetic resonance**, or NMR, the same physics that underlies the MRI scanners in every hospital. In the late 1990s and early 2000s, NMR quantum computers held the lead, produced the field's first algorithmic demonstrations, and culminated in 2001 in the celebrated experiment that factored the number fifteen using Shor's algorithm. It was a landmark, and also, in the end, a technological dead end — a fascinating episode that showed what a quantum computer could do while revealing exactly why a *scalable* one would need a different foundation.

The qubits in an NMR quantum computer are the **spins of atomic nuclei** within the molecules of a liquid. A nucleus like that of hydrogen or carbon behaves like a tiny magnet, and in a strong magnetic field it can point, quantum-mechanically, "up" or "down" — a natural two-state system, a natural qubit. A specially designed molecule, with a handful of distinguishable nuclei, becomes a little quantum computer with a handful of qubits; the different nuclei, sitting in slightly different chemical environments, respond to slightly different radio frequencies, so that carefully shaped radio pulses can address them individually and drive the gates, while the natural interactions between neighboring nuclei in the molecule provide the entangling operations. And because NMR was a mature technology, developed over half a century for chemistry and medicine, the equipment and expertise were readily available. Physicists and chemists at IBM, MIT, Stanford, and Oxford seized on it, and for a few years NMR quantum computing raced ahead of every other approach, notching the first experimental demonstrations of quantum algorithms — small runs of Grover's search, of the Deutsch–Jozsa algorithm, of quantum error-correction primitives.

The crowning achievement came in 2001, when a team at IBM's Almaden Research Center, led by Isaac Chuang — co-author of the field's canonical textbook — used a molecule with seven usable nuclear-spin qubits to run Shor's algorithm and factor the number fifteen into three times five. It was, on one level, an almost comic result: a room-sized apparatus, a custom-synthesized molecule, and years of expertise, deployed to discover a fact that every seven-year-old knows. But the point was never the answer; it was the *method*. For the first time, the algorithm that had launched the field — Shor's period-finding attack on factoring, the thing that had frightened intelligence agencies and opened the funding taps — had actually been executed on physical qubits, start to finish, and had produced the correct factors by genuine quantum period-finding. The uncertain machine had run the world's most famous quantum algorithm. It was a proof that the theory was not merely beautiful but *real*, that these ghostly superpositions and interferences could be marshaled in a laboratory to compute. The image of a quantum computer factoring fifteen became one of the field's founding icons.

And yet, even as they celebrated, the practitioners knew that NMR quantum computing was a road that did not lead where they needed to go, and understanding why is instructive, because it illuminates what a *real* quantum computer requires. The fundamental problem was that NMR did not work with a single molecule in a pure quantum state. It worked with an enormous number of molecules at once — a liquid sample containing something like a billion trillion molecules — all being manipulated together, with the measured signal an average over the whole ensemble. At the ordinary temperatures of the experiments, the individual molecular "computers" were mostly in a jumble of thermal states, not in the clean, cold quantum states an ideal computation demands, and the useful quantum signal had to be teased out of this warm statistical mush by clever tricks. These tricks worked for a handful of qubits, but they came at a devastating cost: the strength of the extractable quantum signal fell off *exponentially* as more qubits were added, so that beyond roughly a dozen qubits the signal would vanish into the noise entirely. There were even pointed arguments that the states produced in these room-temperature NMR experiments contained no genuine entanglement at all — that the "quantum computer" was, in a precise sense, simulable by classical means and not harnessing the true quantum resource. Whatever the interpretation, the scaling verdict was clear and fatal: NMR quantum computing could factor fifteen, but it could never, even in principle, factor anything large, because the approach hit an exponential wall at a handful of qubits. It was a magnificent demonstration and an evolutionary dead end.

In the late 1990s and around the turn of the millennium, the technology that produced the most eye-catching early results was neither trapped ions nor superconducting circuits but something far more surprising: nuclear magnetic resonance, or NMR, the very same physical phenomenon that powers the MRI scanners in hospitals. For a few heady years, NMR quantum computing led the field in the number of qubits demonstrated and in the sheer showmanship of its results, culminating in the single most famous early quantum computation of all — the factoring of the number 15 using Shor's algorithm, in 2001. It is a story of dazzling early success followed by a hard collision with a fundamental wall, and it taught the field a lasting lesson about the difference between a stunt and a scalable machine.

The idea behind NMR quantum computing is ingenious and, at first, almost unbelievable. The qubits are the atomic nuclei within a specially designed molecule — each nucleus behaves like a tiny magnet, a spin that can point "up" (call it 0) or "down" (call it 1) or, being quantum, into a superposition of both. A molecule carefully synthesized with a handful of distinguishable nuclei — different atoms in different chemical positions, each resonating at its own characteristic frequency — becomes a little quantum register, a few qubits bundled into a single molecule. Radio-frequency pulses, exactly like those in an MRI machine and tuned to each nucleus's frequency, rotate the spins to perform single-qubit gates, while the natural magnetic interactions between neighboring nuclei in the molecule provide the coupling needed for two-qubit entangling gates. The molecule computes with its own nuclear spins, steered by radio waves.

But here is the twist that made NMR both powerful and ultimately doomed. You cannot control or measure a single molecule's faint nuclear spins; the signal is far too weak. So NMR quantum computing operates not on one molecule but on an ENORMOUS number of them at once — a whole test tube of liquid, perhaps a billion billion identical molecules dissolved in solvent at room temperature, all being manipulated in parallel by the same radio pulses and all contributing to a single averaged, macroscopic signal that a sensitive detector can read. In effect, you run the same quantum computation simultaneously on a fantastic number of copies and read out the statistical average. This "ensemble" approach is what made NMR so much easier to demonstrate early on: it worked at room temperature, needed no exotic refrigeration or vacuum, and could exploit decades of exquisitely refined NMR technology already sitting in chemistry departments worldwide. It rode in on a mature industry.

The crowning achievement came in 2001, when a team at IBM's Almaden Research Center and Stanford, led by Isaac Chuang, used a custom-synthesized molecule with seven usable nuclear spins to run Shor's algorithm and factor the number 15 into its prime factors, 3 and 5. It was, and remains, iconic: the first time Shor's world-shaking algorithm had actually been executed on physical quantum hardware, a tangible demonstration that the theory ran on real matter. The result made headlines and became a symbol of quantum computing's arrival. For a moment, NMR looked like it might be the road to the machine.

And yet NMR quantum computing turned out to be a technological dead end, for reasons both practical and profound, and understanding why illuminates what a real quantum computer requires. The practical problem is scaling: each additional qubit means finding another distinguishable nucleus in the molecule, and the signal from the ensemble weakens exponentially as qubits are added, so that beyond a dozen or so qubits the approach simply collapses under its own noise. There was no path from seven qubits to seventy, let alone seven thousand. The profound problem is subtler and more interesting: because NMR operates on a warm, thermal ensemble in which the spins are only very slightly biased toward their starting state — the quantum signal is a faint statistical whisper riding atop overwhelming thermal noise — researchers came to doubt whether these room-temperature experiments contained any genuine entanglement at all. Careful analysis suggested that the states involved might be efficiently simulable classically, meaning the "quantum computer" might not have been harnessing true quantum power so much as cleverly mimicking it. The factoring of 15, glorious as a demonstration, may not have been a genuinely quantum computation in the deepest sense.

The NMR episode is worth preserving in the field's memory as a cautionary tale, because it dramatizes, more vividly than any other early chapter, the gap between a headline-grabbing demonstration and a scalable technology. For a few years around the turn of the millennium, NMR quantum computing led the field in the number of qubits demonstrated and in sheer showmanship, culminating in the iconic 2001 factoring of 15 into 3 times 5 using Shor's algorithm

— the first time that world-shaking algorithm had run on physical hardware. It made headlines; it became a symbol of quantum computing's arrival; and for a moment NMR looked like it might be the road to the machine. It was, in fact, a technological dead end, and understanding why teaches a durable lesson about what a real quantum computer requires and how easily a striking result can mislead.

The two reasons for NMR's failure — one practical, one profound — are both illuminating. The practical problem was scaling: each additional qubit meant finding another distinguishable atomic nucleus within the molecule, and the faint signal from the ensemble of countless molecules weakened exponentially as qubits were added, so that beyond a dozen or so qubits the approach simply collapsed under its own noise. There was no path from seven qubits to seventy. The profound problem was subtler and more interesting. Because NMR operates on a warm, thermal ensemble of a fantastic number of molecules — the quantum signal a faint statistical whisper riding atop overwhelming thermal noise — careful analysis suggested that these room-temperature experiments might contain no genuine *entanglement* at all, and might be efficiently simulable classically. The celebrated factoring of 15, glorious as a demonstration, may not have been a genuinely quantum computation in the deepest sense — the machine may have been cleverly *mimicking* quantum power rather than harnessing it.

The two lessons NMR left the field shaped everything that followed, and both are worth carrying forward as tools for judging quantum claims to this day. The first is a caution about hype and appearances: a headline-grabbing result — "quantum computer factors a number!" — can mask a technology with no future, and the number of qubits demonstrated matters far less than whether the machine embodies real, scalable, entangled quantum power. Counting qubits is not the same as counting capability, a lesson the field would have to relearn repeatedly. The second lesson concerns what a serious quantum computer actually demands:

not a warm thermal average over countless molecules, but individually addressable, individually measurable qubits, initialized cleanly into definite states, richly entangled, and controlled one by one. Meeting those demands would require abandoning the comfortable test tube and building qubits from scratch on a chip, cooled to within a whisker of absolute zero — a far harder path, but the only one that led onward. The technology that would ultimately dominate the field took exactly that path, fashioning artificial atoms out of superconducting metal and electric current, and it is to that colder, harder, more scalable road that the story now turns.

Chapter 34: Circuits That Superconduct

The technology that now dominates quantum computing — the one behind Google's Sycamore and Willow, behind IBM's ever-larger processors, behind most of the headlines — began as an unlikely bet. Where the trapped-ion physicists used nature's own perfect atoms, the superconducting-qubit physicists set out to do something that sounded almost perverse: to *build* an artificial atom out of metal and wire, a human-made circuit that would, despite being an object of macroscopic engineering rather than a single particle, behave according to the laws of quantum mechanics, holding superpositions and entangling with its fellows. That such a thing is even possible is a wonder in its own right, and that it became the leading platform in the field, surpassing the purer atomic approaches in raw scale, is a testament to the power of an engineering technology you can design, fabricate, and improve on a chip.

The key to a superconducting qubit is **superconductivity** itself, one of the strangest and most useful quantum phenomena in bulk matter. When certain metals are cooled below a critical temperature — very cold, near absolute zero — their electrical resistance vanishes utterly, and electric current can flow forever without loss. More than that, the electrons in a superconductor pair up and fall into a single collective quantum state, a macroscopic wavefunction in which trillions of electrons march in lockstep as one coherent quantum entity. This is crucial, because it means a superconducting circuit is not a jumble of independently jostling electrons — which would decohere instantly — but a single, unified quantum object, large enough to build and wire, yet coherent enough to serve as a qubit. The superconductor lets quantum behavior scale up from the atomic to the circuit level, and that is the whole trick.

The heart of a superconducting qubit is a peculiar component called a **Josephson junction**, a tiny sandwich in which two superconductors are separated by an ultrathin insulating barrier, so thin that the paired electrons can "tunnel" quantum-mechanically across it. The Josephson junction behaves as a special kind of nonlinear element, and this nonlinearity is precisely what is needed to make a good qubit. An ordinary electrical circuit — a loop with an inductor and a capacitor — oscillates like a pendulum, with a ladder of energy levels all evenly spaced; you cannot use it as a qubit because you cannot address just two levels without accidentally exciting the others, since they are all the same distance apart. The Josephson junction distorts this even ladder, spacing the lowest two energy levels differently from the rest, so that a precisely tuned microwave pulse can drive transitions between just those two levels — the |0> and |1> of the qubit — without disturbing the others. The junction turns a boring linear circuit into an "artificial atom" with an addressable pair of quantum states. The first such qubit was demonstrated in 1999 by a group at NEC in Japan, and though it held its coherence for only a nanosecond — a flicker — it proved that a fabricated circuit could behave as a qubit, and the race was on to make it better.

The great breakthrough in design came in 2007, with the invention of the **transmon** at Yale by a group including Robert Schoelkopf, Michel Devoret, and Steven Girvin. The transmon is a superconducting qubit engineered specifically to be insensitive to one of the dominant sources of noise, electrical charge fluctuations, that had plagued the earlier designs; by operating in a regime that suppressed this sensitivity, the transmon achieved far longer coherence times, and it proved robust and reproducible enough to be fabricated in large numbers. The transmon is, today, the workhorse qubit of the superconducting industry: the basic building block from which Google, IBM, and most others construct their processors. Over the years since 2007, relentless engineering has extended transmon coherence times from nanoseconds to hundreds of microseconds and even approaching milliseconds, improving them by a factor of tens of thousands — a testament to the power of an engineering platform where each fabricated device can be measured, understood, and redesigned, iteration after iteration, the way conventional chips have been improved for decades. Because superconducting qubits are made by the same kind of lithographic fabrication used for ordinary microelectronics — patterned onto a chip with well-established techniques — they can, at least in principle, be manufactured and scaled using the accumulated might of the semiconductor industry, a decisive advantage in the quest for large qubit counts.

The technology that would come to dominate quantum computing — the one behind Google's and IBM's headline machines — is in some ways the strangest of all, because its qubits are not natural objects like atoms but ARTIFICIAL atoms: tiny electrical circuits, fabricated on silicon chips using the same lithographic techniques that make ordinary computer chips, engineered to behave as though they were single quantum particles. That a humble electrical circuit — loops of metal, a capacitor, a junction — could be coaxed into displaying the discrete energy levels and delicate superpositions of an atom is one of the most remarkable facts in the field, and it rests on a phenomenon that itself seemed impossible until it was discovered: superconductivity.

Ordinary metals resist the flow of electricity; their electrons collide with the jostling atoms of the warm crystal lattice, dissipating energy as heat, which is why wires warm up and why a current left running in a loop of copper dies away almost at once. But cool certain metals to within a few degrees of absolute zero and something extraordinary happens: their electrical resistance vanishes, utterly and completely. The electrons, instead of scattering chaotically, pair up (into so-called Cooper pairs) and condense into a single, coherent quantum state that flows without any friction at all — a current started in a superconducting loop will circulate, undiminished, essentially forever. This is superconductivity, discovered in 1911, and its deep significance for our story is that it is a QUANTUM phenomenon manifest on a macroscopic, humanly visible scale. A superconductor is a piece of quantum mechanics you could hold in your hand, if it weren't so cold — a single quantum state spread across billions of billions of electrons, all marching in lockstep.

Because a superconducting circuit hosts a coherent quantum state, it can be engineered to have discrete energy levels, just like an atom, and the two lowest of those levels can serve as the 0 and 1 of a qubit. The crucial ingredient that makes this work is a peculiar component called a Josephson junction — a sandwich in which two superconductors are separated by an ultra-thin insulating barrier, through which the paired electrons can "tunnel" in a purely quantum way. The Josephson junction behaves as a special kind of nonlinear element that spaces the circuit's energy levels UNEVENLY, so that the gap between level 0 and level 1 differs from the gap between 1 and 2. This unevenness is essential: it lets engineers address just the lowest two levels with a precisely tuned pulse without accidentally exciting the qubit up a ladder of higher states. The Josephson junction is the beating heart of the superconducting qubit, the component that transforms a mere circuit into an artificial atom, and it is no exaggeration to say that the modern quantum computing industry is built on these microscopic sandwiches.

The great advantage of superconducting qubits — the reason Google, IBM, and many others bet on them — is that they are MANUFACTURED, not caught. Because they are fabricated on chips using the mature techniques of the semiconductor industry, they can, in principle, be designed to order, patterned in large arrays, wired to their neighbors, and scaled up using the same industrial machinery that already prints billions of transistors on a fingernail of silicon. They are also FAST: superconducting gates take mere tens of nanoseconds, roughly a thousand times quicker than trapped-ion gates, so an enormous number of operations can be crammed into even a short coherence window. Speed and manufacturability are a potent combination, and they are why superconducting circuits pulled ahead in the race to build machines with many qubits.

It is worth pausing on the sheer strangeness of what an "artificial atom" really is, because familiarity can dull the wonder. A natural atom's discrete energy levels come from the quantum confinement of electrons around a nucleus, fixed by the immutable constants of nature. A superconducting qubit's discrete levels come from the collective sloshing of billions of paired electrons around a loop of aluminum or niobium, shaped by an engineer's choice of circuit dimensions — the size of a capacitor, the area of a junction. In other words, physicists *design* the atom's spectrum, tuning the spacing of its energy levels by adjusting the layout on the chip. This is a power natural atoms never grant: you cannot redesign a hydrogen atom, but you can draw a superconducting qubit to specification, choosing its frequency, its sensitivity, its couplings. The artificial atom is quantum mechanics made to order.

The history of the field is, in one sense, a history of coaxing these circuits to hold their quantum states a little longer. The first superconducting qubit to show quantum coherence, demonstrated in Japan in 1999 by Nakamura, Pashkin, and Tsai, held its superposition for barely a nanosecond — long enough to prove the principle, far too short to compute. Two decades of relentless refinement — purer materials, cleaner interfaces, better shielding, smarter designs — have stretched that coherence a hundred-thousand-fold, from a nanosecond to hundreds of microseconds. Every one of those improvements was a small war fought against a specific source of noise: stray two-level defects lurking in the oxide of the junctions, microscopic magnetic impurities, vibrations, radiation leaking in from the warm world above. The engineering is less like building a machine than like gardening in a hostile climate, coaxing a delicate quantum flower to bloom a fraction of a second longer.

A further reflection on the artificial atom illuminates why superconducting circuits, for all their fragility, became the industry's workhorse. The deepest advantage is not merely that the circuits can be manufactured, but that they can be *designed*. A natural atom's properties are fixed by nature; you take what hydrogen or calcium offers. A superconducting qubit's frequency, its sensitivity, its couplings to neighbors, its very energy-level structure are all set by the engineer's choices on the chip — the geometry of a capacitor, the area of a Josephson junction. This tunability lets designers craft qubits to interact in specified ways, to sit at frequencies that avoid crosstalk, to be coupled and decoupled on demand. It is a kind of control that natural atoms never grant, and it is why superconducting circuits could be assembled into the large, wired arrays that carried the field to its first supremacy and error-correction milestones. The artificial atom is not merely a substitute for the real thing; it is a real thing that can be shaped to purpose, and that shapeability is the foundation of the whole superconducting program.

The steep price of this control — extreme fragility and extreme cold — is worth restating as the mirror image of the trapped ion's bargain, because the contrast defines the central strategic choice of the hardware race. A trapped ion is a flawless natural atom, identical to every other, coherent for seconds, but slow and hard to assemble in numbers. A superconducting qubit is a fabricated circuit, each slightly different from its neighbors, coherent for mere microseconds, and requiring temperatures colder than deep space merely to exist — but fast, and printable in large arrays by the same industry that makes the world's chips. It is quality versus quantity, nature versus manufacture, flawlessness versus scalability. The entire superconducting bet is that manufacture will win — that it is easier to build a million flawed-but-improvable qubits and correct their errors than to assemble a million flawless ones. That bet has not yet been settled, and it may not be for years; but it is why, when the public pictures a quantum computer, it almost always pictures superconducting technology, cooled in its iconic golden refrigerator to within a whisker of absolute zero. Managing that cold, it turns out, is one of the most demanding engineering feats in the entire enterprise, and it is the subject to which we now turn.

Chapter 35: The Cold Heart

To understand quantum computing viscerally — to feel in your bones why it is so difficult — there is no better place to look than the temperature at which a superconducting quantum computer must operate. The chip at the bottom of that golden chandelier is held at about ten to fifteen **millikelvin**: ten to fifteen thousandths of a degree above absolute zero, the coldest temperature physics allows, the point at which all thermal motion would cease. This is colder than anything in the natural universe. The cosmic microwave background, the faint afterglow of the Big Bang that fills all of space and sets the temperature of the empty void between the galaxies, sits at about 2.7 kelvin — nearly three degrees above absolute zero, and thus some two hundred times *hotter* than the inside of a quantum computer. To run these machines, human beings routinely create, in laboratory refrigerators, the coldest cubic centimeters in the known cosmos. There is something awe-inspiring in that fact, and something clarifying: the extremity of the cold is a direct measure of the fragility of the quantum states being protected.

Why so cold? The answer goes to the heart of the fragility problem from Part VI. Heat is motion — the random jostling of atoms and the random buzzing of electromagnetic energy — and heat is the mortal enemy of quantum coherence. A superconducting qubit's two states, |0> and |1>, are separated by a minuscule amount of energy, and if the surrounding environment contains thermal energy comparable to that gap, random thermal fluctuations will constantly kick the qubit between its states, scrambling any superposition before a computation can finish. To keep the qubit still and coherent, the thermal energy of its surroundings must be pushed far below the qubit's energy gap — which, given how small that gap is, means pushing the temperature down to millikelvins. Only in that profound cold does the qubit's environment become quiet enough, empty enough of random thermal kicks, for the delicate superposition to survive long enough to be useful. The cold is not a technicality; it is the moat that keeps the warm, noisy, classical world at bay, and the depth of the moat reflects the delicacy of what it protects. Trapped ions, we should note, achieve their isolation differently — the ions themselves are laser-cooled but the apparatus can be warmer — but the superconducting machines that dominate the industry live or die by their refrigeration.

The machine that produces this cold is itself a triumph of engineering, and it is the golden chandelier of the introduction: the **dilution refrigerator**. It cools in stages, each colder than the last, the whole assembly nested like Russian dolls, descending from room temperature at the top to millikelvins at the bottom where the chip sits. The upper stages use liquid helium to reach a few kelvin; the final, coldest stage exploits a remarkable quantum property of a mixture of two isotopes of helium — helium-3 and helium-4 — which, at low temperatures, separate into two phases, and forcing helium-3 atoms across the boundary between them absorbs heat, much as evaporation cools your skin, pulling the temperature down into the millikelvin range. The gold plating that gives the apparatus its jewel-like appearance is functional: gold is an excellent conductor of heat and resists corrosion, helping to carry stray heat away from the delicate lower stages. The coiled tubing, the tiers of plates, the filigree of wiring that carries control signals down to the chip and qubit measurements back up — every element is shaped by the demand to deliver signals to the qubits while letting in as little heat as possible, for every wire that reaches down to the cold chip is also a path by which warmth can leak in. The dilution refrigerator is one of the most beautiful objects in modern science, and its elaborate form is dictated entirely by the war against heat.

The extreme refrigeration is also a sobering reminder of the practical realities that temper the more breathless visions of a quantum future. A quantum computer of this kind is not a device that will ever sit on a desk or in a pocket; it is a room-sized cryogenic installation, drawing significant power to run its refrigeration and its racks of control electronics, requiring specialized facilities and expert staff. The notion, occasionally floated in the more excitable press, of a "quantum laptop" or a quantum chip in your phone is, for the foreseeable future, a fantasy, at least for the superconducting and most other leading approaches; these machines belong in laboratories and data centers, accessed remotely over the internet, as indeed they already are — you can, today, send a program to a real quantum computer in an IBM or Google facility from your web browser and get results back, without ever going near the cold heart that runs it. The cloud, not the desktop, is how quantum computing will reach its users, at least until some far-off breakthrough changes the picture. And the refrigeration imposes its own scaling headaches: every additional qubit needs its own control wires threading down into the cold, and there is only so much room, and only so much heat budget, before the wiring itself becomes a bottleneck — a real and pressing engineering constraint on how large these machines can grow, and a spur to ingenious schemes for controlling many qubits with fewer wires.

If you have seen a photograph of a modern quantum computer, you have almost certainly seen not the quantum computer at all but its refrigerator — the gleaming, golden, chandelier-like tower of tiered plates and cascading copper and brass tubing that hangs suspended in laboratories at Google, IBM, and their rivals. This gorgeous, alien-looking apparatus has become the visual emblem of the entire field, gracing magazine covers and corporate advertisements. It is a dilution refrigerator, and the tiny quantum chip — the actual computer, no bigger than a fingernail — sits bolted to the very bottom, at the coldest point, hidden from view. The glamorous golden tower exists for one purpose: to make that little chip colder than almost anywhere in the known universe, and to keep it there.

The temperatures required are difficult to comprehend. Superconducting qubits must be cooled to around ten to fifteen millikelvin — that is, ten to fifteen THOUSANDTHS of a degree above absolute zero, the coldest temperature physically possible, the point at which all thermal motion ceases. To grasp how cold this is, consider that the vacuum of deep space, far from any star, warmed by the faint afterglow of the Big Bang, sits at about 2.7 kelvin — nearly three full degrees above absolute zero. The chip in a quantum computer is kept roughly two HUNDRED times colder than the emptiness between the galaxies. The coldest place in the universe, so far as we know, is not some remote cosmic void but the bottom plate of a dilution refrigerator in a quantum computing laboratory here on Earth. We have engineered, on purpose, pockets of cold that nature nowhere provides.

Why must it be so absurdly cold? For two intertwined reasons that go to the heart of what a qubit is. First, superconductivity itself requires very low temperatures; warm the circuit and the frictionless flow of paired electrons breaks down, and the artificial atom ceases to exist. Second, and more fundamentally, heat is the enemy of all quantum coherence. Temperature is nothing but random jiggling motion, and that jiggling is a form of noise — a ceaseless buffeting that disturbs delicate superpositions and destroys them through decoherence. The energy gap between a superconducting qubit's 0 and 1 states is minuscule, so even a tiny amount of stray thermal energy, a whisper of heat, is enough to flip the qubit randomly or blur its state. To give the fragile superposition any chance of surviving long enough to compute, the environment must be stilled almost completely, drained of nearly all its thermal jiggling. Cold is not a convenience; it is a precondition for the quantum-ness to exist at all.

The dilution refrigerator achieves this near-perfect stillness through a genuinely ingenious process, and its name comes from the trick at its core. It exploits the peculiar quantum behavior of a mixture of two isotopes of helium — ordinary helium-4 and the rare, lighter helium-3. At temperatures near absolute zero, this mixture spontaneously separates into two distinct phases, one rich in helium-3 and one poor, like oil and water. Forcing helium-3 atoms to cross the boundary from the concentrated phase into the dilute phase absorbs heat — much as a liquid absorbs heat when it evaporates, cooling what it leaves behind — and by continuously pumping helium-3 through this phase boundary in a closed cycle, the refrigerator draws heat relentlessly out of the chip, pushing it colder and colder. The tiered golden structure you see is the cascade of successive cooling stages, each colder than the one above, each plated in copper and gold for the metals' superb thermal conductivity, stepping the temperature down from room warmth at the top to those final few thousandths of a degree at the bottom where the chip lives.

The refrigerator is not a mere accessory; it is one of the central engineering challenges of the whole enterprise, and it shapes the machine's future in ways the public rarely appreciates. Every wire that carries a control signal down to the qubits — and there may be several wires per qubit, for control and readout — also carries heat down from the warm world above, fighting the refrigerator's efforts. As machines grow from hundreds of qubits toward the millions a useful, error-corrected computer will need, the sheer number of wires threatens to overwhelm the cooling: you cannot simply run a million warm cables into a box kept colder than deep space without boiling it. This "wiring bottleneck" is one of the great unsolved practical problems of scaling, and it is driving research into cryogenic control electronics that could live inside the cold, reducing the flood of wires. The beautiful golden chandelier, in other words, is not just an icon but a constraint — a reminder that a quantum computer is as much a triumph of refrigeration and plumbing as of physics.

It is worth pausing on the sheer improbability of what the dilution refrigerator achieves, because familiarity with the iconic golden image can dull the astonishment. To keep its superconducting chip working, a quantum computer must cool that chip to around ten to fifteen millikelvin — ten to fifteen *thousandths* of a degree above absolute zero, the coldest temperature physically possible. Set this against the temperature of deep space: the vacuum between the galaxies, warmed only by the faint afterglow of the Big Bang, sits at about 2.7 kelvin, nearly three full degrees above absolute zero. The chip in a quantum computer is kept roughly *two hundred times colder* than the emptiness between the stars. So far as we know, the coldest places in the entire universe are not remote cosmic voids but the bottom plates of dilution refrigerators in quantum computing laboratories here on Earth. We have engineered, deliberately and routinely, pockets of cold that nature nowhere else provides — and we have done it not for its own sake but because only in such profound stillness can the fragile quantum states of the qubits survive long enough to compute.

The reason such extreme cold is not a luxury but a necessity goes to the heart of what a qubit is, and it is worth spelling out. There are two intertwined reasons. First, superconductivity itself requires very low temperatures; warm the circuit and the frictionless flow of paired electrons breaks down, and the artificial atom simply ceases to exist. Second, and more fundamentally, heat is the mortal enemy of all quantum coherence. Temperature is nothing but random jiggling motion, and that jiggling is a form of noise — a ceaseless buffeting that disturbs delicate superpositions and destroys them through decoherence, the environment's inadvertent measurement of the qubit. The energy gap between a superconducting qubit's 0 and 1 states is minuscule, so even a whisper of stray thermal energy is enough to flip the qubit randomly or blur its state. To give the superposition any chance of surviving long enough to compute, the environment must be stilled almost completely, drained of nearly all its thermal motion. Cold is not a convenience; it is the precondition for the quantum-ness to exist at all — a hush deep enough that the qubit's faint quantum whisper is not drowned by the roar of heat.

The refrigerator is not a mere accessory to the quantum computer but one of its central engineering challenges, and it shapes the machine's future in ways the public rarely appreciates. Every wire that carries a control signal down to the qubits also carries heat down from the warm world above, fighting the refrigerator's efforts, and each qubit may need several such wires. As machines grow from hundreds of qubits toward the millions a useful error-corrected computer will require, the sheer number of wires threatens to overwhelm the cooling: you cannot run a million warm cables into a box kept two hundred times colder than deep space without boiling it. This "wiring bottleneck" is one of the great unsolved practical problems of scaling, driving research into cryogenic control electronics that could live inside the cold and multiplexing schemes that share wires among many qubits. The beautiful golden chandelier, in other words, is not just an icon but a constraint — a standing reminder that a quantum computer is as much a triumph of refrigeration and plumbing as of physics, and that the road to a large machine runs through the unglamorous engineering of cold and wire. With qubits fabricated on chips and kept alive in the coldest place in the universe, the engineers finally had a technology they could build and scale — and the frantic race to count qubits, and to declare victory, could begin.

Chapter 36: Counting Qubits

Before we plunge into the menagerie of rival technologies, it is worth pausing to ask a question that sounds simple and is not: how do you measure the power of a quantum computer? The public conversation fixates on a single number — the count of qubits — and treats the race as a straightforward contest to build machines with ever more of them, as though quantum computing were a horsepower competition. This is one of the most persistent and misleading ideas in the whole subject, and clearing it up is essential before we can judge the contenders sensibly. A quantum computer's usefulness depends on far more than how many qubits it has, and a machine with fewer, better qubits can be vastly more capable than one with more, worse ones. Learning to read past the qubit count is learning to see the field clearly.

Consider why raw qubit count is such a poor measure. A qubit that is noisy, that loses its coherence after a few operations, that makes an error in one gate out of fifty, is nearly worthless for real computation, because any algorithm of interesting length will have dissolved into garbage long before it finishes. A thousand such qubits do not make a useful computer; they make a thousand ways to fail. What matters is the combination of *how many* qubits you have *and how good they are* — how long they stay coherent, how accurately their gates operate, how well they connect to one another, how faithfully they can be measured. A machine with fifty superb qubits, each with a gate error of one in ten thousand and long coherence, can run deeper, more meaningful computations than a machine with a thousand shoddy ones. This is why the announcement of a new "record" qubit count, so beloved of headlines, tells you almost nothing by itself. The right question is always: qubits of *what quality*, connected *how*, and error-corrected or not?

The field has developed several more honest metrics to capture this, and it is worth knowing them, because they recur throughout any serious discussion. **Gate fidelity** measures how accurately a quantum operation is performed — a two-qubit gate fidelity of 99.9 percent means that, on average, the gate does what it should 999 times out of a thousand, and this number, especially for the difficult two-qubit entangling gates, is one of the truest measures of a machine's quality; crossing the surface-code threshold of roughly 99 percent is the critical achievement. **Coherence time** measures how long a qubit retains its quantum state before decohering, setting an upper limit on how long a computation can run. **Connectivity** describes which qubits can directly interact — all-to-all, as in trapped ions, or nearest-neighbor-only, as in superconducting chips — and it strongly affects how efficiently real algorithms can be run, since poorly connected qubits require wasteful extra operations to shuttle information around. IBM introduced a composite metric it calls **quantum volume**, which attempts to roll qubit count, connectivity, and error rates into a single number capturing the size of the most complex circuit a machine can reliably run; others use measures of speed, such as how many quantum operations a machine performs per second. No single number suffices, and anyone who reduces the state of a quantum computer to its qubit count is, wittingly or not, misleading you.

The deepest distinction of all, and the one that will matter most as the field matures, is between **physical qubits** and **logical qubits**, which we met in the discussion of error correction. Every machine of the mid-2020s is, in the terms of Part VI, a *NISQ* machine — Noisy Intermediate-Scale Quantum, John Preskill's phrase — meaning it has an intermediate number of qubits (tens to about a thousand) that are *noisy* and *not* error-corrected. These raw physical qubits are what the headline counts refer to. But the truly transformative machines, the ones that could run Shor's algorithm or long quantum simulations, will be measured in *logical* qubits — error-corrected qubits, each built from many physical ones, reliable enough to run computations of arbitrary length. The transition from counting physical qubits to counting logical qubits marks the passage from the NISQ era to the fault-tolerant era, and it is the great frontier the field is now crossing. A machine with a million physical qubits but no error correction is still a NISQ machine, powerful for some narrow tasks but unable to run the great algorithms; a machine with even a few hundred good *logical* qubits would be a different order of being. When you read, in the chapters ahead, of the latest milestones, the sharpest question to ask is always: are we talking about physical qubits or logical ones? The answer separates incremental progress from historic breakthrough.

As quantum computers moved from single-qubit demonstrations toward machines with dozens and then hundreds of qubits, a number took on outsized importance in the public imagination and in corporate marketing: the qubit count. Press releases trumpeted each new record — 50 qubits, 100, 433, 1,000 — as though the race were a simple matter of piling up more, and the milestones were reported like land-speed records. This chapter is a necessary corrective, for the raw qubit count is one of the most misleading numbers in all of technology, and understanding why is essential to seeing through the hype that perpetually surrounds the field.

The seductive intuition is that more qubits mean a more powerful computer, and there is a kernel of truth in it: because a register of qubits holds a superposition whose size doubles with each qubit added, more qubits do mean an exponentially larger state space, and certain algorithms genuinely need many qubits to run. But this intuition collapses the moment you remember decoherence and error. A qubit is only useful if it is a GOOD qubit — if it holds its quantum state accurately, if the gates that manipulate it are precise, if it can be measured reliably and entangled cleanly with its neighbors. A thousand terrible qubits, each so noisy that a computation dissolves into garbage after a handful of operations, are worth precisely nothing; they cannot run any useful algorithm at all. A dozen superb qubits can do real quantum computations that a thousand bad ones cannot. Quality, not quantity, is what limits what a machine can actually do, and quality is exactly what the headline qubit count omits.

To appreciate the gap, consider what actually determines a quantum computer's power, none of which is captured by counting qubits. There is GATE FIDELITY — the accuracy of each operation, which must be high enough (past the error-correction threshold) that errors do not accumulate faster than a computation can proceed; the difference between 99 percent and 99.9 percent fidelity is the difference between a useless machine and a promising one. There is COHERENCE TIME — how long the qubits survive before decohering, which sets how many operations you can perform before the quantum state evaporates. There is CONNECTIVITY — which qubits can directly interact with which others, since an algorithm may require distant qubits to be entangled, and if they cannot talk directly, expensive extra operations are needed to route information between them. And there is the CIRCUIT DEPTH the machine can sustain — the length of the sequence of operations it can execute before noise overwhelms the signal. A machine's real capability is some fraught combination of all these, and a large qubit count with poor values on any of them is a hollow boast.

The industry, to its credit, has tried to invent honest measures to replace the misleading headline. IBM proposed "quantum volume," a single number that combines qubit count with gate fidelity, connectivity, and achievable circuit depth into one figure of merit that better reflects genuine capability — a machine's quantum volume can only grow if it has many qubits AND those qubits are good AND well-connected. Others speak of "algorithmic qubits" or emphasize the number of reliable logical qubits (error-corrected qubits) rather than raw physical ones, which is arguably the only count that will ultimately matter. These measures are imperfect and themselves subject to gaming, but they reflect a hard-won maturity: the recognition that a quantum computer cannot be reduced to a single number, least of all the crudest one. When a company announces a new record qubit count, the sophisticated reader now asks immediately: yes, but how good are they, and how well connected, and how long do they last?

The distinction between physical and LOGICAL qubits deserves special emphasis, because it is where the qubit-counting hype most badly misleads. Recall the surface code's brutal overhead: protecting a single high-quality logical qubit may require hundreds or thousands of physical qubits. So a machine boasting a thousand physical qubits might, once error correction is switched on, yield only a handful of logical qubits — or, if its physical qubits are below the threshold in quality, none at all. The numbers that make headlines are physical-qubit counts; the numbers that will determine whether a machine can run Shor's algorithm are logical-qubit counts, and the two can differ by a factor of a thousand. A future machine that can factor real encryption keys might be advertised as having a million qubits and possess only a few thousand logical ones. Until that distinction is understood, every qubit-count headline is an invitation to misjudge how close or far the real capability lies.

The seductiveness of the qubit count deserves one more direct assault, because it is the single most persistent source of public misunderstanding about quantum computing, endlessly renewed with each new record announcement. The intuition that "more qubits equals more powerful" contains a kernel of truth — a register of qubits holds a superposition whose size doubles with each qubit added, so more qubits do mean an exponentially larger state space — but it collapses the instant you remember decoherence and error. A qubit is only useful if it is a *good* qubit: if it holds its state accurately, if the gates manipulating it are precise, if it can be measured reliably and entangled cleanly with its neighbors. A thousand terrible qubits, each so noisy that a computation dissolves into garbage after a handful of operations, cannot run any useful algorithm at all, while a dozen superb qubits can do real quantum computations the thousand bad ones cannot. Quality, not quantity, limits what a machine can actually do — and quality is exactly what the headline count omits.

The specific quantities that a raw qubit count ignores are worth enumerating, because knowing them turns a naive reader into a discerning one. There is gate fidelity — the accuracy of each operation, which must clear the error-correction threshold or errors accumulate faster than a computation can proceed; the difference between 99 percent and 99.9 percent fidelity is the difference between a useless machine and a promising one. There is coherence time — how long the qubits survive before decohering, which sets how many operations you can perform. There is connectivity — which qubits can directly interact with which others, since an algorithm may need distant qubits entangled, and poor connectivity forces expensive routing operations. And there is the circuit depth the machine can sustain — the length of the operation sequence it can execute before noise overwhelms the signal. A machine's real capability is a fraught combination of all these, and a large qubit count paired with poor values on any of them is a hollow boast. The industry's own honest measures — IBM's "quantum volume," which folds count, fidelity, connectivity, and depth into one figure, and the emphasis on logical rather than physical qubits — reflect a hard-won maturity: the recognition that a quantum computer cannot be reduced to a single number, least of all the crudest one.

The physical-versus-logical distinction is where qubit-counting hype most badly misleads, and it is worth restating as the crux of the whole chapter. Because of the surface code's brutal overhead, protecting a single high-quality logical qubit may require hundreds or thousands of physical qubits, so a machine boasting a thousand physical qubits might, once error correction is switched on, yield only a handful of logical qubits — or, if its physical qubits are too noisy to be below threshold, none at all. The numbers that make headlines are physical-qubit counts; the numbers that will determine whether a machine can run Shor's algorithm are logical-qubit counts, and the two can differ by a factor of a thousand. A future machine capable of factoring real encryption keys might be advertised as having a million qubits and possess only a few thousand logical ones. Until this distinction is understood, every qubit-count headline is an invitation to misjudge how near or far the real capability lies — which is why the sophisticated reader, confronted with any new record, asks immediately: yes, but how good are they, how well connected, how long do they last, and how many logical qubits can they form? Keep that skepticism firmly in hand as we enter the modern hardware race, where the temptation to trumpet qubit counts has never been greater.

Part VIII — The Qubit Zoo

Chapter 37: Superconducting — the Industrial Path

If quantum computing has a front-runner in the popular imagination, it is the superconducting qubit, because it is the technology behind the most famous milestones and the deepest corporate pockets. When Google announced quantum supremacy in 2019, it was with a superconducting chip. When IBM unveils its latest processor with a record number of qubits, they are superconducting. The superconducting approach, whose physics we met in Part VII, has become the industrial mainstream of quantum computing, and understanding why — its particular blend of advantages and liabilities, and the corporate strategies built around it — is essential to reading the field's news.

The central strategic advantage of superconducting qubits is that they are *manufactured*. Unlike a trapped ion, which is a gift of nature identical to every other ion of its species, a superconducting qubit is a fabricated circuit, patterned onto a silicon or sapphire chip by the same lithographic techniques that produce conventional microprocessors. This means the vast, mature, staggeringly capable infrastructure of the semiconductor industry — the fabrication plants, the design tools, the accumulated know-how of packing ever more components onto a chip — can, at least in principle, be brought to bear on scaling up quantum processors. It means qubits can be designed and tuned, their properties engineered rather than accepted, and improved iteration after iteration. And it means the qubits can be packed together and wired on a chip, allowing relatively large numbers to be integrated. This manufacturability is why superconducting machines have led the race in raw qubit count, climbing from a handful in the 2000s to Google's 53-qubit Sycamore in 2019, IBM's 127-qubit Eagle in 2021, its 433-qubit Osprey in 2022, and its 1,121-qubit Condor in 2023 — a steep ascent that no other platform has matched in sheer numbers.

The strategic liabilities are the price of using a fabricated, solid-state object rather than a perfect atom. Superconducting qubits are, individually, less pristine than trapped ions: their coherence times, though improved by tens of thousands of times since 1999, are still shorter, measured in hundreds of microseconds, and their gate error rates, while now past the crucial threshold, are generally higher than the best ions'. Because they are manufactured, no two are exactly alike, so each chip must be painstakingly characterized and calibrated, its qubits' individual quirks measured and compensated — a laborious process that grows harder as chips grow larger. Their connectivity is limited: a superconducting qubit interacts mainly with its immediate neighbors on the chip, so algorithms requiring distant qubits to interact must shuffle information across the grid through chains of operations, wasting gates and accumulating errors. And they demand the extreme refrigeration of Part VII, with all its bulk, expense, and wiring constraints. These liabilities are real, and they are why the raw qubit-count lead of superconducting machines does not translate directly into a lead in useful computational power.

The two giants of the superconducting approach, Google and IBM, have pursued notably different strategies, and the contrast illuminates the field. Google, through its Quantum AI division founded by Hartmut Neven and long driven by the physicist John Martinis (who led the supremacy effort before departing in 2020), has focused on quality and on the specific goal of demonstrating error correction — building smaller, exquisitely characterized chips optimized to prove scientific milestones, culminating in the 2024 Willow processor's below-threshold error-correction demonstration. IBM, through a long-published and closely watched roadmap, has emphasized scale, connectivity, and building an accessible ecosystem — pushing qubit counts upward, introducing modular "Heron" chips designed to be linked together, developing the software stack and cloud access that let thousands of outside researchers run programs on IBM machines, and pivoting, as the field matured, from raw qubit count toward error mitigation, quantum "utility," and a declared path to fault tolerance later in the decade. Other companies and national laboratories — Rigetti in the United States, IQM in Europe, and numerous academic groups — pursue superconducting qubits as well, but Google and IBM define the platform's frontier.

Superconducting qubits are the workhorses of the modern quantum computing industry, the technology behind the machines that have generated the most headlines, attracted the most investment, and reached the largest qubit counts. Google and IBM, the two giants whose rivalry has defined the public face of the field, both built their programs on superconducting circuits, and it is worth drawing together, in one place, why this technology pulled ahead and what its prospects and problems are as the machines grow toward the scales that useful computation will demand.

The case for superconducting qubits rests on the twin pillars we have already met: manufacturability and speed. Because the qubits are lithographically fabricated on chips using techniques adapted from the semiconductor industry, they inherit that industry's staggering capacity for miniaturization and mass production; you can pattern a hundred or a thousand qubits on a single chip and wire them to their neighbors in one fabrication run, and the roadmap to still larger arrays uses machinery the world already knows how to build. And because superconducting gates operate in tens of nanoseconds, roughly a thousand times faster than trapped ions, an enormous number of operations can be executed within even the brief coherence window, and computations run quickly. For a field where the goal is eventually millions of qubits performing billions of operations, the marriage of chip-scale manufacturing with blistering gate speed is a compelling foundation, and it is why the largest machines to date are superconducting.

The dominant design is the transmon qubit, a refinement introduced in the late 2000s that solved an early plague of superconducting circuits: their extreme sensitivity to electrical charge noise, tiny stray fluctuations in the environment that scrambled the qubit's state. The transmon, by cleverly adjusting the circuit's parameters — increasing a certain ratio so that the qubit becomes insensitive to charge fluctuations while retaining just enough nonlinearity to be addressable — dramatically extended coherence times and made superconducting qubits stable enough to be practical. The transmon is now the near-universal building block of superconducting quantum computers, and successive generations of it, together with better materials, cleaner fabrication, and improved control, have pushed coherence times from mere nanoseconds in the earliest devices to hundreds of microseconds today — a thousandfold improvement won through two decades of painstaking engineering.

The qubit-count milestones have accordingly been dominated by superconducting machines, and the pace has been striking. Google's Sycamore processor of 2019, with 53 qubits, made the first quantum-supremacy claim (the subject of coming chapters). IBM embarked on an aggressive public roadmap, unveiling processors with playful bird names in a deliberate escalation — Eagle with 127 qubits in 2021, Osprey with 433 in 2022, and Condor with over

1,000 in 2023 — while simultaneously, and more importantly, shifting its emphasis from raw counts toward quality, modularity, and error correction, having learned the lesson that bigger is not automatically better. The trajectory demonstrated that superconducting technology could indeed scale in numbers; the harder question, still being answered, was whether it could scale while keeping the qubits good enough, and well enough connected, to be worth having.

For the problems of superconducting qubits are real and stubborn, and honesty requires facing them. Their coherence times, though vastly improved, remain far shorter than trapped ions' — microseconds versus seconds — leaving a much narrower window in which to compute. Because each qubit is individually fabricated, no two are identical, and large chips require exhausting, continual calibration to keep every qubit and every gate tuned. Connectivity is limited: a qubit typically talks only to its immediate neighbors on the chip, so entangling distant qubits demands chains of intermediary operations that consume time and inject error. And looming over everything is the wiring and refrigeration bottleneck — every qubit needs control lines threading down into the coldest place in the universe, and it is not yet clear how to route the millions of such lines a full-scale machine would require without overwhelming the refrigerator. Solving that may demand a shift to modular architectures, in which many smaller chips are linked together rather than one impossibly large chip built, and much current effort goes into exactly this.

The rivalry between Google and IBM, both building on superconducting circuits, has been one of the most productive competitions in modern science, and it is worth appreciating how differently the two giants have approached the same technology. Google, under John Martinis and his successors, pursued a strategy of deep, focused demonstrations — building the specific chip needed to prove a specific point, whether quantum supremacy (Sycamore, 2019) or below-threshold error correction (Willow, 2024). IBM, by contrast, pursued breadth and openness: a public, aggressive roadmap with named processors, free cloud access that put real quantum hardware into the hands of hundreds of thousands of users worldwide, and a deliberate cultivation of an ecosystem of programmers, educators, and startups. The two philosophies — the sharp demonstration and the broad platform — have pushed the whole field forward faster than either could alone, and the competition between them has been a gift to the science.

A recurring theme worth dwelling on is the way the superconducting community learned, painfully, that qubit count is a snare. IBM's early roadmap emphasized raw numbers — 127, then 433, then over 1,000 qubits — and the milestones drew headlines. But the company's own engineers understood that a thousand noisy qubits could not run a single useful error-corrected computation, and IBM's more recent strategy has visibly shifted toward quality, connectivity, modularity, and error correction, culminating in a roadmap reorganized around efficient qLDPC codes and a fault-tolerant target near the end of the decade. This maturation — from counting qubits to building *good* qubits and welding them into logical ones — mirrors the field's broader passage from the hype of supremacy to the sober work of fault tolerance, and it is a story the superconducting camp has lived out in public.

A closing reflection on the superconducting era captures both its genuine triumphs and its unresolved anxieties. By most measures of scale, superconducting circuits pulled decisively ahead in the race to build machines with many qubits, and the milestones they produced — Google's Sycamore reaching quantum supremacy in 2019, IBM's processors climbing past a thousand qubits, and above all Google's Willow demonstrating error correction below threshold in 2024 — are among the defining achievements of the whole field. These were not incremental steps but genuine landmarks, and they were possible because superconducting technology married chip-scale manufacturing to blistering gate speed, allowing large arrays of qubits to be built and operated quickly. When the history of the first fault-tolerant quantum computers is written, superconducting circuits will occupy a central place, for they carried the field from tabletop demonstrations to machines that entered regimes beyond classical simulation and, crucially, proved that quantum errors can be systematically beaten.

Yet the honest verdict is that front-runner is not the same as sure winner, and the superconducting platform's unresolved problems are precisely the ones that could yet let a rival overtake it. Its coherence times, though improved a thousandfold over two decades, remain far shorter than trapped ions'; each individually-fabricated qubit differs slightly from its neighbors and demands exhausting continual calibration; connectivity is limited to nearest neighbors, forcing costly routing to entangle distant qubits; and above all looms the wiring and refrigeration bottleneck — the seemingly intractable problem of threading millions of control lines into a box kept colder than deep space without boiling it. Solving that may require abandoning the dream of one impossibly large chip in favor of modular architectures, in which many smaller chips are linked by short quantum interconnects, and much current effort goes into exactly this. The field is emphatically not settled on a single technology, and the wise observer treats the hardware race as genuinely open. Several rivals — trapped ions with their pristine quality, neutral atoms with their surging scale and mobility, photonics with its room-temperature robustness, silicon with its manufacturing promise, topological qubits with their dream of built-in protection — each have strengths exactly where superconductors are weak. Superconducting circuits made quantum computing a large-scale industrial endeavor rather than a tabletop science, and they lead by most measures of scale today; but the machine that finally delivers useful, error-corrected computation may yet be built from something else, and to see why, we turn to the chief challenger.

Chapter 38: Trapped Ions — the Fidelity Kings

If the superconducting qubit is the industrial mainstream, the trapped ion is the connoisseur's choice — the platform that consistently posts the highest quality numbers, the cleanest gates, the longest coherence, the closest approach to the textbook ideal of a qubit. We met the trapped ion's physics in Part VII; here we cast it as the chief rival to the superconducting approach, and examine the strategy of quality-over-quantity on which its champions have staked their bet. The rivalry between these two platforms is the central drama of the modality wars, and it turns on a genuine and unresolved question: in the race to useful, fault-tolerant quantum computing, does it matter more to have *many* qubits or *good* ones?

The trapped ion's case rests on its extraordinary fidelity. Because the qubit is encoded in the internal states of a single atom — one of nature's own flawless, identical objects, isolated in vacuum and shielded from the world — trapped-ion qubits achieve the best gate accuracies and longest coherence times of any technology. The finest trapped-ion two-qubit gates reach fidelities of 99.9 percent and beyond, meaning fewer than one error per thousand entangling operations, comfortably past the surface-code threshold and generally better than the best superconducting qubits. Trapped ions also enjoy **all-to-all connectivity**: through the shared vibrational bus of the ion string, any ion can be made to interact directly with any other, without the shuffling that superconducting chips require, so algorithms map onto the hardware more efficiently and waste fewer operations. And because all ions of a given species are perfectly identical, there is no chip-to-chip variability to calibrate away. On the honest metrics of Part VII — fidelity, coherence, connectivity — trapped ions lead.

The trapped ion's burden is scale, and it is a heavy one. A single trap can hold only a limited number of ions in a well-behaved line before the shared vibrational modes that enable the gates become too numerous and crowded to control cleanly, and the gates themselves are relatively *slow* — microseconds per operation, far longer than the nanosecond gates of superconducting circuits, because they rely on the physical motion of atoms. Building a large trapped-ion computer therefore requires escaping the single-trap limit, and the field has developed two main strategies for doing so, each demanding. One, the **quantum charge-coupled device** or QCCD architecture, physically *shuttles* ions between multiple connected trapping zones on a chip, moving them around like beads on an abacus to bring the right pairs together for gates — an approach that works but adds great complexity in controlling the ions' motion. The other links separate ion-trap modules through *photons*, using light to entangle ions in different traps and so building a larger machine out of networked small ones — a beautiful idea, still maturing. Either way, scaling trapped ions to the thousands or millions of qubits that fault tolerance demands is a formidable engineering challenge, and it is why trapped-ion machines have lagged behind superconducting ones in raw qubit count even as they lead in quality.

The two flagship companies of the trapped-ion approach embody the platform's strengths and its corporate seriousness. **IonQ**, spun out of research at the University of Maryland and Duke by the physicists Christopher Monroe and Jungsang Kim, became one of the first pure-play quantum-computing companies to go public, and it emphasizes the high fidelity and all-to-all connectivity of its ions, arguing that its qubits' quality lets it achieve more with fewer of them. **Quantinuum**, formed by the merger of Honeywell's quantum division with Cambridge Quantum, has pursued the QCCD shuttling architecture and has posted some of the field's most impressive results — record quantum volumes, high-fidelity logical qubits, and early demonstrations of error-corrected operations that outperform the underlying physical qubits, a crucial milestone on the road to fault tolerance. When, in the mid-2020s, teams began demonstrating small numbers of genuine *logical* qubits — error-corrected qubits performing better than their physical constituents — trapped-ion machines, with their superior fidelity, were among the leaders, precisely because high physical fidelity makes error correction cheaper and easier.

Having surveyed the superconducting front-runner, we return to its chief rival, the trapped ions we first met as the field's earliest triumph — for in the modern era they have grown from a laboratory technique into a serious industrial technology, embodied by companies whose machines, though smaller in qubit count, are in crucial respects the best in the world. The trapped-ion story is the great counterexample to the superconducting emphasis on scale, a standing argument that quality may matter more than quantity, and it is championed most prominently by two companies: IonQ, spun out of the University of Maryland and the pioneering work of Christopher Monroe, and Quantinuum, the trapped-ion venture that grew out of Honeywell's deep expertise in precision engineering and later merged with Cambridge Quantum.

The case for trapped ions, restated for the modern race, comes down to fidelity, coherence, and connectivity — the three qualities that the raw qubit count ignores and that ultimately determine what a machine can do. Trapped-ion gate fidelities are the highest of any technology, routinely exceeding 99.9 percent and in the best two-qubit demonstrations pushing toward 99.99 percent, comfortably beyond the error-correction threshold; their coherence times are measured in seconds or longer, a millionfold advantage over superconducting circuits; and, because the ions share collective modes of motion, they can offer all-to-all connectivity, so that any qubit can be entangled directly with any other, without the routing overhead that plagues chip-based rivals. On the metrics that measure genuine capability rather than mere size, trapped-ion machines have repeatedly led the field, and Quantinuum in particular has posted some of the highest quantum-volume figures and the most advanced demonstrations of error correction and logical qubits anywhere.

Those error-correction demonstrations deserve emphasis, because they are where trapped ions have most dramatically shown the value of quality. Quantinuum's machines have been used to create logical qubits — error-corrected qubits encoded across several physical qubits — whose logical error rates are actually LOWER than the physical error rates of their constituent parts, the crucial crossover that proves error correction is working as intended. In one striking result, a small number of physical qubits were used to build logical qubits whose error rates were suppressed hundreds of times below the physical rate. Because trapped-ion qubits are so clean to begin with and so flexibly connected, they can implement sophisticated codes with modest overhead, and they achieved some of the first convincing demonstrations that the whole error-correction program actually delivers on its promise. Where superconducting machines have led in qubit count, trapped ions have often led in the far more meaningful achievement of making qubits BETTER through error correction.

The Achilles' heel, as ever, is scaling, and it is the mirror image of the superconducting predicament. The very features that make trapped ions so good — their reliance on shared vibrational modes and on precisely aimed laser beams — become obstacles as numbers grow. A single trap can hold only a few dozen ions before their collective motion becomes too complex to control cleanly, and the laser systems needed to address many ions individually grow forbiddingly intricate. The gates are also intrinsically slow, microseconds rather than nanoseconds, so that although each operation is exquisitely accurate, the machine runs at a more deliberate pace. The path to thousands or millions of trapped-ion qubits therefore cannot be simply "build a bigger trap"; it requires clever architectures.

The leading architectural idea is the quantum charge-coupled device, or QCCD, in which ions are physically shuttled around a chip between separate zones — some for storage, some for gates, some for measurement — like tiny charged marbles moved through a microscopic switchyard by electric fields, so that only manageable handfuls of ions ever share a vibrational mode at once. Quantinuum's machines use exactly this shuttling approach, and it has proven powerful, though it is intricate and demanding. A complementary long-range idea is to link separate ion traps together using photons — flying qubits of light that carry entanglement between distant modules — knitting many small, high-quality traps into a larger networked machine. Both approaches are actively advancing, and both aim to preserve the trapped ion's superb per-qubit quality while escaping the ceiling that a single trap imposes.

To feel the trapped ion's advantage in something like human terms, consider the numbers side by side. A superconducting qubit holds its quantum state for perhaps a few hundred microseconds; a trapped-ion qubit can hold its for seconds, and in specialized "clock" states for minutes — a difference of a millionfold or more. In that longer window an ion can undergo vastly more operations before its coherence decays, which is why, despite their slow gates, trapped ions can execute deep, complex circuits that noisier platforms cannot sustain. It is the difference between a sprinter who tires in seconds and a marathon runner who paces steadily for hours; for computations that require many sequential steps — and error correction requires enormous numbers of them — endurance can matter more than raw speed.

The all-to-all connectivity of trapped ions is a subtler advantage that repays a closer look, because connectivity turns out to be one of the most underappreciated determinants of a quantum computer's real power. On a superconducting chip, a qubit can typically talk only to its immediate neighbors, so entangling two distant qubits requires a chain of intermediary operations — a game of telephone that consumes time and injects error at every hop. In a trapped-ion register, any ion can be entangled directly with any other through their shared motion, as if every qubit sat next to every other qubit at once. For many algorithms, and especially for the intricate connectivity that efficient error-correcting codes demand, this flexibility is worth a great deal, effectively multiplying the usefulness of each qubit. It is one reason trapped-ion machines punch so far above their modest qubit counts on measures of genuine capability like quantum volume.

The human story behind trapped ions is worth telling, because it illustrates how academic science becomes industry. The technology grew from decades of painstaking work in atomic physics laboratories — the same tradition that produced the world's best atomic clocks — and its migration to computing owes much to a handful of pioneers, notably David Wineland, whose exquisite control of single ions earned a share of the 2012 Nobel Prize, and Christopher Monroe, who carried the science out of the university and into the marketplace by co-founding IonQ. When Honeywell, a company better known for thermostats and aerospace, brought its formidable precision-engineering expertise to bear and produced trapped-ion machines of exceptional quality — later spun into Quantinuum — it signaled that the quantum industry would be built not only by software giants and startups but by old-line engineering firms with deep skills in vacuum systems, lasers, and control. The trapped ion, the field's first qubit, thus remains its quality benchmark, a standing proof that in quantum computing, as in much else, the best is not always the biggest.

Chapter 39: Neutral Atoms — the Dark-Horse Surge

Every technological race has its surprise, the contender that comes from behind to upend the settled order, and in quantum computing that role has been played, in the early 2020s, by the neutral-atom platform. A few years earlier it had been a promising laboratory curiosity; by the middle of the decade it was posting results that rivaled the best trapped-ion and superconducting machines and offered a more natural path to large numbers, and it had become one of the most closely watched approaches in the field. The neutral-atom surge is a reminder of how open the hardware race remains, and how quickly the landscape can shift.

The neutral-atom qubit is, like the trapped ion, encoded in the internal states of a single atom — but a *neutral* atom, one that has not been stripped of an electron and carries no net charge. This is both the platform's central challenge and, once solved, its great advantage. Because a neutral atom has no charge, you cannot hold it with the electric fields that trap ions; instead you hold it with light. A tightly focused laser beam creates an **optical tweezer**, a microscopic well of light in which a single atom can be caught and suspended, and — crucially — arrays of hundreds or thousands of such tweezers can be created at once, using holographic techniques to split a laser into a whole grid of tiny traps, each holding one atom. The atoms can be arranged in essentially arbitrary two- or three-dimensional patterns, and, remarkably, *rearranged* on the fly: the tweezers can be moved, shuttling atoms into position, sorting a randomly loaded array into a perfect lattice, reconfiguring the machine's very geometry between operations. This reconfigurability — the ability to move qubits around and connect distant ones by physically relocating them — gives neutral atoms a flexible connectivity that neither superconducting chips nor fixed ion strings possess.

The trick that turns an array of trapped neutral atoms into a *computer* — that provides the entangling interaction — is the **Rydberg state**. Ordinarily, neutral atoms sitting a few micrometers apart barely interact at all, which is good for storing quantum information without unwanted disturbance but useless for performing gates. The solution is to briefly excite an atom into a **Rydberg state**, in which one of its electrons is boosted to a very high energy level and orbits far from the nucleus, swelling the atom to enormous size. In this bloated Rydberg state the atom interacts strongly with its neighbors — the so-called Rydberg blockade, in which one excited atom prevents a nearby one from being excited too — and this strong, controllable interaction is used to perform entangling gates between chosen atoms. By exciting atoms into Rydberg states at will, the machine can turn interactions on and off, entangling selected pairs, running the gates that quantum computation requires. The atoms rest quietly, well-isolated, in their ground states, and spring into interaction only when briefly promoted to Rydberg states — a clean separation of storage and interaction that is one of the platform's elegant features.

The neutral-atom approach offers a compelling blend of the virtues of its rivals. Like trapped ions, it uses nature's perfect, identical atoms, and it has rapidly closed the fidelity gap, achieving two-qubit gate accuracies approaching those of the best ions. Like superconducting chips, it offers a natural path to large numbers — arrays of hundreds or even thousands of atoms have been assembled, far more than any single ion trap holds — and its reconfigurable geometry gives it a flexibility all its own. The atoms are cooled and held in vacuum, but the apparatus does not require the extreme dilution-refrigerator temperatures of superconducting circuits, simplifying some of the engineering. These advantages drove a burst of progress that startled the field: in the early 2020s, academic groups at Harvard and MIT (associated with the physicist Mikhail Lukin) and companies such as **QuEra**, **Pasqal**, and **Atom Computing** demonstrated neutral-atom machines of unprecedented scale and, increasingly, quality — including some of the most impressive early demonstrations of error correction with genuine logical qubits, marshaling hundreds of physical atoms into dozens of protected logical qubits, results that put neutral atoms at the very forefront of the fault-tolerance race.

Among the rival hardware platforms, one has surged from relative obscurity to the front rank with startling speed:

neutral atoms, held motionless not by the electric fields that cage charged ions but by finely focused beams of laser light called optical tweezers. In just the past several years, neutral-atom quantum computers have gone from a promising laboratory curiosity to one of the most exciting and closely watched technologies in the field, posting results that have forced everyone to take them seriously. Championed by companies such as QuEra (spun out of Harvard and MIT), Pasqal in France, and Atom Computing, the neutral-atom approach combines some of the best features of its rivals and has lately achieved milestones that superconducting and trapped-ion machines could not.

The core idea is elegant. A neutral atom — rubidium or cesium or a similar species — is captured at the focus of a tightly concentrated laser beam, which acts like a tiny bowl of light, holding the atom trapped at its center. Because these optical tweezers can be created and steered with holographic techniques, experimentalists can arrange hundreds or thousands of atoms into arbitrary two- or three-dimensional patterns — grids, lattices, arbitrary shapes — each atom a qubit, each held in its own beam of light. Like trapped ions, these qubits are perfect and identical, because they are natural atoms manufactured to universal specification, and they enjoy long coherence times; unlike trapped ions, they need not share a single vibrational mode, so the number in a single system can be far larger. The atoms are placed, individually and precisely, by light itself.

The entangling mechanism is the platform's masterstroke, and it exploits a wonderful piece of atomic physics called the Rydberg blockade. To make two neutral atoms interact — normally they barely notice one another — you excite them into "Rydberg states," in which an outer electron is promoted to a vast, bloated orbit far from the nucleus, swelling the atom to thousands of times its ordinary size. So enlarged, the atoms exert strong forces on each other across surprising distances, and the "blockade" effect means that once one atom is excited into its giant Rydberg state, it prevents a nearby atom from being excited too — a conditional interaction that can be harnessed to build entangling gates. By flashing the right laser pulses, experimentalists switch this strong interaction on and off at will, entangling chosen pairs of atoms. The atoms sit inert and isolated most of the time, then briefly balloon into giants to shake hands, then shrink back to quiet solitude.

The neutral-atom platform's virtues are a compelling blend. It scales readily to large numbers — machines with hundreds and even over a thousand atoms have been demonstrated, rivaling or exceeding superconducting qubit counts, while retaining atomic-quality coherence. It offers flexible, reconfigurable connectivity, because the tweezers can physically MOVE atoms around during a computation, bringing distant qubits together to interact and then returning them, a mobility neither fixed superconducting chips nor (as easily) trapped ions possess. And it operates without the millikelvin refrigeration that superconducting circuits demand — the atoms are cold, but cooled by laser light rather than by a dilution refrigerator, sidestepping some of the cryogenic wiring nightmare. For a technology that was a minor player only a decade ago, this combination of scale, quality, and flexibility is remarkable.

The most electrifying recent results have come in error correction, and they are why neutral atoms leapt to prominence. In late 2023, a Harvard-led team working with QuEra demonstrated logical qubits and error-correcting operations at a scale and quality that stunned the field — creating dozens of logical qubits and performing entangling operations between them, exploiting the atoms' mobility to shuttle them into the patterns the codes require. It was one of the most advanced demonstrations of error correction on any platform, and it announced neutral atoms as a genuine contender for the endgame, not merely a promising newcomer. The ability to physically rearrange qubits mid-computation turns out to be a powerful asset for implementing the intricate connectivity that error-correcting codes demand.

The speed of the neutral-atom platform's rise is worth dwelling on, because it is a vivid lesson in how quickly the quantum hardware landscape can shift. A decade ago, neutral atoms were regarded mainly as a tool for quantum simulation and precision measurement, not as a serious contender to build a general-purpose quantum computer; the entangling gates were too slow and too error-prone. Then a cluster of advances — better optical tweezers, sharper control of the Rydberg blockade, and above all the realization that atoms could be physically shuttled around during a computation — transformed the field's prospects almost overnight. By the mid-2020s, neutral-atom machines were demonstrating hundreds of qubits and some of the most advanced error-correction results anywhere. Anyone who had written the platform off would have been badly wrong, and the episode stands as a warning against confident predictions about which technology will win.

The mobility of neutral atoms — the ability to pick up qubits and move them with beams of light — is a genuinely novel capability with no real counterpart among the fixed qubits of superconducting chips, and its importance for error correction can hardly be overstated. Recall that the most efficient error-correcting codes, the qLDPC codes that promise to slash the crushing overhead of fault tolerance, require qubits that are not merely nearest neighbors to interact — a demand that flat, fixed hardware struggles to meet. A neutral-atom machine can simply *carry* distant qubits together, entangle them, and carry them back, implementing the long-range connectivity that these powerful codes need. This marriage of a fast-scaling platform with the connectivity that efficient codes demand is a large part of why neutral atoms have surged to the front rank, and why some observers now rate them among the likeliest routes to a large-scale fault-tolerant machine.

There is a pleasing scientific lineage here, too, that connects the newest hardware to the oldest ideas in this book. Neutral atoms cooled and trapped by light are the same systems that gave us the world's most precise atomic clocks and the first Bose-Einstein condensates — exotic states of matter in which thousands of atoms merge into a single quantum entity. The techniques of laser cooling and optical trapping, developed over decades of fundamental atomic physics and honored with multiple Nobel Prizes, turned out to be exactly the tools needed to build a quantum computer, just as they had earlier built the finest clocks and sensors. It is another instance of the deep unity running through this whole story: the same mastery of preparing, holding, and measuring delicate quantum states underlies clocks, sensors, and computers alike, so that an advance in one quietly enriches the others. The atom held motionless in a bowl of light is at once humanity's finest timekeeper, its most sensitive probe, and now, perhaps, a building block of its most powerful computer.

Chapter 40: Photons — Computing with Light

There is a certain poetic fitness in computing with light, for it was light — Planck's glowing oven, Einstein's photons — that opened the quantum age in the first place. The photonic approach to quantum computing encodes qubits in individual particles of light, photons, and manipulates them with the optical components familiar from any physics laboratory: beam splitters, mirrors, phase shifters, and detectors. It is one of the more distinctive and, in some ways, most radical of the platforms, and it comes with a bundle of advantages and disadvantages so different from the matter-based approaches that it almost belongs to a separate category of machine.

The photon's great virtue is its aloofness. A photon barely interacts with its environment as it flies — it can travel through kilometers of optical fiber or the vacuum of space with its quantum state largely intact — which means photonic qubits are remarkably resistant to decoherence, the enemy that forces every other platform into extreme isolation and cold. Photonic quantum computers can operate at or near *room temperature* (though their single-photon detectors often still need cooling), sidestepping the elaborate dilution refrigerators of the superconducting approach. And because photons *are* light, the natural carrier of information over distance, the photonic platform is uniquely suited to quantum *communication* — to linking quantum processors over networks, to building the quantum internet, to distributing entanglement between distant nodes. In a future of networked quantum machines, photons are the natural connective tissue, whatever the processors themselves are made of, and this dual role — as both a computing platform and the medium of quantum communication — gives photonics a special strategic importance.

The photon's aloofness is also its curse, and here is the deep difficulty of the approach. The very fact that photons barely interact with their environment means they barely interact *with each other* — and interaction is exactly what entangling two-qubit gates require. Getting one photon to conditionally affect another, as a CNOT gate demands, is enormously hard, because photons simply pass through one another without noticing. The resolution, worked out in a landmark theoretical result by Emanuel Knill, Raymond Laflamme, and Gerard Milburn in 2001, is both ingenious and demanding: you can perform effective gates between photons using only beam splitters, phase shifters, and *measurement*, exploiting the interference of photons and the act of detection to induce the needed interactions indirectly. But this measurement-based approach makes the gates **probabilistic** — they succeed only some of the time, at random — and building a reliable computer out of unreliable, sometimes-failing gates requires either heroic amounts of redundancy or clever architectures that turn the probabilistic successes into deterministic computation. This is the central engineering burden of photonic quantum computing, and much of the field's cleverness goes into managing it, through schemes such as "fusion-based" and "measurement- based" quantum computing that are built around the probabilistic nature of photonic gates rather than fighting it. Photon loss — photons simply going missing, absorbed or scattered before they reach a detector — is a further relentless enemy, since a lost photon is a lost qubit.

The photonic approach has produced two very different kinds of landmark. On the scientific side, photonic machines have been used for some of the field's most striking demonstrations of quantum computational advantage through a task called **boson sampling** — a specialized problem, proposed by Scott Aaronson and Alex Arkhipov, that asks for the statistics of many photons passing through a complex optical network, a task believed to be exponentially hard for classical computers. In 2020, a group at the University of Science and Technology of China led by Jian-Wei Pan reported a photonic machine named **Jiuzhang** that performed a boson-sampling task said to be far beyond classical reach, a Chinese counterpart to Google's superconducting supremacy claim and a demonstration that photonics could reach the frontier of quantum advantage. On the commercial side, companies have bet heavily on photonics as a path to *scalable* fault-tolerant computing: **PsiQuantum**, notably, has staked its strategy on building a large-scale photonic quantum computer using **silicon photonics** — fabricating the optical components on chips using semiconductor-industry techniques, and pursuing the millions of components a fault-tolerant photonic machine would need — while **Xanadu** in Canada pursues its own photonic architecture with an emphasis on networkability and continuous-variable encodings.

Every other qubit technology we have surveyed shares a common feature: it holds its qubits STILL — atoms suspended in traps, circuits fixed on chips, all pinned in place and shielded from disturbance. The photonic approach to quantum computing does something radically different and, at first hearing, almost paradoxical: it computes with qubits that never stop moving, that fly through the apparatus at the speed of light and can never be held still at all. Its qubits are individual particles of light — photons — and the appeal and the difficulty both stem from this restless, luminous nature. Championed by companies such as PsiQuantum, Xanadu, and others, photonic quantum computing offers a strikingly different bet on the future, with its own distinctive advantages and formidable obstacles.

The great advantage of photons is that they barely interact with their environment — and this makes them nearly immune to decoherence. The reason atoms and circuits must be so exquisitely isolated and, in the superconducting case, cooled colder than deep space, is that they couple readily to their surroundings, which lets the environment disturb and destroy their quantum states. Photons are the opposite: a photon can travel across the galaxy, or through miles of optical fiber, scarcely touched by the matter it passes, its quantum state preserved. This means photonic qubits can maintain coherence at ROOM TEMPERATURE, with no dilution refrigerator, no ultra-high vacuum, no millikelvin cold — an enormous practical simplification. Photonic quantum computers can, in large part, be built and operated warm, using components adapted from the vast, mature telecommunications and integrated-optics industries, which already manufacture the lasers, waveguides, and detectors that move light through the world's fiber networks. A photonic quantum chip can look like a piece of patterned glass on silicon.

But the very property that makes photons wonderfully robust — their reluctance to interact — is also their central curse, and it goes to the heart of what computing requires. To build entangling two-qubit gates, you need your qubits to INTERACT, to influence one another, and photons, gliding past each other like ghosts, refuse to do so directly. Two beams of light simply pass through one another without effect. This makes deterministic two-photon gates extraordinarily hard, and it is the fundamental obstacle photonic quantum computing has wrestled with for decades. The ingenious workaround, developed in a landmark 2001 scheme by Emanuel Knill, Raymond Laflamme, and Gerard Milburn (the "KLM" protocol), is to induce the needed interactions INDIRECTLY, through measurement: by combining photons at beam-splitters and detecting some of them, one can, through the strange logic of quantum measurement and interference, coax the surviving photons into an entangled state. The catch is that this trick is probabilistic — it only succeeds some of the time, and you must try repeatedly, keeping the successes and discarding the failures, which demands generating and shuttling vast numbers of photons and heralding the rare lucky events.

This measurement-based, probabilistic character has pushed photonic quantum computing toward a distinctive architecture, and it is a bold and clarifying bet. Rather than fight to build reliable individual gates, some photonic companies — PsiQuantum most prominently — have embraced a "fusion-based" or measurement-based model in which the computation is driven by generating enormous streams of photons, entangling them in a large resource state, and performing cascades of measurements, accepting the probabilistic failures and compensating with sheer volume and error correction. PsiQuantum has staked its strategy explicitly on manufacturability, partnering with a semiconductor foundry to fabricate photonic chips at industrial scale, wagering that the path to a useful machine runs through mass-producing millions of optical components on silicon wafers rather than perfecting a few exotic qubits. It is a very different theory of how to win the race — leaning on the manufacturing might of the existing photonics industry.

Photons also possess a unique and increasingly important virtue: they are nature's own carriers of quantum information over distance. Because a photon can travel through optical fiber without losing its quantum state, photons are the ideal medium for a quantum NETWORK — for linking separate quantum processors together, for quantum communication, and for the quantum internet we will discuss in later chapters. Even quantum computers built from other technologies, like trapped ions or neutral atoms, increasingly plan to use photons as the "flying qubits" that carry entanglement between modules. So even if photonic machines do not win the race to a standalone general-purpose computer, photonics is likely to be an indispensable part of the broader quantum ecosystem, the connective tissue of any distributed or networked quantum future.

The paradox at the heart of photonic quantum computing — that photons are wonderful because they barely interact, and terrible for the same reason — is worth turning over once more, because it shapes the entire strategy of the companies pursuing it. Every other platform fights a losing battle against the environment: its qubits interact too much, and the environment reaches in to destroy their coherence. Photons have the opposite problem and the opposite virtue. They ignore the environment, gliding through glass and air unperturbed, so their coherence is superb and they need no refrigeration — but they also ignore *each other*, so making them interact to build the two-qubit gates that computation requires is agonizingly hard. The whole art of photonic quantum computing is to manufacture interactions that nature is reluctant to provide, and the cleverness required has made the field a showcase of ingenuity.

The measurement-induced interaction at the heart of the KLM scheme deserves a second, more intuitive pass, because it is one of the subtler ideas in the field. When two photons meet at a beam-splitter and you detect where they go, the very act of detection — through the strange logic of quantum interference and the indistinguishability of photons — can force the photons that survive into an entangled relationship they did not have before. The interaction is not a force one photon exerts on another; it is a correlation conjured by measurement itself. The price is that it works only probabilistically, succeeding at random some fraction of the time, which is why photonic machines must generate torrents of photons and sift the rare successes from the failures. This has pushed the leading photonic effort, at PsiQuantum, toward a strategy of overwhelming volume — mass-producing millions of optical components on silicon wafers in a commercial semiconductor foundry, and accepting the probabilistic failures as a cost to be paid in sheer scale rather than fought with exotic hardware.

The dual role of photons — as a candidate computing platform and as the indispensable carrier of quantum information between machines — is perhaps their most important feature for the field's future, and it is easy to overlook amid the focus on standalone computers. Even if photonic machines never win the race to build a universal quantum computer, photons will almost certainly be the wires of any quantum network, the flying qubits that carry entanglement between trapped-ion or superconducting or neutral-atom processors, knitting many small machines into one larger modular whole and eventually linking distant machines into a quantum internet. In this sense photonics is less a rival to the other platforms than a connective partner to them all — the medium through which quantum information travels when it must move. A future large-scale quantum computer may well be a hybrid: matter qubits doing the computation, photonic qubits carrying information between the modules, each technology doing what it does best. Photons, the qubits that cannot hold still, may find their destiny not in staying put to compute but in ferrying quantum information across the room, across the city, and eventually across the world.

Chapter 41: Spin in Silicon — the Chip-Fab Bet

Of all the ways to build a qubit, one has a special allure for anyone who remembers how the classical computer revolution actually happened: encode the qubit in a single electron's spin, held in a tiny structure carved from **silicon**, the very material on which the entire digital age was built. The silicon spin-qubit approach makes a seductive strategic bet — that the winning quantum platform will be the one that can ride the coattails of the trillion-dollar semiconductor industry, fabricating qubits by the million using the same fabrication plants, materials, and techniques that already produce the processors in every phone and laptop. If quantum computing's central problem is scaling to enormous numbers of qubits, what better foundation than the one technology humanity has already perfected for putting billions of components on a chip?

The qubit, in this approach, is the **spin** of a single electron (or sometimes an atomic nucleus) confined in a **quantum dot** — a minuscule pocket in a semiconductor, engineered to hold one electron in isolation, so small it behaves as an artificial atom. The electron's spin, pointing "up" or "down" in a magnetic field, is a natural two-state system, a natural qubit, and it can be controlled with electrical and magnetic signals and read out by sensitive charge detectors. Silicon is a particularly good host: ordinary silicon can be purified to a special isotope, silicon-28, whose nuclei have no spin and therefore create a magnetically "quiet" environment in which the electron spin can retain its coherence remarkably well, shielded from the magnetic noise that would otherwise scramble it. Silicon spin qubits have achieved impressive fidelities and long coherence times, and they are extraordinarily *small* — far tinier than superconducting circuits or the apparatus around trapped atoms — which is exactly what you want if you dream of packing millions onto a chip.

The strategic appeal is obvious and much emphasized by the approach's backers, chief among them **Intel**, whose entire business is high-volume silicon fabrication, along with academic and commercial groups in Delft (the QuTech institute), Australia (Silicon Quantum Computing, and the company Diraq), and elsewhere. Their argument runs: every other platform must invent its manufacturing from scratch, but silicon spin qubits can, in principle, be made in existing semiconductor foundries, inheriting the industry's unmatched ability to produce identical, tiny components in astronomical numbers with high yield. If that promise is realized, the silicon approach could leapfrog the scaling struggles that bedevil its rivals, going from hundreds to millions of qubits on the strength of infrastructure that already exists. It is the "if we can just make it a chip problem, we win" bet, and given the history of classical computing it is not one to dismiss lightly.

The challenges are equally real, and they explain why silicon spin qubits, despite their promise, have lagged behind superconducting and trapped-ion machines in demonstrated qubit counts. Controlling single electron spins in quantum dots is exquisitely delicate: the dots must be fabricated with near-atomic precision, the electrons coaxed one at a time into their pockets, and the control signals delivered with great accuracy to structures a few tens of nanometers across. Making two spin qubits interact to perform entangling gates requires bringing their electrons close enough to feel one another, a finicky business. And the very smallness that makes the qubits attractive for scaling makes them hard to wire and control individually, raising, in acute form, the same "wiring bottleneck" that haunts all solid-state approaches — how do you deliver control signals to a million qubits packed onto a chip? For years, silicon spin qubits could be demonstrated only in ones and twos, far behind the qubit counts of other platforms, and the field wondered whether the approach's manufacturing promise would ever be realized or would remain perpetually just over the horizon.

Of all the approaches to building a quantum computer, one carries a special strategic allure: making qubits out of silicon, the very material on which the entire trillion-dollar semiconductor industry — and thus the entire modern digital world — is already built. The dream of silicon spin qubits is seductive in the extreme: if quantum information could be encoded in tiny structures fabricated on silicon chips, then the staggering manufacturing infrastructure that produces billions of transistors on a fingernail could, in principle, be turned to producing billions of qubits, and the scaling problem that torments every other technology might be solved by the industry that already knows how to make chips at planetary scale. Championed by Intel, by academic groups in Delft, Sydney, and elsewhere, and by companies like Silicon Quantum Computing and Quantum Motion, this approach bets on the ultimate manufacturing advantage.

The qubit in this scheme is, in most versions, the SPIN of a single electron — or sometimes of a single atomic nucleus — trapped in a tiny structure on a silicon chip called a quantum dot, an artificial "box" a few tens of nanometers across, defined by carefully placed electrodes, that can hold one electron in isolation. An electron behaves like a minuscule magnet, a spin that can point "up" (call it 0) or "down" (call it 1) or into a superposition of both, and this spin is the qubit. It is manipulated with magnetic fields or with electrical pulses applied to the surrounding electrodes, and entangled with its neighbors through the natural interaction between adjacent electron spins. The information lives in the orientation of a single electron's spin, held in a cage of silicon and voltage. These are, in a sense, the smallest possible qubits — a single particle's spin — and their tininess is part of the promise, since more of them can be packed into a given area than any rival technology.

The advantages, beyond the overwhelming one of manufacturability, are genuine and growing. Silicon spin qubits are extraordinarily small, so a future chip could in principle host millions or billions of them in the same area that holds a few hundred large superconducting qubits — the density is potentially enormous. They can boast long coherence times, especially when built from isotopically purified silicon (silicon-28, from which the magnetic silicon-29 nuclei that disturb the spins have been removed), because a purified silicon crystal is a remarkably quiet, non-magnetic home for a spin. And, tantalizingly, they may be able to operate at slightly higher temperatures than superconducting qubits — around one kelvin rather than a few thousandths — which, though still very cold, would ease the crushing refrigeration and wiring burden and allow control electronics to sit closer to the qubits.

But silicon spin qubits have lagged the leading technologies in qubit count, and honesty requires explaining why the dream has been so slow to materialize. The very smallness that makes them promising makes them fiendishly difficult to fabricate and control: placing and wiring structures at the scale of single electrons, with the precision quantum coherence demands, pushes even the semiconductor industry's formidable techniques to their limits, and tiny variations from one quantum dot to the next make every qubit slightly different and hard to calibrate. For years, groups struggled to build more than a handful of spin qubits that worked well together, while superconducting and atomic machines raced ahead into the hundreds. Demonstrating high-fidelity two-qubit gates — getting neighboring electron spins to entangle reliably — proved especially hard, though fidelities have now been pushed above the error-correction threshold in the best devices, a crucial milestone reached only recently.

The recent progress has been encouraging enough to keep the dream alive and, some believe, to position silicon for a comeback. Groups have demonstrated small arrays of spin qubits with gate fidelities exceeding 99 percent, operation of qubits at the "hot" temperature of around one kelvin, and the integration of qubits with conventional control electronics on the same chip — each a step toward the manufacturing-driven scaling that is the whole point of the approach. Intel has fabricated spin-qubit test chips on its standard production lines, demonstrating that the qubits can be made with industrial tools. The bet is a long one: silicon spin qubits are behind today, but IF their manufacturing advantage can be realized, they could scale past every rival with breathtaking speed, leapfrogging from hundreds to millions by riding the same exponential that carried classical chips from thousands of transistors to trillions. Whether that promise is redeemed remains one of the field's great open wagers.

The strategic logic behind silicon spin qubits is worth restating starkly, because it explains why the approach commands such patience despite its slow start. Every other quantum computing platform must build its manufacturing base essentially from scratch — new fabrication techniques, new materials, new supply chains. Silicon spin qubits, alone among the contenders, could in principle inherit the entire, staggering apparatus of the semiconductor industry: the trillion-dollar fabs, the decades of accumulated expertise in patterning structures at the nanometer scale, the supply chains, the yield-management know-how, the sheer institutional capacity to manufacture complex devices by the billion. If quantum information could be encoded in structures that ordinary chip factories already know how to make, then the scaling problem that torments every rival — how to go from hundreds of qubits to millions — might be solved by an industry that has already gone from thousands of transistors to trillions. It is the ultimate manufacturing bet, and its potential payoff is why Intel and others have pursued it for years despite its lagging qubit counts.

The reason the payoff has been so slow to arrive lies in a cruel irony: the very smallness that makes spin qubits so promising for scaling makes them punishingly hard to build and control today. A spin qubit is a single electron confined in a quantum dot a few tens of nanometers across — smaller than the wavelength of visible light, at the very limit of what even the most advanced lithography can pattern with the precision quantum coherence demands. Tiny variations from one dot to the next, unavoidable at this scale, make every qubit slightly different and fiendishly hard to calibrate in concert; getting neighboring electron spins to entangle reliably, with the high fidelity error correction requires, proved a years-long struggle. The field spent much of the 2010s able to build only a handful of spin qubits that worked well together, while atomic and superconducting machines raced into the hundreds — a frustrating position for a technology whose whole promise was eventual mass production.

Yet the recent progress has been real enough to keep the dream alive and to suggest that silicon's moment may still come. Groups have demonstrated two-qubit gate fidelities above the error-correction threshold, operation of qubits at the relatively balmy temperature of around one kelvin — which, though still cryogenic, is enormously easier to achieve than the millikelvin cold superconducting qubits demand, and which would allow control electronics to sit close to the qubits — and the fabrication of spin-qubit test chips on Intel's standard commercial production lines, proving the qubits can be made with industrial tools. The bet remains a long one, and silicon starts the race well behind. But if its manufacturing advantage can finally be realized, silicon spin qubits could scale with the same breathtaking exponential that carried classical chips from the first integrated circuits to the billions-of-transistors processors in every modern phone — a latecomer that, riding the mightiest manufacturing engine ever built, leapfrogs the field. Whether that promise is kept is one of the great open wagers of quantum computing.

Chapter 42: Topological — Microsoft's Long, Contested Wager

The most intellectually beautiful approach to quantum computing is also the one that, after two decades and a vast investment, has the least to show for itself — a gap between promise and delivery that makes the topological quantum computer both the field's most alluring dream and its most cautionary tale. The idea, which traces to Alexei Kitaev (whose surface code we met in Part VI) and to the mathematician Michael Freedman, is to solve the error problem not by the laborious active correction of Part VI but by encoding quantum information in a form so intrinsically robust that noise cannot corrupt it in the first place. If it worked, it would change everything. Whether it can be made to work at all is, as of the mid-2020s, still genuinely unsettled.

The dream rests on a exotic idea from physics: that in certain rare states of matter, there can exist strange quasiparticles called **anyons** — not ordinary particles but collective excitations, disturbances in a material that behave as if they were particles with bizarre properties found nowhere else in nature. In particular, some anyons have the remarkable feature that when you move them around one another — *braid* them, in the language of the field, weaving their paths through space and time like strands of a plait — the quantum state of the system changes in a way that depends only on the *topology* of the braid: on how the strands are knotted and linked, not on the precise details of the paths. And topology, by its nature, is robust: you can wiggle and deform a knot without changing what knot it is, just as a coffee cup and a doughnut remain topologically the same however you stretch them. If quantum information is encoded in the braiding of anyons, then local noise — a random jiggle here, a stray disturbance there — cannot corrupt it, because such local perturbations do not change the topology of the braid. The information is protected not by watchful correction but by the deep, global structure of the physical system itself. Error resistance would be built into the hardware, "for free," at the level of physics. It is one of the most elegant ideas in all of quantum computing.

The catch — and it is an enormous catch — is that the specific anyons needed, called **non-abelian anyons**, and in particular the **Majorana zero modes** that Microsoft's effort has pursued, are extraordinarily difficult to create and to detect, and it has never been unambiguously established that a working topological qubit can be built from them at all. Majorana modes are predicted to arise at the ends of certain specially engineered superconducting nanowires under exacting conditions, and creating those conditions, and then *proving* that what you have created really is a Majorana mode and not some mundane look-alike, has consumed thousands of researcher- years. The history is chastening: promising experimental claims of Majorana detection have repeatedly been made, celebrated, and then cast into doubt or retracted, most notoriously a high-profile 2018 paper that a leading group was compelled to withdraw in 2021 after problems with the data came to light. The very robustness that makes topological qubits attractive — their reluctance to interact with the environment — also makes them maddeningly hard to observe and manipulate, and the field has been repeatedly burned by results that did not hold up.

Microsoft has pursued this approach with singular determination, staking a large part of its quantum program on topological qubits for the better part of two decades, on the bet that if it could just get the physics to work, its qubits would be so intrinsically stable that they would leapfrog the painstaking error-correction overhead that burdens every rival platform — that a topological quantum computer might reach useful fault tolerance with dramatically fewer resources. In February 2025 the company announced a chip it called **Majorana 1**, claiming to have created and controlled the elusive topological qubits at last, and presenting it as a breakthrough toward a scalable topological machine. The announcement drew intense interest and intense skepticism in equal measure: some physicists welcomed it as a genuine milestone, while others argued the evidence fell short of proving that true topological qubits had been demonstrated, reviving the field's long-standing wariness about Majorana claims. As of this writing the matter remains contested, a fitting emblem of an approach that has always hovered between revolutionary promise and unproven physics.

The most radical and intellectually beautiful approach to quantum computing is topological quantum computing, an idea that would solve the fragility problem not by correcting errors after they happen but by building qubits so intrinsically robust that errors are suppressed at the deepest physical level — protection woven into the very fabric of the qubit rather than bolted on afterward. It is the boldest bet in the field, pursued most prominently and expensively by Microsoft, and after more than two decades of effort it remains the great unfulfilled promise of quantum computing, a technology of sublime theoretical elegance and punishing experimental difficulty. Its intellectual father is Alexei Kitaev, whose surface code we met earlier; topological quantum computing is his other, more audacious, dream.

The idea builds on the topological protection we encountered with the surface code, but pushes it to the hardware itself. Recall that topology concerns global properties of shape that survive local deformation — the hole in a doughnut cannot be removed by any amount of poking or squeezing. Topological quantum computing would encode quantum information in exactly such global, topological properties of an exotic quantum system, so that local disturbances — the stray photons and thermal jiggles that destroy ordinary qubits — simply cannot corrupt it, any more than a dent can remove a doughnut's hole. The information would be stored non-locally, smeared across the system in a way no local noise could touch, making the qubit inherently, physically stable. Error correction, in effect, would be a free gift of the physics rather than an expensive overhead of thousands of extra qubits.

The exotic ingredients this requires are extraordinary entities called non-abelian anyons, and in particular Majorana zero modes — strange quasiparticles predicted to arise at the ends of special superconducting nanowires under exactly the right conditions. A Majorana is a peculiar object: in a sense it is "half" of an electron, and a pair of them separated across a wire jointly stores one qubit of information in a way that is delocalized and therefore protected — neither end alone holds the information, so disturbing one end cannot destroy it. To perform quantum gates, one would physically move these anyons around one another, BRAIDING their paths through space and time like strands of a plait, and the remarkable mathematical fact is that the result of a computation depends only on the topology of the braid — how the strands wind around each other — and not on the precise details of the paths, which is exactly why it is robust. You compute by braiding, and the answer cares only about the knot, not the wiggle. It is one of the most beautiful ideas in all of physics.

Microsoft made an enormous, sustained bet on this vision, investing heavily over many years in the belief that a topological qubit, once achieved, would be so intrinsically stable that it would leap past the noisy qubits of every rival and reach useful, error-corrected computation with far less overhead. The strategic logic is compelling: why build thousands of mediocre qubits and spend most of them on error correction, when you could build a few superb ones that correct themselves? If topological qubits worked, they might render the whole brutal overhead of the surface code unnecessary. The prize justified the patience, and Microsoft assembled a formidable team and pursued the approach with singular focus while its competitors piled up conventional qubits.

But the experimental road has been brutal, and the story is a cautionary one about the gap between beautiful theory and stubborn reality. Majorana zero modes are exceedingly difficult to create and, worse, exceedingly difficult to identify unambiguously — the subtle experimental signatures that would prove their existence can be mimicked by more mundane effects, and the field has been repeatedly burned by premature claims. Most painfully, a prominent 2018 paper claiming clear evidence for Majorana particles, published in a leading journal by researchers associated with Microsoft's effort, was RETRACTED in 2021 after other scientists raised serious concerns about the data analysis — a humbling and public setback that cast a shadow over the whole program and fueled skepticism about whether topological qubits would ever arrive. It was a vivid reminder that in frontier science, the desire to see a long-sought result can outrun the evidence.

The intellectual seduction of topological quantum computing is worth understanding on its own terms, because it explains why so many brilliant people have devoted decades and fortunes to a technology that has yet to produce a single unambiguous qubit. Every other platform accepts the fragility of quantum states as a fact of life and fights it with the elaborate, expensive machinery of error correction, spending hundreds or thousands of physical qubits to protect a single logical one. Topological quantum computing proposes to make that whole struggle largely unnecessary by building qubits that are *intrinsically* protected — encoding information in global, topological properties that local noise simply cannot disturb, so that error correction is baked into the physics rather than bolted on at ruinous cost. If it worked, a handful of topological qubits might do the work of thousands of ordinary ones, collapsing the crushing overhead that makes fault-tolerant machines seem so remote. The prize is not a better qubit but a fundamentally cheaper road to the whole fault-tolerant future.

The physics is as beautiful as the promise. The information would be stored in pairs of Majorana zero modes — strange objects, each in a sense "half" an electron, arising at the ends of specially engineered superconducting nanowires — with the information smeared non-locally across the pair, so that no local disturbance to either end can destroy it. Computation would proceed by physically braiding the world-lines of these objects around one another through space and time, and the astonishing mathematical fact is that the result depends only on the topology of the braid — which strand crossed over which, how many times they wound — and not on the precise wiggling paths, which is exactly why the computation is robust against the small perturbations that ruin ordinary qubits. You would compute by tying knots in the histories of quasiparticles, and the answer would care only about the knot. Few ideas in physics so perfectly marry deep mathematics to practical purpose.

But the experimental story is a hard lesson in the gap between beautiful theory and stubborn nature, and it is worth telling honestly. Majorana zero modes are not only extraordinarily difficult to create — requiring exquisitely clean nanowires, strong magnetic fields, and temperatures near absolute zero — but extraordinarily difficult to *identify* unambiguously, because the subtle experimental signatures that would prove their existence can be counterfeited by more mundane effects lurking in imperfect devices. The field has been burned repeatedly by premature claims, most painfully when a prominent 2018 paper reporting clear evidence of Majorana particles was retracted in 2021 after independent scientists raised serious concerns about how the data had been analyzed — a public and humbling setback that deepened skepticism about whether topological qubits would ever arrive. Microsoft, the approach's chief backer, has pressed on and reported renewed progress in the years since, claiming steps toward creating and controlling topological qubits; whether those claims will withstand scrutiny remains genuinely uncertain. Topological quantum computing thus stands as the field's grandest gamble: a technology of sublime elegance that could leapfrog every rival if it works, and could consume decades and fortunes chasing a mirage if it does not. Nature has not yet revealed which.

Chapter 43: The Annealers — D-Wave and the Road Not Taken

While the mainstream of quantum computing pursued the universal, gate-model machine we have described — the programmable device that runs Shor's algorithm, Grover's search, and quantum simulations by composing quantum gates — a Canadian company took a different road entirely, and reached the market long before anyone else. D-Wave Systems, founded in 1999, built and sold **quantum annealers**: specialized machines that are not universal quantum computers at all, that cannot run Shor's algorithm or Grover's, but that are designed to do one thing — solve a certain class of optimization problems — and that were commercially available, with thousands of qubits, years before the gate-model machines reached even a hundred. The D-Wave story is a fascinating and controversial counterpoint to the main narrative, a lesson in the difference between building a quantum computer and building a *useful* one, and a cautionary tale about how hard it is to prove that a quantum machine is doing anything a classical one could not.

Quantum annealing is a restricted form of quantum computing, related to the "adiabatic" model mentioned back in Part IV, and it is aimed squarely at **optimization problems** — the ubiquitous task of finding the best arrangement among an astronomical number of possibilities, such as the most efficient delivery route, the lowest- energy configuration of a molecule, or the optimal schedule for a fleet of aircraft. Many such problems can be recast as finding the lowest point in a vast, rugged "energy landscape" of possible solutions, and a quantum annealer attacks them by a clever physical process. It sets up a quantum system whose energy landscape encodes the problem, starts it in an easily prepared quantum state, and then slowly, gently transforms the system, allowing it to settle — with the help of quantum effects such as "tunneling" through barriers rather than climbing over them — toward the lowest-energy configuration, which encodes the optimal solution. It is a beautiful idea, and it is far easier to build than a universal gate-model machine, because it does not require the precise, high-fidelity gate operations and error correction that the universal machines demand. This is why D-Wave could field machines with thousands of qubits — reaching into the thousands while the gate-model machines were still counting in the tens — and sell them to customers including Google, NASA, Lockheed Martin, and Volkswagen.

The controversy — and it has been fierce and long-running — is whether D-Wave's machines actually deliver any **quantum speedup**, any advantage over classical optimization algorithms running on ordinary computers. This turns out to be devilishly hard to establish, and it illustrates a theme that runs through the whole field: it is far easier to build a machine that *might* be exploiting quantum effects than to *prove* that those effects are providing a genuine computational benefit. For years, independent researchers subjected D-Wave's machines to careful benchmarking, and the results were sobering: while the machines demonstrably worked and solved the optimization problems posed to them, it proved extremely difficult to show that they solved them any *faster* than good classical algorithms running on conventional hardware. Some studies found no quantum advantage; others found evidence of quantum tunneling at work but no overall speedup once classical methods were properly tuned; the debate went back and forth, with claims and counterclaims, and no clean, decisive demonstration that the annealers beat the best classical competition on a practically important problem. D-Wave's qubits, moreover, were of much lower quality than the gate-model machines' — noisier, less coherent, more limited in connectivity — and critics argued that whatever quantum effects were present were too weak and too swamped by noise to yield real advantage.

D-Wave's defenders countered, with some justice, that the company had built real, large, commercially available quantum hardware while the gate-model camp was still struggling to reach double digits, that its machines were being used by real customers to explore real optimization problems, and that quantum annealing might yet prove its worth as the hardware improved and as the right problems were found — problems whose landscapes are shaped in just the way that quantum tunneling can exploit. The company continued to develop its technology, improving its machines and pursuing both annealing and, later, gate-model systems, and it remained a genuine and pioneering presence in the field. But the mainstream of quantum computing, and the great milestones this book tracks, moved along the universal gate-model road, because it is the gate-model machine, not the annealer, that can run the algorithms — Shor's, Grover's, quantum simulation — that give quantum computing its world-changing promise. The annealer is a specialized instrument of uncertain advantage; the universal machine is the real prize.

While most of the quantum computing world pursued the universal, gate-based machine — the kind that runs Shor's and Grover's algorithms and everything else — one company took a dramatically different path that got quantum processors into customers' hands years before anyone else, and in doing so ignited one of the fiercest and most instructive controversies in the field's history. The company was D-Wave Systems, a Canadian firm founded in 1999, and its machines are quantum ANNEALERS: special-purpose devices designed to solve one particular kind of problem — optimization — rather than to run arbitrary quantum algorithms. The D-Wave story is a case study in the difference between a general-purpose quantum computer and a specialized quantum gadget, and in how hard it can be to prove that a quantum machine is actually delivering quantum advantage.

Quantum annealing is built for optimization problems — the ubiquitous task of finding the best configuration among an astronomical number of possibilities, the lowest-cost route, the most efficient schedule, the arrangement of least energy. The technique reframes such a problem as a landscape of hills and valleys, where the solution you seek is the very lowest valley, and the challenge is to find that deepest point without getting stuck in some shallower dip along the way. A classical optimizer is like a ball rolling downhill: it easily falls into the nearest valley but can get trapped there, unable to climb out and over a ridge to find a deeper valley beyond. Quantum annealing exploits a genuinely quantum trick — quantum tunneling — to let the system pass THROUGH energy barriers rather than having to climb over them, potentially escaping shallow traps and finding the true lowest valley more reliably. The machine is initialized in a simple quantum state and then slowly, gently evolved so that, if the evolution is slow enough, it settles into the lowest-energy configuration, which encodes the answer.

D-Wave's achievement in engineering was real and should not be dismissed: the company built machines with hundreds and then thousands of superconducting qubits — far more than any gate-based machine at the time — and actually sold them to serious customers including Lockheed Martin, Google, NASA, and Los Alamos National Laboratory, at prices reported in the millions of dollars. By raw qubit count, D-Wave was for years the runaway leader, boasting thousands of qubits when gate-based machines had mere dozens. To a public trained to equate qubit count with power, D-Wave looked like the front-runner by a wide margin, and it garnered enormous, often breathless, media attention.

But here the controversy erupted, and it is deeply instructive. Skeptical scientists asked the hard question that qubit counts always beg: were D-Wave's machines actually delivering a QUANTUM SPEEDUP? Was the quantum annealing genuinely solving problems faster than the best classical algorithms running on ordinary computers — and was quantum tunneling really doing the work, rather than the machine behaving like an expensive classical optimizer? For years the evidence was murky and hotly contested. Careful benchmarking studies repeatedly found that, on the problems tested, D-Wave's machines did NOT clearly outperform good classical algorithms running on conventional hardware; whenever the quantum machine seemed to win, a cleverer classical method would soon match or beat it. Whether the devices exhibited genuine, useful quantum effects at all was fiercely debated, and the annealer's qubits, being noisy and specialized, were of a very different and less pristine character than the gate-based qubits elsewhere. The company's bold marketing collided with the scientific community's demand for rigorous proof.

The D-Wave saga crystallized several lessons that echo throughout this book. First, it hammered home, yet again, that qubit count is a treacherous measure: D-Wave had thousands of qubits and no clearly demonstrated advantage, while gate-based machines with far fewer qubits were doing more fundamentally quantum things. Second, it illustrated the profound difficulty of PROVING quantum advantage — of showing convincingly that a quantum machine beats not just some classical method but the very best classical method, a moving target that improves the moment it is challenged. Third, it drew a bright line between a universal, gate-based quantum computer, which can in principle run any quantum algorithm, and a special-purpose quantum device like an annealer, which is limited to one class of problem and whose quantum credentials are harder to establish. A quantum annealer is not a lesser universal computer; it is a different kind of machine altogether.

The D-Wave saga is worth lingering on as a parable, because it dramatizes, more sharply than any other episode, the single hardest question in the whole field: how do you *prove* that a quantum machine is actually doing something quantum, and doing it better than a classical computer could? D-Wave built machines with thousands of qubits — dwarfing every gate-based rival — and sold them to serious customers for millions of dollars, and by the crude measure of qubit count it looked, for years, like the runaway leader of the entire industry. Yet the deeper the scientists probed, the harder it became to show that all those qubits were delivering any genuine advantage. Whenever the annealer seemed to win a benchmark, a cleverer classical algorithm would soon match or beat it, and whether the machine was truly harnessing quantum tunneling or merely behaving like an expensive, specialized classical optimizer remained fiercely contested. A thousand qubits, it turned out, could add up to no demonstrated advantage at all.

The lesson generalizes far beyond D-Wave, and it is one of the most important in this book. Proving quantum advantage is extraordinarily hard because it requires beating not some classical method but the *very best* classical method — a moving target that improves, often dramatically, the moment it is challenged, and frequently improves *because* the quantum claim inspired classical researchers to try harder. This pattern, first sharply visible with D-Wave, would recur throughout the field's history: Google's Sycamore supremacy claim was chipped away by improved classical simulations; IBM's quantum-utility demonstration was quickly reproduced classically; claim after claim of near-term advantage narrowed or dissolved under the assault of a motivated, resourceful classical-computing community. The annealer was simply the first arena in which this drama played out in full, and it taught the field a durable humility about the difference between a machine that looks powerful and one that is provably, usefully so.

It is only fair, though, to resist the temptation to write D-Wave off entirely, because the story is more nuanced than "hype exposed." Quantum annealing is a legitimate approach with genuine research merit; the company built real, working, large-scale superconducting hardware and got quantum processors into the hands of real users years before anyone else, seeding practical experimentation across industries from logistics to finance. Some evidence for genuine quantum effects in its machines has accumulated over time, and the technology continues to develop. What the annealer is *not* is a universal, gate-based quantum computer — the kind that can run Shor's and Grover's algorithms and everything else — and much of the confusion around D-Wave came from conflating the two very different kinds of machine. A quantum annealer is not a lesser universal computer; it is a different device entirely, built for one class of problem, and judged by different standards. Its greatest legacy may be the sharpness with which it posed the field's defining challenge — the challenge of demonstrating real quantum advantage against the full might of classical computing — a challenge that would soon move from the specialized world of annealers to the center of the world's attention, in the fierce and famous battle over quantum supremacy.

Part IX — Supremacy and Its Discontents

Chapter 44: Naming the Prize

Before a race can be won, its finish line must be drawn, and in quantum computing the drawing of the finish line was itself a subtle and consequential act. What, exactly, would count as proof that a quantum computer had done something no classical computer could? The question is harder than it looks, and the two men who did most to frame it — coining the terms that now structure the entire debate — were not hardware builders but theorists, and one of them coined *both* of the field's defining phrases. His name is John Preskill, and to understand the supremacy wars, and indeed the whole current era of quantum computing, you must understand the vocabulary he gave it.

John Preskill is a theoretical physicist at Caltech, a figure of great authority and unusual thoughtfulness, who came to quantum information from particle physics and cosmology and brought to it a gift for seeing the field whole and naming its features clearly. In 2012 he coined the term **quantum supremacy** to denote the milestone at which a quantum computer performs some task — *any* task, however useless — that is beyond the practical reach of any classical computer. Note carefully what the term does and does not claim.

Quantum supremacy does not require that the task be *useful*; it is a purely scientific milestone, a demonstration that quantum hardware has crossed into a regime of computation that classical machines cannot follow, even if what it computes is of no practical value whatever. The point of the milestone is to prove, once and for all, that quantum computers are not merely a theoretical possibility but a physical reality capable of outrunning the classical world — to silence the skeptics who doubted that controllable quantum computation could ever be built at scale. It is a flag planted on a summit, not a treasure delivered to a customer.

The term "supremacy" later became controversial for reasons unrelated to physics — some researchers found its connotations unfortunate and proposed alternatives such as "quantum advantage" — and the two phrases are now used somewhat interchangeably, though purists distinguish them, reserving "advantage" for the more demanding milestone of a *useful* quantum speedup on a problem someone actually cares about. This distinction matters and will recur: "supremacy" (or advantage in the weak sense) means beating classical computers on *some* task, however contrived; "quantum advantage" in the strong sense means beating them on a *useful* one. The first is a scientific milestone; the second is the real economic prize, and, as we will see, it remains, in the mid-2020s, still unclaimed on any clear-cut, practically important problem — a crucial and often-obscured fact about the true state of the field.

Preskill's second and even more influential coinage came in 2018, in an essay that named and defined the era we have been living in ever since. He called it the **NISQ** era — Noisy Intermediate-Scale Quantum — and the acronym has become the single most useful label in the field. Each word is doing work. *Noisy*: the qubits are error-prone, and there is no full error correction, so computations are limited in length before noise overwhelms them. *Intermediate-Scale*: the machines have an intermediate number of qubits — tens to a few thousand — enough to be interesting and beyond easy classical simulation, but far short of the millions that fault-tolerant computing will require. *Quantum*: they are genuine quantum machines, exploiting superposition and entanglement. The NISQ era is the awkward adolescence of quantum computing: past the infancy of a handful of qubits, but not yet arrived at the fault-tolerant maturity where error-corrected logical qubits run arbitrarily long computations. Preskill's essay was clear-eyed and influential precisely because it was *honest* about the limitations — it explicitly cautioned that NISQ machines might or might not find useful applications, that full fault tolerance remained "a rather distant dream," and that the field should not overpromise. It was a theorist's sober framing of a moment that the marketing departments were prone to hype, and it gave the community a shared, realistic vocabulary for where things actually stood.

Before the dramatic experiments that would claim it, the goal needed a name, and the name it received — "quantum supremacy" — was both brilliant and unfortunate, capturing the public imagination while sowing years of confusion and controversy. It was coined in 2012 by John Preskill, a distinguished theoretical physicist at Caltech and one of the field's most thoughtful elder statesmen, and understanding exactly what he meant by it — and what he did NOT mean — is essential to making sense of the fierce battles that followed.

Preskill introduced the term to mark a specific, well-defined milestone: the moment when a quantum computer performs some task — ANY task, however useless — that no classical computer could accomplish in a reasonable time. The word "supremacy" was chosen to convey that at this threshold the quantum machine would demonstrably surpass all classical computers combined, crossing from the realm of "quantum computers are interesting in principle" into "quantum computers can do something classical computers provably cannot." It was meant as a scientific landmark, a proof of concept, a decisive demonstration that the exponential power latent in quantum mechanics could be made real and measurable. Preskill was careful and precise about it: quantum supremacy did not mean the quantum computer was USEFUL, did not mean it could break codes or simulate molecules or solve any practical problem, and did not mean quantum computers were ready to replace classical ones. It meant only that a clear, quantifiable line had been crossed — that for at least one task, the quantum machine had left every classical computer behind.

The distinction between this milestone and practical usefulness cannot be overstated, and the failure to grasp it is the source of endless public confusion. The task chosen to demonstrate supremacy could be — and, as it turned out, was — completely contrived and pointless, a problem designed specifically to be hard for classical computers and natural for quantum ones, with no application whatsoever beyond serving as a benchmark. Achieving supremacy would be like proving a new kind of engine could momentarily produce more power than any existing engine, by having it spin a flywheel to no purpose: a genuine and important demonstration of capability, but not yet a vehicle that could take you anywhere. Preskill's milestone was the flywheel, not the journey. He himself later expressed some regret about the term, and much of the field, uncomfortable with the word's political and hierarchical overtones, has increasingly preferred alternatives like "quantum advantage" or "quantum computational advantage."

That discomfort with the word itself became a genuine controversy, and it is worth recording because it reflects real tensions in science. The term "supremacy" carries unavoidable and ugly connotations — echoes of "white supremacy" and other odious hierarchies — and many researchers argued it was an unfortunate, even harmful, choice for a scientific milestone, one that could alienate people and cheapen the achievement. In 2019, around the time of the first supremacy claim, a group of scientists published a letter in the journal Nature urging the community to abandon the term in favor of "quantum advantage," arguing that language matters and that the field should not casually adopt a word freighted with such associations. Others countered that "advantage" was too weak and vague, failing to convey the sharp, provable, all-or-nothing nature of Preskill's original milestone, and that renaming established terminology mid-stream sows its own confusion. The debate has never been fully resolved, and both terms remain in use, sometimes interchangeably and sometimes with subtly different meanings, a small linguistic muddle layered atop the scientific one.

There is a deeper reason the naming matters, beyond politics, and it goes to the intellectual substance of the milestone. To claim quantum supremacy, you must do two things, and the second is far harder than the first. You must run your quantum machine on the chosen task — and you must PROVE that no classical computer could have done the same in a reasonable time. That second requirement is treacherous, because it demands a claim about the limits of ALL possible classical algorithms, including ones no one has yet invented. You cannot simply show that your machine beat today's best classical method; you must argue, from complexity theory and careful estimation, that no feasible classical method could keep up, that the classical cost genuinely explodes. And classical algorithms and hardware keep improving, so a supremacy claim is always vulnerable to being overturned by a cleverer classical simulation devised after the fact. This is exactly what would happen — repeatedly — turning the supremacy milestone from a clean finish line into a moving, contested boundary, fought over inch by inch between the quantum experimentalists and a motivated, resourceful classical-computing community determined to prove the quantum machines had not really won.

The confusion sown by the word "supremacy" is worth examining as a case study in how language shapes public understanding of science, for few technical terms have been so consistently misread. To a physicist, "quantum supremacy" meant something precise and modest: a demonstration, on some contrived task, that a quantum machine had outrun all classical computers — a scientific proof of principle, nothing more. To the public and much of the press, "supremacy" sounded like triumph, dominance, the arrival of machines that had surpassed ordinary computers in general — a wholesale revolution. The gap between these readings has generated a decade of misunderstanding, with people believing that powerful, useful quantum computers had arrived in 2019, when what had actually happened was that a machine had performed one useless benchmark task no classical computer could match. The lesson is old but perpetually relevant: in science, the choice of a vivid word can mislead as much as it illuminates, and "supremacy" oversold a real but narrow achievement.

The deeper reason the milestone is so slippery — and this is the point that repays real attention — is that proving quantum supremacy requires proving a claim about *all possible classical algorithms*, including ones not yet invented. It is not enough to show that your quantum machine beat today's best classical method; you must argue that no feasible classical method, however clever, could ever keep up, that the classical cost genuinely explodes exponentially. This is a claim about the ultimate limits of classical computation, and such claims are treacherous, because classical algorithms keep improving, often specifically in response to a supremacy challenge. A supremacy claim is therefore always provisional, always vulnerable to being undermined by a cleverer classical simulation devised after the fact — which is exactly what would happen, repeatedly, turning the clean finish line into a contested, shifting boundary fought over inch by inch.

This is why some thoughtful researchers argued from the start that "advantage" was the wiser word, and why the terminology remains unsettled to this day. "Quantum advantage" carries less triumphalist baggage and better conveys the provisional, relative, hard-won nature of the achievement; but critics countered that it was too weak and vague, failing to capture the sharp, provable, all-or-nothing character of Preskill's original milestone. Layered atop this technical debate was the genuine discomfort many felt with the word "supremacy" itself, freighted as it is with ugly historical associations, prompting a 2019 letter in the journal *Nature* urging the community to abandon it. The upshot is a small but telling muddle: two terms in use, sometimes interchangeably and sometimes with subtly different meanings, a linguistic uncertainty mirroring the scientific one. For the reader, the practical lesson is simply to look past the word to the substance — to ask, of any such claim, exactly what task was performed, whether it was useful, and how the classical competition responded.

Chapter 45: Sycamore — Google's 2019 Claim

In October 2019, the journal *Nature* published a paper from Google's quantum-computing group with a title of studied understatement — "Quantum supremacy using a programmable superconducting processor" — and the world's press erupted. Google, the paper announced, had used a 53-qubit superconducting processor named **Sycamore** to perform a computation in about two hundred seconds that, it claimed, would take the world's most powerful classical supercomputer approximately ten thousand years. If true, this was the milestone Preskill had named seven years earlier: quantum supremacy, the first demonstration that a quantum machine could do something beyond the practical reach of any classical computer. It was hailed as a historic moment, compared by some to the Wright brothers' first flight, and it thrust quantum computing into the global spotlight as never before. It also ignited an immediate and instructive controversy that reveals, better than any success story could, how genuinely difficult it is to know when a quantum computer has truly won.

Understanding what Google actually did requires setting aside the natural assumption that Sycamore solved some important problem. It did not, and was never meant to. The task Google chose, called **random circuit sampling**, was selected precisely because it is hard for classical computers and natural for a quantum processor, not because its answer is of any use to anyone. The idea is this: program the quantum processor with a random sequence of gates, run it, and measure all the qubits, producing a random-looking string of bits; repeat many times to build up a statistical distribution of which strings appear. Because of quantum interference, some strings are more likely than others in a complicated, structured way, and the resulting distribution is a kind of quantum fingerprint of the specific random circuit. To *predict* this distribution classically — to compute, for a given random circuit, which output strings should appear and how often — requires, in effect, simulating the full 53-qubit quantum state, and that, as we know from Part II, means tracking an astronomical number of amplitudes: two to the fifty-third power, roughly nine quadrillion complex numbers, evolving through the circuit. This is at or beyond the edge of what the largest classical supercomputers can do. The quantum processor, by contrast, produces samples from the distribution effortlessly, simply by *being* the quantum system and being measured. Sampling from the distribution is easy for the quantum machine and desperately hard for the classical one — which is exactly the asymmetry a supremacy demonstration needs.

Sycamore itself was a considerable engineering achievement, embodying everything Part VII described: 53 functioning superconducting transmon qubits (one of the intended 54 failed) arranged on a chip, chilled in a dilution refrigerator to near absolute zero, with gate fidelities high enough — just barely — to run circuits deep enough to reach the supremacy regime before noise overwhelmed the computation. Google's team, led at the hardware level by John Martinis, had spent years improving qubit quality, reducing errors, and learning to calibrate the chip's many imperfect qubits into a working whole. Running its random circuits and collecting a few million samples in about two hundred seconds, Sycamore produced a distribution that bore the expected quantum fingerprint, verified by careful statistical tests, at a scale that Google argued no classical computer could match in any reasonable time. On its face, the finish line had been crossed.

In October 2019, Google announced to the world that it had achieved quantum supremacy, and the moment became the most famous single event in the history of quantum computing — a genuine scientific milestone wrapped in corporate triumph, drama, and immediate, ferocious controversy. The claim was published in the journal Nature, the venue of record for the most important scientific results, and it declared that Google's quantum processor had performed in about two hundred seconds a task that would take the world's fastest supercomputer, Google estimated, roughly ten thousand years. If true, it meant that a line first imagined by Preskill had at last been crossed: a quantum machine had done something no classical computer on Earth could match.

The machine that did it was called Sycamore, a superconducting quantum processor with 53 working qubits (one of a planned 54 had failed), fabricated as a chip of transmon qubits arranged in a two-dimensional grid, cooled in a dilution refrigerator to the now-familiar temperature colder than deep space. The task Sycamore performed was, as Preskill's framework had anticipated, utterly useless in itself: a contrived benchmark called random circuit sampling. The idea is to apply a long sequence of randomly chosen quantum gates to the 53 qubits, producing an immensely complex, deeply entangled quantum state — a state that is a superposition across the roughly nine quadrillion possible outcomes of measuring 53 qubits — and then to measure the qubits many times, collecting samples from the fantastically intricate probability distribution that the random circuit produces. Because the qubits interfere in a hopelessly complicated pattern, some outcomes are far more likely than others, and reproducing the correct distribution of samples is exactly the kind of task that grows exponentially hard for a classical computer, which would have to track all nine quadrillion amplitudes at once. Sycamore, being itself a quantum system, simply embodied the state and sampled from it directly, in seconds.

The choice of task was deliberate and clever, and it embodied the whole logic of the supremacy milestone. Random circuit sampling has no use — no one needs samples from a random quantum circuit — but it was chosen precisely because it plays to a quantum computer's deepest strength (holding and sampling a vast entangled superposition) while being provably, or at least very plausibly, hard for classical machines. It is the flywheel, spun to no purpose except to prove the engine's power. Google's team, led by John Martinis, who had spent years building the hardware, framed the result carefully in the scientific paper as a demonstration of a milestone, not a useful application, and many physicists hailed it, rightly, as a landmark — the first concrete, experimental proof that a programmable quantum machine could outpace all classical computation on a well-defined task. The dream of decades had, in this narrow but real sense, been realized.

But the announcement was immediately and fiercely contested, and the controversy revealed everything about how hard the supremacy milestone is to pin down. The challenger was IBM — Google's chief rival in the field — and its objection struck at the very heart of the claim. Recall that to prove supremacy you must show not only that your quantum machine did the task but that no classical computer could have done it in reasonable time; Google's "ten thousand years" was an ESTIMATE of the best classical approach. IBM's researchers argued that Google had overestimated the classical difficulty — that with a cleverer algorithm exploiting the vast disk storage of a supercomputer to hold the enormous quantum state, a classical machine could perform the same task not in ten thousand years but in a matter of days. If IBM was right, then Sycamore had done in two hundred seconds something a classical computer could do in days — impressive, a real speedup, but NOT the decisive, exponential, "no classical computer could ever" supremacy that had been proclaimed. The finish line, it turned out, depended entirely on how good your opponent's classical algorithm was.

There was a further, poignant human dimension to the Sycamore triumph. The result was the culmination of years of grinding effort by John Martinis and his team, who had been recruited to Google to build exactly this machine, and it should have been the crowning moment of a career. Instead, within months of the supremacy announcement, Martinis found himself sidelined in a reorganization of Google's quantum effort and departed the company in 2020, an abrupt and jarring coda to the field's greatest technical achievement — a reminder that even the most brilliant scientific triumphs are embedded in the messy realities of corporate ambition and internal politics. The man who had led the building of the first supreme quantum computer left the company shortly after building it.

It is worth savoring just how vast, and how meaningless, the "ten thousand years" figure was, because it captures the strange character of the whole supremacy enterprise. Google's claim was that Sycamore had done in about two hundred seconds a computation that the world's fastest supercomputer would need roughly ten thousand years to reproduce — a ratio so enormous it beggars intuition. And yet the task itself, random circuit sampling, was of no use to anyone: it produced a list of random-looking bit strings, sampled from the fantastically complex probability distribution that a random quantum circuit generates, with no application beyond serving as a benchmark. Here was a machine performing, in minutes, a feat that would outlast civilizations on a classical computer — and the feat was worthless. This is the paradox of quantum supremacy in its purest form: an epochal demonstration of raw capability, spun to absolutely no purpose, like an engine roaring at full power while connected to nothing.

The technical achievement embodied in Sycamore, setting aside the useless task, was nonetheless genuinely formidable and deserves recognition. To perform random circuit sampling on 53 qubits, Google's engineers had to fabricate a chip whose qubits were good enough — coherent enough, accurate enough in their gates, reliable enough in their measurement — to apply around twenty layers of carefully calibrated operations and still produce a signal that rose detectably above the noise. This demanded years of relentless engineering: pushing gate fidelities high enough, calibrating each of the individually-fabricated qubits, taming the countless sources of error that afflict superconducting circuits. Reaching the regime where a quantum machine could enter territory beyond easy classical simulation was not a matter of simply building more qubits but of building qubits *good enough* to sustain a deep computation — a milestone in quantum engineering quite apart from the philosophical debate over what "supremacy" meant.

The bittersweet human coda to Sycamore bears repeating, because it reveals the human realities behind the gleaming press releases. The triumph was the culmination of years of grinding effort by John Martinis and his team, who had been recruited to Google expressly to build such a machine; it should have crowned a distinguished career. Instead, within months of the supremacy announcement, Martinis found himself sidelined in a reorganization of Google's quantum program and departed the company in 2020 — an abrupt and jarring end to the greatest technical achievement in the field's history. That the man who led the building of the first "supreme" quantum computer left the company so soon after building it is a reminder that even the most brilliant scientific triumphs are embedded in the messy, sometimes brutal realities of corporate ambition, internal politics, and personality. The chandelier hummed on in its cold room; the architect had gone. And even as he left, the classical-computing community was already at work proving that Sycamore's ten-thousand-year gap was far smaller than Google had claimed — a rebuttal that would reveal how contested the very idea of a finish line really was.

Chapter 46: The IBM Rebuttal

The ink was scarcely dry on Google's *Nature* paper when IBM struck back. In a blog post and technical paper released just before the Google result was officially published — the two companies' quantum groups being fierce rivals, and IBM having caught wind of the coming announcement — IBM's researchers disputed the headline claim, not by denying that Sycamore worked, but by attacking the ten-thousand-year figure at its foundation. Google's estimate, IBM argued, rested on an inefficient method of classical simulation, one that failed to exploit the enormous *disk storage* available to a supercomputer like Summit. Using a smarter classical approach that traded some computation for the clever use of Summit's vast memory and storage, IBM contended, the same random-circuit- sampling task could be simulated classically in about two and a half *days* — not ten thousand years. Two and a half days is a long time, and Sycamore's two hundred seconds was still far faster; but two and a half days is not ten thousand years, and if IBM was right, then Sycamore had not performed a task "beyond the practical reach of any classical computer" at all. The classical computer could, with effort, keep up. The supremacy claim, IBM suggested, was premature.

The exchange was pointed, and it had an edge of corporate rivalry — IBM and Google were competing for leadership of the field, and IBM had reasons of pride and positioning to puncture its rival's balloon. But strip away the competitive theater and the episode illuminates something genuinely deep about the nature of the quantum supremacy milestone, something that no amount of better hardware can make go away. A quantum supremacy claim is not a statement about the quantum computer *alone*; it is a statement about the quantum computer *relative to the best possible classical alternative*. And the best possible classical alternative is a moving target, because classical algorithms can always be improved. To claim supremacy, you must argue that *no* classical method, however clever, could reasonably reproduce your result — and that is a claim about the entire space of possible classical algorithms, present and future, which is fiendishly hard to establish and impossible to prove definitively. Every supremacy demonstration is therefore provisional, vulnerable to a cleverer classical algorithm coming along and closing the gap. The finish line is not fixed; it recedes or advances depending on how good the classical competition gets, and the classical competition is a large, motivated community that treats every quantum supremacy claim as a challenge to be met.

And meet it they did, repeatedly, in the years after Sycamore. Google's claim touched off a productive arms race between quantum experimentalists and classical algorithm designers. Spurred by the supremacy claim, classical-computing researchers devised ever-cleverer methods for simulating random circuit sampling, using new algorithms, clever mathematical techniques, and the raw power of the largest supercomputers and GPU clusters. Over the following few years, a series of papers steadily eroded Google's ten-thousand-year gap, showing that Sycamore's specific task could be classically simulated in days, then hours, then — by some accounts, using new algorithmic insights and major computing resources — in a matter of minutes or seconds, at least for the particular circuits Google had run. The classical community had, in effect, largely caught up to the original Sycamore demonstration. This did not mean Sycamore had failed or that quantum computing was a mirage; it meant that the *specific* task, at the *specific* scale of 53 qubits and that circuit depth, had turned out to be closer to the classical frontier than Google's first estimate suggested, and that the classical frontier, once challenged, could be pushed further than anyone had bothered to push it before. Google, for its part, responded by running larger, deeper circuits on improved processors, re-establishing a gap that classical methods would then again attack — an ongoing back-and-forth that continues to this day and that has, on balance, been *healthy* for the science, since it forces honest, quantitative comparison rather than breathless claims.

IBM's response to Google's supremacy claim was swift, pointed, and deeply revealing about the nature of the whole enterprise, and it deserves a chapter of its own because it exposed the fragile foundations on which any supremacy claim must rest. Rather than dispute that Sycamore had run its random-circuit-sampling task, IBM attacked the second, harder half of the claim: the assertion that no classical computer could match it in reasonable time. This was the vulnerable point, and IBM struck it precisely.

Google had estimated that reproducing Sycamore's two-hundred-second computation would take the world's most powerful supercomputer — Summit, at Oak Ridge National Laboratory — about ten thousand years. IBM's researchers, who happened to have built Summit, argued that this estimate was badly inflated because it assumed a particular, memory-limited classical method. With a smarter approach that exploited the machine's enormous supply of disk storage — using hard drives to hold the gigantic quantum state that would not fit in fast memory — IBM calculated that Summit could perform the same task in about two and a half DAYS. Days, not millennia. IBM never actually ran the full computation, but its argument was credible and detailed, and its conclusion was pointed: Google had not, strictly speaking, achieved quantum supremacy in the fullest sense, because a classical computer COULD do the task in a feasible, if lengthy, time. The word "supremacy," IBM implied, had been claimed prematurely.

Notice carefully what this dispute was and was not. IBM did not claim that classical computers were as fast as Sycamore — two hundred seconds versus two and a half days is still a speedup of roughly a thousandfold in Sycamore's favor. Nor did IBM deny that quantum computers held enormous promise. The disagreement was about a definition and a boundary: whether "a classical computer would take two and a half days" counts as the quantum machine being SUPREME, or merely faster. It was, in part, a semantic and rhetorical battle, but it rested on a genuine and important scientific truth that the whole affair drove home with unusual force: quantum supremacy is not a fixed, absolute achievement but a MOVING TARGET, defined relative to the best available classical algorithms and hardware, both of which keep improving — sometimes dramatically, and often specifically in response to a supremacy claim.

This is the crucial and lasting lesson of the IBM rebuttal, and it recurs throughout the years that followed: every time a quantum machine claims to have done something no classical computer can match, the enormous, motivated, and resourceful classical-computing community immediately sets to work trying to prove it wrong by inventing better classical algorithms — and remarkably often, they succeed, at least partially. The Sycamore result itself was progressively chipped away in the years after 2019, as researchers devised ever-cleverer classical methods (exploiting mathematical structures called tensor networks, and harnessing large clusters of ordinary processors and graphics chips) that could simulate Google's specific random circuits faster and faster, until some groups claimed they could match or even beat Sycamore's original task on classical hardware. The ten-thousand-year gap did not merely shrink; in the eyes of some classical experts, it very nearly closed. Each such advance forced the quantum experimentalists to run larger, harder instances to stay ahead, and the classical community would then attack those in turn. It became a genuine, ongoing arms race.

Some observers found this dispiriting, reading it as evidence that quantum supremacy was hollow, a mirage that dissolved on close inspection. But that reading misses the deeper significance, and the truer interpretation is more interesting. The back-and-forth is exactly how science is supposed to work: an extraordinary claim provokes extraordinary scrutiny, rival groups push each other to ever-greater rigor, and the boundary between quantum and classical capability is mapped with increasing precision through fierce, productive competition. The fact that classical algorithms improved dramatically in response to the quantum challenge is itself a valuable scientific outcome — the quantum machines spurred genuine advances in classical computation, sharpening our understanding of what is and is not classically hard. And crucially, the quantum machines kept improving too, and the fundamental point stood: for these sampling tasks, the classical cost genuinely does grow exponentially with the size of the quantum system, so that as quantum machines scale up even modestly, the classical competitors must eventually, inevitably, fall hopelessly behind. The arms race has a winner in the long run, and the exponential is on the quantum side.

There is a delicious irony in the identity of the challenger that is worth savoring, because it reveals how the competition among quantum computing's giants sharpens the science. The rebuttal to Google's supremacy claim came from IBM — Google's chief rival in quantum computing — and it concerned the classical simulation of Google's quantum machine on IBM's own supercomputer, Summit, at Oak Ridge National Laboratory. IBM, in other words, was arguing that its *classical* machine could nearly keep pace with Google's *quantum* one, using clever software rather than exotic qubits. The company that was itself racing to build quantum computers had a strong motive to deflate a rival's quantum triumph by showing that classical computing was not so easily beaten — and in doing so it performed a genuine service to the science, forcing a more honest accounting of exactly what Sycamore had and had not proven. Rivalry, here, was a spur to rigor.

The precise nature of IBM's argument repays attention, because it illustrates how a supremacy claim can be attacked without disputing that the quantum machine did anything at all. IBM did not deny that Sycamore had performed its random-circuit-sampling task in two hundred seconds. It attacked only Google's *estimate* of the classical difficulty — the "ten thousand years" figure — arguing that Google had assumed a particular, memory-limited classical method, and that a smarter approach exploiting Summit's enormous disk storage to hold the giant quantum state could reproduce the task in about two and a half days rather than ten thousand years. IBM never actually ran the full computation, but its argument was detailed and credible. And notice what it did and did not concede: two hundred seconds versus two and a half days is still a thousandfold speedup in Sycamore's favor — a real and impressive advantage — but it is emphatically not the decisive, exponential, "no classical computer could ever" supremacy that had been proclaimed. The disagreement was about where, precisely, to draw the line between "faster" and "supreme."

The lasting lesson, which the years since have hammered home again and again, is that quantum supremacy is not a fixed achievement but a *moving target*, defined relative to the best available classical algorithms and hardware, both of which keep improving — and often improve precisely in response to a quantum claim. In the years after 2019, Sycamore's specific result was progressively chipped away as researchers devised ever-cleverer classical methods — exploiting mathematical structures called tensor networks, harnessing clusters of graphics processors — that simulated Google's circuits faster and faster, until some groups claimed they could nearly match Sycamore on classical hardware. Each such advance forced the quantum experimentalists to run larger, harder instances to stay ahead, and the classical community would then attack those in turn: a genuine, ongoing arms race. Far from undermining quantum computing, this contest is exactly how the boundary between quantum and classical power gets mapped — through fierce, productive competition — and the fundamental mathematics ensures the quantum side wins in the end, because the classical cost of these tasks grows exponentially with the size of the quantum machine, so that even modest scaling eventually leaves every classical competitor hopelessly behind.

Chapter 47: Jiuzhang — China, Photons, and Boson Sampling

Quantum computing is not solely an American enterprise, and any honest history must reckon with the fact that one of the world's most formidable quantum-computing efforts is China's, backed by enormous state investment and driven by scientists of the first rank. In December 2020, just over a year after Google's Sycamore, a group at the University of Science and Technology of China in Hefei, led by the eminent physicist Jian-Wei Pan — often called, in the Western press, "the father of quantum" in China — announced its own quantum supremacy demonstration, achieved by an entirely different physical route. Their machine, named **Jiuzhang** after an ancient Chinese mathematical text, used not superconducting qubits but *photons*, and not random circuit sampling but a task called **boson sampling**. It was a striking demonstration that the supremacy frontier could be reached by multiple, independent hardware paths, and a signal that the quantum race was global.

Boson sampling, which we touched on in the chapter on photonic computing, is a task tailor-made to be hard for classical computers and natural for photons. The idea, proposed by Scott Aaronson and Alex Arkhipov in 2011, is to send many identical photons into a large, complicated network of beam splitters and mirrors, let them interfere in the bewildering quantum way that identical particles of light do, and then detect where they emerge. The pattern of where the photons come out is governed by quantum interference among all the myriad paths they could have taken, and computing the probabilities of the various output patterns turns out to be connected to a notoriously hard mathematical quantity (the "permanent" of a matrix) that classical computers cannot efficiently calculate. So, as with random circuit sampling, you have the crucial asymmetry: the photonic machine produces samples from the hard distribution effortlessly, simply by sending the photons through and detecting them, while a classical computer would have to grind through an intractable calculation to predict the same statistics. Jiuzhang sent dozens of photons through a large optical network and collected samples from a distribution that, Pan's group argued, would take classical supercomputers an impractically long time to reproduce.

Jiuzhang's demonstration had certain features that made it, in some respects, a stronger supremacy claim than Sycamore's, and certain features that made it weaker, and the comparison is instructive. On the stronger side, boson sampling with photons operates at or near room temperature (no dilution refrigerator), and the specific mathematical hardness underlying it is, some argued, better understood and more robust than that of random circuit sampling, making the classical-simulation gap harder to close. Pan's group also pressed the advantage, following the first Jiuzhang with improved versions using more photons, widening the claimed gap. On the weaker side, the original Jiuzhang was not a *programmable* machine in the way Sycamore was — its optical network was essentially fixed, set up to perform one particular boson-sampling task, not a general-purpose device that could be loaded with arbitrary circuits — so it was even further than Sycamore from being a universal quantum computer, more a special-purpose demonstration of quantum computational hardness than a step toward a practical machine. And, as with Sycamore, classical algorithm designers rose to the challenge, developing improved classical methods that chipped away at the claimed supremacy gap, in the same healthy back-and-forth that followed Google's claim.

The deeper significance of Jiuzhang, beyond the specifics of boson sampling, is what it revealed about the shape of the global quantum race, and this is worth dwelling on because it frames much of the geopolitics we will reach in the final part. China had, by 2020, mounted a quantum-computing and quantum-communication effort of the very highest caliber, competitive with anything in the West, and in some areas — notably quantum communication, where China had earlier launched the Micius satellite to demonstrate space-based quantum key distribution, a genuine world first — arguably ahead. Pan's group demonstrated supremacy by both photonic *and*, later, superconducting routes, showing mastery of multiple platforms. The Chinese state had recognized quantum technology as a strategic priority and was investing accordingly, on a scale that alarmed policymakers in Washington and elsewhere and helped drive the West's own national quantum initiatives. Quantum computing had become, by the early 2020s, not merely a scientific competition among laboratories and companies but a geopolitical one among nations, freighted with implications for economic power and, above all, for the security of secrets — because the same Shor's algorithm that hangs over Western encryption hangs over everyone's, and the nation that first builds a cryptographically relevant quantum computer gains an extraordinary and perhaps decisive intelligence advantage.

Just over a year after Google's Sycamore made headlines, a team in China announced its own claim to quantum computational supremacy, using a machine and a method entirely different from Google's — and in doing so both strengthened the case that quantum machines could genuinely outpace classical ones and announced, unmistakably, that quantum computing had become an international race with the world's two superpowers at its center. The machine was called Jiuzhang, named for an ancient Chinese mathematical text, and it computed not with superconducting circuits but with light.

Jiuzhang, built by a team led by the physicist Jian-Wei Pan

— often called the father of Chinese quantum research — at the University of Science and Technology of China, was a PHOTONIC quantum machine, and it performed a task called boson sampling, a problem tailor-made for photonic systems and, like Google's random circuit sampling, chosen precisely because it is believed to be extraordinarily hard for classical computers while natural for the quantum hardware. Boson sampling, first proposed by Scott Aaronson and Alex Arkhipov in 2011, involves sending many photons through a large, intricate network of beam-splitters and mirrors and then measuring where they emerge; because photons are quantum particles that interfere in fantastically complex ways as they traverse the network, predicting the distribution of where they come out requires a classical computer to evaluate a mathematical quantity (the "permanent" of a large matrix) that is notoriously, provably hard to compute. The photonic machine, by simply letting the photons do what photons do, samples from that distribution directly. In 2020, Jiuzhang was reported to have performed a boson-sampling task that Google's benchmark classical methods would supposedly take some six hundred million years to match.

The Jiuzhang result was scientifically important for several reasons beyond national pride, and it is worth being clear about them. First, it provided INDEPENDENT confirmation, on a completely different physical platform and a completely different task, that quantum machines could achieve computational supremacy — no longer a one-off from a single company on a single kind of hardware, but a phenomenon reproduced across radically different technologies, which greatly strengthened the case that the achievement was real and not an artifact of one machine or one task. Second, unlike Google's programmable but superconducting Sycamore, Jiuzhang in its original form was not a programmable computer at all but a fixed optical apparatus built to perform one specific task — a distinction that mattered, since a fixed special-purpose device is less flexible than a universal machine, though the demonstration of supremacy was no less valid for it. Third, and most consequentially for the world beyond physics, it demonstrated in the most public possible way that China was at the absolute forefront of quantum technology, capable of matching and in some respects exceeding the achievements of American giants like Google and IBM.

That last point transformed quantum computing from a scientific and commercial contest into a matter of geopolitics and national security. Jian-Wei Pan's group had already made China a leader in related quantum technologies — the same team had earlier launched the world's first quantum-communication satellite, Micius, and demonstrated quantum-encrypted links across thousands of kilometers — and Jiuzhang extended that leadership into computation. Governments took notice. The realization that quantum computers could eventually break the encryption protecting military, financial, and diplomatic secrets, combined with the demonstration that China could build world-class quantum machines, helped spur enormous national investments in quantum technology around the world, framed increasingly in the language of strategic competition. Quantum computing had become an arena of great-power rivalry, and Jiuzhang was a shot heard in every capital.

The Jiuzhang claims, like Google's, were subjected to the now-familiar cycle of classical challenge, and honesty requires noting it. Skeptics and classical algorithm developers scrutinized the boson-sampling results, and in the following years, improved classical algorithms — again exploiting clever mathematical structures and large computing clusters — narrowed the claimed gap, arguing that the original supremacy margins were overstated and that classical machines could come closer than first thought. The Chinese team responded, as Google had, by building larger and more capable versions of Jiuzhang, using more photons and more complex networks to push the task further beyond classical reach, and the arms race continued on the photonic front just as it had on the superconducting one. The pattern was universal: claim, classical counterattack, larger quantum machine, repeat.

The significance of a *second* supremacy demonstration, on entirely different hardware, is easy to underrate and worth making explicit. When a single company achieves a startling result on a single kind of machine performing a single kind of task, a reasonable skeptic may wonder whether the achievement is an artifact — of that particular hardware, that particular benchmark, that particular company's assumptions. Jiuzhang answered that skepticism decisively. Here was quantum computational supremacy reproduced on a completely different physical platform (photons rather than superconducting circuits), on a completely different task (boson sampling rather than random circuit sampling), by a completely different team on the other side of the world. The phenomenon was not a fluke of one machine or one benchmark; it was a robust, reproducible feature of what sufficiently controlled quantum systems can do. This independent confirmation, on radically different foundations, greatly strengthened the case that quantum supremacy was genuine and not an illusion of one group's optimistic bookkeeping.

The transformation of quantum computing into a matter of geopolitics dates, more than to any other single event, to Jiuzhang, and the shift deserves emphasis. Before it, quantum computing was widely seen, especially in the West, as a contest among American technology giants — Google, IBM, and their kin. Jiuzhang announced, in the most public and dramatic way possible, that China was at the absolute forefront of quantum technology, capable of matching and in some respects exceeding the achievements of those giants. Jian-Wei Pan's group had already made China a leader in quantum communication — launching the Micius satellite and building the world's most extensive quantum-secured networks — and Jiuzhang extended that leadership into computation. Governments took notice, and quantum technology moved swiftly up the agenda of national-security planners, framed increasingly in the language of great-power competition, strategic advantage, and the specter of an adversary who might one day break the world's encryption. A photonic experiment in Hefei had become a shot heard in every capital.

Jiuzhang's claims, of course, were subjected to the same relentless cycle of classical challenge that had greeted Sycamore, and the pattern is by now familiar enough to be almost reassuring. Classical algorithm developers scrutinized the boson-sampling results and, in the years that followed, devised improved classical methods that narrowed the claimed gap, arguing that the original supremacy margins had been overstated. The Chinese team responded, exactly as Google had, by building larger and more capable versions of the machine — more photons, more complex optical networks — to push the task further beyond classical reach, and the arms race continued on the photonic front just as it had on the superconducting one. Claim, classical counterattack, larger quantum machine, repeat: the rhythm was universal, transcending platform, task, and nation. Between them, Sycamore and Jiuzhang established beyond reasonable doubt that quantum computational supremacy was a real, reproducible phenomenon — demonstrated on two continents with two utterly different technologies — and equally that it was a contested, shifting milestone, perpetually harried by an improving classical adversary and widely misunderstood by a public told it meant more than it did.

Chapter 48: What Supremacy Did and Did Not Prove

It is time to take stock, soberly and without either the boosterism of the press releases or the sourness of the debunkers, of what the great supremacy demonstrations actually established — because the gap between what quantum computers have *done* and what they are popularly *believed* to have done is, in the mid-2020s, the single most important thing for an honest observer to understand. This chapter is the reckoning that the two previous ones set up, and its conclusions are more modest, and more interesting, than the headlines suggest.

Begin with what supremacy genuinely proved, for it is real and should not be minimized. Before Sycamore and Jiuzhang, there remained room for a determined skeptic to doubt whether controllable quantum computation could ever be scaled to the point of outrunning classical machines — whether the accumulating noise and the difficulty of controlling many qubits would forever keep quantum computers penned in a regime that classical computers could easily simulate, making them at best a curiosity. The supremacy demonstrations closed off that pessimism. They showed, with real hardware, that quantum processors could be built and controlled well enough to reach a regime where classical simulation becomes extraordinarily difficult — that the exponential resources of entangled quantum states could be marshaled, imperfectly but genuinely, at a scale beyond easy classical reach. This is a profound scientific achievement, a validation of decades of theory and engineering, and a genuine milestone in the history of computation. Quantum computers are real, they work, and they can do things that strain the largest classical machines. That much is now beyond serious doubt.

Now the essential caveats, which the hype systematically obscures. First, the tasks were *useless*. Random circuit sampling and boson sampling were chosen precisely because they are hard for classical computers and natural for quantum hardware, not because their answers matter to anyone; no scientific, commercial, or practical benefit flows from knowing the output distribution of a random quantum circuit. The supremacy machines did not cure a disease, break a code, design a material, or optimize a supply chain; they performed elaborate, verifiable, but pointless statistical demonstrations. Second, the claims were *contested and eroded*: as we saw, cleverer classical algorithms repeatedly narrowed the gaps, and the specific "ten thousand years" kind of figure did not survive scrutiny. Third, and most importantly, these were *NISQ* demonstrations — noisy, not error-corrected, limited in depth — and they did not, and could not, run the great algorithms that give quantum computing its world-changing promise. Sycamore cannot run Shor's algorithm to factor a meaningful number; it does not have the qubits, the fidelity, or the error correction. The supremacy machines proved that quantum hardware had reached a frontier, but the frontier they reached is a long way from the summit where useful quantum computing lives.

This brings us to the crucial distinction the whole part has been building toward, and it deserves to be stated as starkly as possible, because getting it wrong is the source of most public confusion about quantum computing. There are two quite different milestones, and they may be separated by a decade or more. **Quantum supremacy** (or "advantage" in the weak sense) means outperforming classical computers on *some* task, however contrived — and it has, arguably, been achieved, on useless problems, by NISQ machines, subject to ongoing classical challenge. **Quantum advantage in the strong sense** — sometimes called "practical quantum advantage" — means outperforming classical computers on a *useful* problem, one whose answer someone actually wants: a valuable molecule simulated, a real optimization solved faster, a genuine speedup on a task with economic or scientific worth. And this second milestone, the one that actually matters for the technology's promised revolution, *has not yet been clearly achieved on any practically important problem, by any quantum computer, anywhere,* as of the mid-2020s. This is the single most important and least understood fact about the state of quantum computing. The machines have achieved supremacy on toy problems; they have not achieved useful advantage on real ones. The revolution has been demonstrated in principle and not yet delivered in practice.

Why the long gap between the two milestones? Because useful advantage requires either fault-tolerant machines with error-corrected logical qubits (still being built, as Part XII will describe) or the discovery of NISQ algorithms that extract real value from noisy, uncorrected hardware (a hope that has proved harder to realize than many expected). The supremacy tasks were, in a sense, cheats: chosen to be exactly what NISQ machines can do and classical computers cannot, with no requirement of usefulness. Useful problems do not come so conveniently tailored; they demand more qubits, more depth, more reliability than NISQ machines possess, and closing that gap is the central work of the field's current era. Some sober observers worry that the gap could prove long indeed, that the NISQ era might yield few or no useful applications and that real value awaits full fault tolerance still years away — a concern we will weigh honestly in the final part. Others are more optimistic, pointing to rapid recent progress in error correction. But the honest present-tense answer is that quantum computers, for all their genuine and remarkable achievements, have not yet done anything *useful* that a classical computer cannot do, and anyone who tells you otherwise is selling something.

After the triumphant announcements and the fierce rebuttals, after the moving finish lines and the international rivalry, it is worth stepping back to ask the simple, essential question calmly: what did quantum supremacy actually PROVE? The answer is genuinely important, and it lies carefully between the breathless hype that greeted the announcements and the dismissive skepticism of the debunkers. Both extremes miss the truth, and getting it right is the best inoculation against a lifetime of misunderstanding quantum computing news.

Start with what supremacy did NOT prove, because the misconceptions are so widespread and so tenacious. It did not prove that quantum computers are useful — the tasks performed, random circuit sampling and boson sampling, are deliberately, completely useless, chosen only as benchmarks. It did not prove that quantum computers can break encryption, simulate molecules, solve optimization problems, or do anything of practical value; those capabilities require far larger, error-corrected machines that do not yet exist. It did not prove that quantum computers are ready to replace or even complement classical computers for any real-world task. And it did not prove that quantum computers are, in general, faster than classical ones — for the overwhelming majority of computational problems, quantum machines offer no advantage at all. Anyone who came away from the 2019 headlines believing that powerful, useful quantum computers had arrived was badly misled, and much of the subsequent public confusion and disappointment traces to exactly that overreading.

Now consider what supremacy DID prove, which is real and profound and should not be diminished by the debunkers. It proved that a programmable quantum machine could be built, controlled, and operated with enough precision — enough qubits, enough coherence, enough gate fidelity — to reach a genuinely new regime, producing and sampling from quantum states of a complexity that strains or exceeds what classical computers can handle. This was not guaranteed. For decades it had been an open question whether decoherence and error would forever prevent quantum machines from doing anything beyond trivial demonstrations; the pessimists had warned that the fragility might be insurmountable in practice. Quantum supremacy answered that question: the machines work, at least well enough to enter the regime where quantum effects give real computational leverage. It was a decisive proof of principle that the exponential resource latent in quantum mechanics could be harnessed by an engineered, programmable device — the experimental vindication of everything from Feynman's vision to Deutsch's universal machine to the whole edifice of quantum algorithms.

The honest framing, then, is that quantum supremacy was

a crucial SCIENTIFIC milestone but not yet a PRACTICAL one — a proof of capability, not of utility. A useful analogy is the first controlled nuclear chain reaction, achieved by Enrico Fermi's team beneath a Chicago squash court in 1942. That reaction produced only about half a watt of power, enough to light a small bulb dimly, and was utterly useless as a source of energy. But it proved, decisively, that a controlled chain reaction was possible, and everything that followed — reactors, and yes, weapons — flowed from that proof of principle. Quantum supremacy is quantum computing's Chicago Pile: a small, useless, epochal demonstration that the thing can be done at all, from which, if the engineering can be pushed far enough, world-changing applications may eventually flow. The dim bulb is not the point; the proof that the reaction sustains itself is.

The debate over the moving finish line — the classical community's repeated success in narrowing or closing the claimed gaps — does not undermine this core achievement, and it is important to see why. Yes, cleverer classical algorithms repeatedly ate into the specific supremacy margins, and for some early demonstrations the classical machines nearly caught up. But this reflects the difficulty of proving a NEGATIVE (that no classical method could keep up) far more than any weakness in the quantum machines themselves. And the fundamental mathematics is unforgiving in the quantum machine's favor: the classical cost of these sampling tasks grows EXPONENTIALLY with the size of the quantum system, so while classical algorithms can win skirmishes against a fixed small machine, they cannot win the war — each modest increase in the number of quantum qubits multiplies the classical burden until no conceivable classical computer, however clever, can follow. The arms race is real, but the exponential guarantees which side wins as the machines grow. Supremacy proved that the exponential lever exists and can be pulled; scaling will make the lever irresistible.

The Chicago Pile analogy deserves to be pressed a little further, because it captures, better than any other, the precise nature of what quantum supremacy achieved. When Enrico Fermi's team achieved the first controlled nuclear chain reaction beneath the stands of a Chicago squash court in December 1942, the reactor produced about half a watt of power — enough to light a single small bulb dimly, and utterly useless as a source of energy. By the crude measure of practical output, it was nothing. But it proved, decisively and for the first time, that a controlled chain reaction was possible, and from that proof of principle flowed everything that followed: the reactors that would one day power cities, and the weapons that would reshape geopolitics. The dim bulb was not the point; the self-sustaining reaction was. Quantum supremacy is quantum computing's Chicago Pile — a small, useless, epochal demonstration that the fundamental thing can be done at all, from which, if the engineering can be pushed far enough, world-changing capabilities may eventually flow.

The debate over the moving finish line — the classical community's repeated success in narrowing the claimed gaps — should not, on reflection, diminish the core achievement, and understanding why requires holding two ideas together. Yes, cleverer classical algorithms repeatedly ate into the specific supremacy margins, and for some early demonstrations the classical machines nearly caught up; this reflects the genuine difficulty of proving that *no* classical method could keep pace. But the fundamental mathematics is unforgiving in the quantum machine's favor: the classical cost of these sampling tasks grows *exponentially* with the size of the quantum system, so while classical algorithms can win skirmishes against a fixed, small machine, they cannot win the war. Each modest increase in the number of quantum qubits multiplies the classical burden until no conceivable classical computer, however clever, can follow. The arms race is real and productive, but the exponential guarantees which side prevails as the machines grow. Supremacy proved that the exponential lever exists and can be pulled; scaling will make the lever irresistible.

The balanced verdict, then — the one worth carrying away — is that quantum supremacy was a genuine and historic *scientific* milestone but not yet a *practical* one, a proof of capability rather than of utility, and one widely and badly misunderstood as more than it was. It proved that engineered, programmable quantum machines can be built and controlled precisely enough to enter a regime beyond easy classical simulation — the experimental vindication of everything from Feynman's founding vision to Deutsch's universal machine to the whole edifice of quantum algorithms. It did not prove that quantum computers are useful, that they can break codes or simulate molecules or solve any practical problem, or that they are ready to replace classical machines for anything. Anyone who came away from the 2019 headlines believing that powerful, useful quantum computers had arrived was badly misled — and much of the subsequent public confusion and eventual disappointment traces to exactly that overreading. Holding the balance — real achievement, real remaining distance — is the mark of understanding. And there was one arena where the mere *prospect* of the future machine already reshaped the present with real and urgent stakes: the codes that guard the secrets of the world.

Part X — The Cryptography Reckoning

Chapter 49: The Locks of the Internet

Every time you see a little padlock icon in your web browser, an invisible act of mathematical magic has just taken place. In the fraction of a second before a secure web page loads, your computer and a distant server have performed a cryptographic ritual that lets them agree on a secret key — a key known only to the two of them, even though every byte they exchanged to establish it traveled across the open internet, visible to anyone who cared to look. This ritual, repeated billions of times a day, is the foundation of essentially all security online, and it depends on a family of mathematical problems believed to be hard for classical computers. Quantum computers, when large enough, will make those problems easy. To understand the magnitude of the threat, and the frantic effort to defend against it, you must first understand what, exactly, quantum computers threaten — the locks of the internet and how they work.

The central invention that makes secure communication possible among strangers is **public-key cryptography**, and it is worth appreciating what a startling idea it is, because its very cleverness is what makes it vulnerable. Before public-key cryptography, encryption required that sender and receiver *already share a secret key* — a code book, a password, something agreed in advance through some secure channel. This is fine for spies who can meet in person, but hopeless for the internet, where your computer must establish secure communication with a server it has never contacted before and with which it shares no prior secret. Public-key cryptography, invented in the 1970s (publicly by Diffie, Hellman, Rivest, Shamir, and Adleman; secretly, it later emerged, somewhat earlier by researchers at Britain's GCHQ), solved this seemingly impossible problem with a beautiful asymmetry. Each party has *two* keys: a **public key**, which can be shared with the whole world, and a **private key**, kept secret. The keys are mathematically linked so that a message scrambled with the public key can be unscrambled only with the matching private key. Anyone can use your public key to encrypt a message to you, but only you, with your private key, can decrypt it. No prior shared secret is needed; the public key can be shouted from the rooftops.

The security of this scheme rests entirely on a **one-way** mathematical relationship between the public and private keys — a computation that is easy to perform in one direction but believed to be practically impossible to reverse. In the RSA system, as we saw in the chapter on Shor's algorithm, the one-way function is multiplication: the public key is built from the product of two large secret primes, and recovering the primes from the product — *factoring* — is believed to be intractable for classical computers when the numbers are large enough. In the Diffie–Hellman and elliptic-curve systems, which underlie much of the key exchange that actually happens when you load a secure web page, the one-way function is different — it involves the **discrete logarithm** problem, a kind of arithmetic that is easy forward and believed hard to reverse — but the principle is identical: security rests on a mathematical operation easy to do and, classically, practically impossible to undo. These systems — RSA, Diffie–Hellman, elliptic-curve cryptography — are the locks of the internet, the mechanisms that establish the secret keys, verify identities, and create the digital signatures that authenticate software, certificates, and transactions. They are everywhere, invisibly, in every secure connection, and the entire edifice of digital trust is built upon the presumed difficulty of the mathematical problems they rely on.

Here is the vulnerability, stated plainly: *every one of these systems is broken by Shor's algorithm.* Recall from Part V that Shor's algorithm efficiently solves both the factoring problem (breaking RSA) and the discrete logarithm problem (breaking Diffie–Hellman and elliptic-curve cryptography). The one-way functions that secure the internet are one-way only for *classical* computers; for a sufficiently large quantum computer running Shor's algorithm, they are two-way, easily reversible, no locks at all. A quantum computer that could run Shor's algorithm on cryptographically sized numbers would be able to derive private keys from public ones — to pick every public-key lock on the internet. It could decrypt communications, forge digital signatures, impersonate any party, and undermine the authentication that the entire online world depends on. This is not a weakening, as Grover's algorithm merely weakens symmetric encryption; it is a *total break* of the public-key layer. The locks do not become harder to pick; they cease to be locks at all.

To understand why governments and corporations are pouring billions into quantum computing — and into defending against it — you must understand the invisible locks that secure the entire digital world, because those locks are exactly what a large enough quantum computer would pick. Every time you see a little padlock icon in your web browser, every time you send a message, make a payment, log into an account, or transmit anything private over the internet, you are relying on cryptography — mathematical locks that keep your information secret from anyone who might intercept it. These locks are the silent, ceaseless machinery that makes digital life possible, and almost no one thinks about them. Quantum computing threatens to break the most important of them, and this chapter explains what they are and why they are vulnerable.

The most consequential of these locks is public-key cryptography, and the ingenious problem it solves is worth appreciating, because its solution is also its weakness. Imagine you want to receive secret messages from someone you have never met and with whom you have never privately exchanged a password. How can they encrypt a message that only you can read, when everyone — including eavesdroppers — can see all the communication between you? For most of history this was impossible; secure communication required the two parties to first share a secret key in private, face to face or through a trusted courier. Public-key cryptography, invented in the 1970s, performed what seemed like magic: it let two strangers establish secure communication over a completely public channel, with no prior shared secret. It works using a pair of mathematically linked keys — a PUBLIC key that anyone can see and use to encrypt a message TO you, and a PRIVATE key that only you possess and that alone can decrypt those messages. You publish your public key to the world; anyone can lock a message with it; but only your private key can unlock it. It is as if you handed out open padlocks to everyone, keeping the only key to yourself.

The security of this scheme rests entirely on certain mathematical problems being easy to perform in one direction but practically impossible to reverse — the "one-way functions" we met in the discussion of Shor's algorithm. The dominant system, RSA (named for Rivest, Shamir, and Adleman, who published it in 1977), rests on the difficulty of factoring large numbers: your public key is essentially a very large number that is the product of two enormous secret primes, and your private key is those primes. Anyone can multiply the primes together — that is easy — but recovering the primes from their product is, for numbers of the sizes used in practice (hundreds of digits), so staggeringly hard for classical computers that it would take longer than the age of the universe. A closely related system, elliptic-curve cryptography, rests on a different but analogously hard problem (the "discrete logarithm"), and is increasingly used because it achieves the same security with smaller keys. Together, RSA and elliptic-curve cryptography secure essentially all of the encrypted communication on the internet: your banking, your messaging, your shopping, the software updates that keep your devices safe, the very certificates that let your browser trust that a website is genuine.

The vulnerability is now obvious, and it is total. Shor's algorithm, as we have seen, factors large numbers efficiently on a sufficiently powerful quantum computer — and it solves the elliptic-curve discrete-logarithm problem too. This means a large-scale, error-corrected quantum computer running Shor's algorithm would completely break both RSA and elliptic-curve cryptography, rendering them useless. Every secret protected by these systems — which is to say, nearly every secret transmitted over the internet — would lie open to anyone with such a machine. The padlocks that secure the digital world would spring open at a touch. This is not a partial weakness or a matter of using longer keys; it is a fundamental, catastrophic break of the entire mathematical foundation on which public-key cryptography stands. The one-way function is one-way no longer.

It is essential to be precise about the scope of the threat, because precision is what separates informed concern from panic, and both extremes are common. Not all cryptography is equally doomed. SYMMETRIC cryptography — the kind that uses a single shared secret key to both encrypt and decrypt, such as the widely used AES standard that protects data once a connection is established — is only mildly threatened: Grover's algorithm can speed up the search for a symmetric key, but only quadratically, which effectively halves the key's strength, a threat easily neutralized by simply doubling the key length (using 256-bit keys instead of 128-bit).

Symmetric cryptography survives the quantum era with a minor adjustment. It is PUBLIC-KEY cryptography — RSA and elliptic curves — that faces total destruction from Shor's algorithm, and because public-key cryptography is what allows strangers to establish secure connections in the first place (and to distribute the symmetric keys that then do the bulk encryption), its collapse would unravel the whole system. The lock that lets everything else work is the one that breaks.

It is worth appreciating just how quietly miraculous public-key cryptography is, because we rely on it constantly without a thought, and its cleverness is the very thing quantum computers threaten to undo. Before the 1970s, secure communication required the two parties to share a secret key in advance, exchanged in private — face to face, or through a trusted courier — which made secure communication among strangers, at scale, essentially impossible. Public-key cryptography performed what seemed like magic: it let two people who had never met, and who shared no prior secret, establish perfectly secure communication over a completely public channel that eavesdroppers could freely observe. It works through a pair of mathematically linked keys — a public key anyone can use to encrypt a message *to* you, and a private key that only you possess and that alone can decrypt those messages. You publish the open padlock to the world; anyone can lock a message with it; but only you hold the key. Every padlock icon in your browser, every secure login, every encrypted payment, every software update your devices trust — all of it rests on this quiet miracle.

The mathematical foundation of that miracle — and its Achilles' heel — is the one-way function, and the distinction it embodies is worth grasping precisely. Certain mathematical operations are easy to perform in one direction but practically impossible to reverse. Multiplying two enormous prime numbers together is easy — a pocket calculator does it instantly — but recovering those primes from their product (factoring) is, for numbers hundreds of digits long, so staggeringly hard for classical computers that it would take longer than the age of the universe. RSA encryption builds its keys from exactly this asymmetry: your public key is essentially a huge number that is the product of two secret primes, and your private key is those primes. A closely related system, elliptic-curve cryptography, rests on a different but analogous one-way problem (the discrete logarithm) and achieves the same security with smaller keys. Together these two systems secure essentially all of the encrypted communication on the internet — the difficulty of reversing their one-way functions is the wall behind which the digital world shelters.

Shor's algorithm, as we have seen, tears that wall down, and the completeness of the destruction bears emphasizing, along with its precise limits. A large-scale quantum computer running Shor's algorithm would factor large numbers and solve the discrete-logarithm problem efficiently, completely breaking both RSA and elliptic-curve cryptography — rendering them useless, exposing every secret they protect. This is not a partial weakness patchable with longer keys; it is a total, catastrophic break of the mathematical foundation of public-key cryptography. But precision matters, for it separates informed concern from panic. *Symmetric* cryptography — the kind using a single shared key to both encrypt and decrypt, such as the AES standard that does the bulk of data encryption once a connection is established — is only mildly threatened: Grover's algorithm speeds the search for a symmetric key merely quadratically, effectively halving its strength, a threat neutralized simply by doubling the key length. Symmetric cryptography survives the quantum era with a minor adjustment. It is *public-key* cryptography — the lock that lets strangers establish secure connections and distribute those symmetric keys in the first place — that faces total destruction, and because it is the lock that lets everything else work, its collapse would unravel the whole system. The threat is specific, profound, and not yet realized — but already urgent, because of the harvest-now-decrypt-later logic that gives Q-Day its long backward shadow.

Chapter 50: Q-Day — When the Locks Break

Cryptographers have a name for the hypothetical day on which a quantum computer capable of breaking current public-key cryptography first comes into existence: **Q-Day**. It is a deliberately dramatic term for a genuinely momentous prospect — the day the locks of the internet, described in the previous chapter, cease to be locks. On Q-Day, or in the period following it, whoever possesses the machine gains the ability to break RSA, Diffie–Hellman, and elliptic-curve cryptography, and with them the confidentiality and authenticity of vast swaths of the world's digital communications and stored data. This chapter asks the two questions everyone wants answered — how bad would Q-Day be, and when will it come — and gives the honest answer to the second, which is that nobody knows, and anyone who claims certainty is guessing.

Consider first the stakes, to understand why Q-Day commands the attention of governments and security professionals worldwide. The immediate consequence of a cryptographically relevant quantum computer would be the compromise of public-key cryptography everywhere it is used, which is nearly everywhere: the secure web (every padlock), encrypted email and messaging, virtual private networks, the authentication of software updates and digital certificates, the security of financial transactions and banking systems, the protection of government and military communications, the integrity of blockchains and cryptocurrencies (whose signatures rely on elliptic-curve cryptography), and much more. An adversary with such a machine could, in principle, decrypt intercepted communications, forge the digital signatures that certify software and identities, impersonate trusted parties, and undermine the chains of trust on which digital society runs. It would be, if it arrived suddenly and without preparation, a security catastrophe of the first order — which is precisely why the world is not waiting for it to arrive suddenly, but preparing, as the following chapters describe. The threat is severe enough that even a modest probability of its near-term arrival justifies serious, expensive precaution.

Now the timing, and here honesty requires humility. Estimates for when a cryptographically relevant quantum computer might exist vary enormously, from optimists (or pessimists, depending on your point of view) who suggest the early 2030s, to those who think it decades away, to a minority who suspect it may never be practical at all. The uncertainty is genuine and irreducible, because it depends on engineering breakthroughs that have not yet happened and whose difficulty is hard to forecast. Recall the daunting arithmetic: breaking a 2048-bit RSA key with Shor's algorithm requires not merely thousands of qubits but thousands of *error-corrected logical* qubits, each built from perhaps a thousand or more noisy physical qubits, implying a machine with on the order of a *million* physical qubits running coherently for hours — while the best machines of the mid-2020s have hundreds to about a thousand noisy physical qubits and no large-scale error correction. The gap is vast. Closing it requires sustained progress on qubit quality, qubit count, and above all error correction, and while that progress has been real and, lately, encouraging, extrapolating it to a firm date is guesswork. Adding to the uncertainty, the resource estimates themselves keep changing: better algorithms and error-correction schemes have repeatedly *lowered* the number of qubits thought necessary, pulling the threat closer, even as the engineering reality of building those qubits keeps it far. The result is a forecast with enormous error bars, and responsible experts express it as a probability distribution over decades, not a date.

The uncertainty itself has become a subject of formal study and polling. Surveys of quantum-computing experts, conducted periodically, reveal a wide spread of opinion but a discernible center of mass: a substantial fraction of experts estimate a meaningful probability — often cited around one in seven, or higher over longer horizons — that a quantum computer capable of breaking RSA-2048 will exist within a decade or so, and a majority think it more likely than not within two to three decades. These are not predictions but aggregated expert judgments under deep uncertainty, and they should be read as such. But their practical import is clear and sobering: the probability of Q-Day within the planning horizon of long-lived secrets and long-lived infrastructure is high enough that ignoring it would be reckless. A government protecting secrets that must stay secret for thirty years, or an industry deploying infrastructure meant to last decades, cannot treat a one-in-seven-within-ten-years threat as negligible. Prudence demands action well before Q-Day, not after.

"Q-Day" is the name given, half-seriously and half in genuine dread, to the hypothetical future day when a quantum computer becomes powerful enough to break the public-key cryptography that secures the world — the day RSA and elliptic-curve encryption fail, and the digital locks protecting banking, government, military, and personal secrets all spring open at once. It is a concept that has moved from science-fiction speculation into the sober planning documents of intelligence agencies, central banks, and standards bodies, and understanding both why it is taken so seriously and why it has not yet arrived is essential to a clear-eyed view of quantum computing's real-world stakes.

The most unsettling feature of Q-Day is that its threat operates BACKWARD in time, reaching into the present through a strategy chillingly known as "harvest now, decrypt later." An adversary — a hostile nation's intelligence service, say — does not need to possess a quantum computer today to begin exploiting the quantum threat today. They need only intercept and record encrypted communications and data NOW, storing the scrambled ciphertext in vast archives, and wait. Whenever Q-Day arrives — in five years, or fifteen, or thirty — they simply run Shor's algorithm on their hoard and decrypt everything they collected in the intervening decades. This means that any secret which must remain secret for a long time is arguably ALREADY compromised, right now, even though the quantum computer does not yet exist. State secrets, diplomatic cables, military plans, intelligence sources, medical records, the design of long-lived infrastructure, biometric data that can never be changed — anything with a shelf life of a decade or more is potentially being harvested this very moment for future decryption. The quantum computer of 2040 threatens the secrets of 2026. The future reaches back and steals from the present.

This backward-reaching logic is precisely why Q-Day is treated with such urgency despite lying somewhere in the uncertain future. If the only risk were to communications sent AFTER a quantum computer was built, there would be time to relax and prepare at leisure. But because secrets sent today can be captured today and cracked tomorrow, the deadline for protecting long-lived information is not Q-Day itself but NOW — the migration to quantum-resistant cryptography must be well underway before Q-Day, or an enormous trove of already-harvested secrets will be exposed the instant the machine switches on. Security officials speak of this in terms of a simple, alarming inequality attributed to the cryptographer Michele Mosca: if the length of time your secrets must stay secret, plus the number of years it will take to migrate all your systems to new cryptography, exceeds the number of years until Q-Day, then you are already too late. For many organizations, with sprawling legacy systems that take a decade or more to upgrade and secrets that must last decades, the math is deeply uncomfortable.

So when will Q-Day come? Here honesty demands humility, because no one knows, and the estimates span a vast range. Breaking a real 2,048-bit RSA key with Shor's algorithm is thought to require a quantum computer with thousands of high-quality LOGICAL qubits — which, given the brutal overhead of error correction, translates to perhaps millions of PHYSICAL qubits, all working in concert with error rates below the threshold, executing billions of gate operations. Today's best machines have a few hundred noisy physical qubits and no more than a handful of logical ones. The gap is enormous — a factor of thousands or more in every dimension — and bridging it requires not just incremental scaling but sustained breakthroughs in error correction, qubit quality, and system engineering. Expert estimates for when a cryptographically relevant quantum computer will exist range from optimistic guesses of a decade to pessimistic ones of several decades to a quiet suspicion, held by some, that it may never be practical at all. Anyone who tells you they know the date with confidence is selling something.

Yet the uncertainty about timing has not produced complacency; on the contrary, it has produced a global scramble, because the "harvest now, decrypt later" logic means the prudent response to an uncertain deadline is to act NOW. Governments have issued directives ordering their agencies to inventory their cryptography and begin migrating to quantum-resistant systems on aggressive timelines; in the United States, national security memoranda and legislation have mandated the transition, with target dates in the 2030s. Intelligence agencies treat the harvesting threat as active and ongoing. And an entire field of "post-quantum cryptography" — new encryption schemes designed to resist even a quantum computer — has moved from academic research to urgent standardization and deployment, a story we will take up in the coming chapters. The looming, dateless certainty of Q-Day has, in effect, already arrived as a force reshaping how the world thinks about securing its secrets, long before the machine that defines it has been built.

The strange temporality of Q-Day — a future capability that reaches backward to endanger the present — is worth restating because it is so counterintuitive and so consequential. Ordinarily, a future threat gives you until it arrives to prepare; a lock-pick that will exist in fifteen years leaves fifteen years to install better locks, and whatever you protect in the meantime stays safe. Q-Day shatters this logic through the "harvest now, decrypt later" strategy: because encrypted data, once transmitted, can be copied and stored indefinitely, an adversary can record your secrets today and simply wait for the machine that will one day decrypt them. The theft happens now; the reading happens later. This means the effective deadline for protecting any secret with a long shelf life is not Q-Day but the present moment — a deeply unsettling inversion that transforms an uncertain future milestone into a present-tense emergency for anyone whose secrets must endure.

The range of expert estimates for when Q-Day will arrive is worth laying out plainly, because the honest answer is genuine uncertainty, and both false confidence and false panic are common. Breaking a real 2,048-bit RSA key with Shor's algorithm is thought to require a quantum computer with thousands of high-quality logical qubits, which — given the brutal overhead of error correction — translates to perhaps a million or more physical qubits, all working in concert below the error-correction threshold, executing billions of gate operations with the fragile superposition held intact throughout. Today's best machines have a few hundred noisy physical qubits and a handful of logical ones. The gap is enormous — a factor of thousands in every dimension — and bridging it requires not just incremental scaling but sustained breakthroughs. Expert estimates range from optimistic guesses of a decade, through sober assessments of two or three decades, to a quiet suspicion held by some that a cryptographically relevant machine may never be practical at all. Anyone who states the date with confidence is selling something.

Yet the uncertainty has produced not complacency but a global scramble, and the reason returns us to the harvest-now logic. Because long-lived secrets sent today can be captured today and cracked whenever Q-Day comes, the prudent response to an uncertain deadline is to act *now*. Governments have issued directives ordering their agencies to inventory their cryptography and begin migrating to quantum-resistant systems on aggressive timelines, with target dates in the 2030s; intelligence services treat the harvesting threat as active and ongoing; and an entire field of post-quantum cryptography has moved from academic research to urgent standardization and deployment. The dateless, possibly-distant, possibly-never certainty of Q-Day has, in effect, already arrived as a force reshaping how the world secures its secrets — a future machine casting a long shadow backward over the present, forcing the world to defend today against a computer that does not yet exist. The defense takes two forms: replacing the vulnerable codes with new mathematical locks even a quantum computer cannot pick, and building security from the laws of quantum physics themselves. To the first — the frantic, worldwide effort to armor the internet before Q-Day dawns — we turn next.

Chapter 51: Harvest Now, Decrypt Later — the Threat That Is Already Here

Of all the reasons quantum computing matters, the one that makes it a *present* concern rather than a future one is a piece of cold strategic logic so simple that anyone can grasp it in a sentence, yet so consequential that it has reshaped the security planning of governments and corporations worldwide. It is called **harvest now, decrypt later** — sometimes "store now, decrypt later," or "retrospective decryption" — and it means exactly what it says: an adversary intercepts and stores encrypted data today, while it cannot yet read it, and simply waits until a future quantum computer can decrypt the hoard. This chapter dwells on this idea because it is the hinge on which the entire practical urgency of the quantum threat turns, and because it is so often overlooked by those who reassure themselves that Q-Day is decades away.

The logic is inescapable once stated. Encrypted data traveling across the internet — or stored on servers, or flowing through undersea cables — can be copied and archived by anyone with access to the traffic, most obviously by nation-state intelligence agencies with the reach to tap the internet's backbone. The encryption protecting that data is, today, unbreakable; the intercepting adversary cannot read it *now*. But encryption does not decay gracefully — it is either secure or, once the key can be recovered, completely broken — and so an adversary who patiently stores the encrypted data can, on the day a capable quantum computer arrives, retroactively decrypt everything it has been hoarding: years, even decades, of communications, laid bare all at once. The secret you send today, believing it safe behind RSA or elliptic-curve encryption, is safe only for as long as no quantum computer exists — and if a copy of it is sitting in an adversary's data warehouse, it will be read the moment one does. Your secret's true lifespan is not "until Q-Day"; it is "from the moment you send it," if the adversary is already listening.

This transforms the quantum threat's timeline in a way that is essential to understand. The relevant question is not "when will Q-Day arrive?" but "how long does my data need to stay secret, and could someone be storing it now?" Cryptographers formalize this with what is sometimes called **Mosca's inequality**, after the researcher Michele Mosca: you have a problem if the length of time your data must remain secure, *plus* the time it will take to migrate your systems to quantum-resistant cryptography, is *greater than* the time until a quantum computer arrives. Rearranged, this says: if your secrets must last, say, twenty-five years, and migrating your infrastructure to new cryptography will take ten years, then you must complete that migration before a quantum computer is within thirty-five years — which, given the expert estimates of the previous chapter, means you are already late, or nearly so. For any organization holding long-lived secrets, the harvest-now-decrypt-later logic, combined with the slow pace of cryptographic migration, means the deadline for action is not Q-Day but *years before* it — and for the most sensitive, longest-lived secrets, that deadline may already have passed.

Which secrets are actually at risk from this strategy? Not the ephemeral ones — your lunch order, a one-time password, a message whose contents are worthless in a year, are safe enough, because by the time they can be decrypted no one will care. The danger is to information with a *long shelf life*: state and military secrets that remain sensitive for decades; intelligence sources and methods; the designs of weapons systems and critical infrastructure; diplomatic communications; medical records and genetic data, which are sensitive for a lifetime; trade secrets and intellectual property; the personal data of individuals; and the long-term keys and root certificates that anchor digital trust. For all of these, the possibility that an adversary is recording encrypted traffic today with the intent of reading it after Q-Day is a live and serious concern, and it is widely assumed — though by its nature hard to prove — that the major intelligence powers are already engaged in exactly this kind of large-scale harvesting, patiently filling data warehouses against the day their quantum computers come online. The absence of public proof is cold comfort; the strategy is cheap, obvious, and deniable, and prudence must assume it is underway.

The single most important idea for understanding why the quantum threat is urgent TODAY, even though the threatening machine does not yet exist, deserves a chapter of its own, because it inverts our normal intuitions about security and deadlines. It is the strategy called "harvest now, decrypt later," sometimes rendered "store now, decrypt later," and it transforms Q-Day from a distant future problem into a present-tense emergency for anyone whose secrets must endure.

The ordinary way to think about a future threat is that you have until it arrives to prepare. If a new kind of lock-pick will exist in fifteen years, you have fifteen years to install better locks; anything locked up after the pick appears is at risk, but what you protect in the meantime is safe. Harvest now, decrypt later shatters this comfortable logic for one simple reason: encrypted data, once transmitted, can be COPIED and STORED indefinitely by anyone who can intercept it. Your encrypted message travels across the internet through many hands — routers, cables, undersea links, internet exchanges — and any adversary positioned along that path, or tapping the fiber, can silently make a perfect copy of the scrambled ciphertext and file it away. They cannot read it today; the encryption holds. But they do not need to read it today. They need only keep it, patiently, until the day a quantum computer can run Shor's algorithm and unlock their entire archive at once. The theft happens now; the reading happens later.

This means the relevant deadline for protecting long-lived secrets is not Q-Day but the present moment. Consider what has a long secret shelf life. Government and military secrets may need to stay secret for decades — the identities of intelligence sources, the designs of weapons systems, diplomatic communications, strategic plans. Personal data that cannot be changed is even more starkly exposed: your genome, your biometric fingerprints and iris scans, your medical history — these are secret for LIFE, and if harvested today and decrypted in twenty years, they are exposed for the rest of your life and beyond, unchangeable. The blueprints of infrastructure meant to last fifty years, the trade secrets underpinning a company's value, the financial records of individuals and institutions — all of these carry the kind of long shelf life that makes harvest-now-decrypt-later not a hypothetical but a live danger. Anything you send encrypted today that still matters in fifteen or twenty years may already be sitting in an adversary's archive, awaiting Q-Day.

There is strong reason to believe this harvesting is not merely theoretical but actively underway. Well-resourced intelligence agencies have both the technical means — the ability to tap major communication links at scale — and the obvious motive to be collecting and storing encrypted traffic in anticipation of future decryption capability. Building vast data-storage facilities to hoard intercepted communications is well within the capacity and known practice of major state actors, and the logic is compelling enough that any competent intelligence service would be negligent NOT to be doing it. Security professionals largely assume, as a prudent baseline, that adversaries are harvesting encrypted data today with quantum decryption in mind. You should assume that a copy of your important encrypted communications may already exist in someone's archive, patiently waiting.

The response this demands is captured in a simple, sobering piece of reasoning known as Mosca's inequality, after the cryptographer Michele Mosca, and it is worth internalizing because it turns a vague anxiety into a concrete decision rule. Let X be the number of years your secrets must remain secure. Let Y be the number of years it will take your organization to migrate all its systems to quantum-resistant cryptography — and for large institutions with tangled legacy systems, Y is often surprisingly large, five or ten years or more. Let Z be the number of years until Q-Day. Mosca's point is stark: if X plus Y is greater than Z — if the time your secrets must last plus the time it takes you to upgrade exceeds the time until the quantum computer arrives — then you are already in trouble, because some of your long-lived secrets, sent during or before the migration, will still be sensitive when Q-Day cracks them. Since Z is uncertain and possibly short, and X and Y are often large and fixed, the only safe response is to start migrating immediately.

It is worth making concrete just how many kinds of secret carry the long shelf life that makes harvest-now-decrypt-later so dangerous, because the abstraction can obscure the stakes. Some secrets expire quickly — the location of a surprise party, next week's meeting agenda — and for these the quantum threat is irrelevant; by the time Q-Day arrives, no one cares. But an astonishing range of information must stay secret for a decade, two decades, or a lifetime. State and military secrets — the identities of intelligence sources, the designs of weapons systems meant to serve for forty years, diplomatic communications, strategic plans — must often remain confidential for generations. Personal data that cannot be changed is even more starkly exposed: your genome, your fingerprints, your iris scans are secret for life, and if harvested today and decrypted in twenty years, they are exposed for the rest of your days, forever unchangeable, unlike a password you can reset. Trade secrets, financial records, the blueprints of long-lived infrastructure, the private medical histories of individuals — all carry the kind of endurance that turns today's interception into tomorrow's catastrophe.

There is strong reason to believe the harvesting is not hypothetical but actively underway, and the logic is worth spelling out because it justifies the urgency. Well-resourced intelligence agencies possess both the technical means — the ability to tap major communication links, including the undersea cables and internet exchanges through which the world's traffic flows — and the obvious motive to be collecting and storing encrypted data in anticipation of future decryption. Building vast facilities to hoard intercepted communications is well within the known capacity and practice of major state actors, and the reasoning is compelling enough that any competent intelligence service would be negligent *not* to be doing it: the cost of storage is modest, and the potential payoff, when Q-Day comes, is the retroactive exposure of decades of an adversary's secrets. Security professionals therefore largely assume, as a prudent baseline, that hostile services are harvesting encrypted data today with quantum decryption in mind. You should assume that a copy of your important encrypted communications may already sit in someone's archive, waiting patiently for a machine that does not yet exist.

Mosca's inequality, the simple rule that captures all of this, deserves to be stated once more as a decision tool, because it converts a vague dread into a concrete imperative. Let X be the number of years your secrets must stay secret; let Y be the number of years it will take to migrate all your systems to quantum-resistant cryptography — often surprisingly large, five or ten years or more, for organizations with tangled legacy systems; and let Z be the number of years until Q-Day. Mosca's point is stark: if X plus Y exceeds Z, you are already too late, because some of your long-lived secrets, sent during or before the migration, will still be sensitive when the quantum computer cracks them. Since Z is uncertain and possibly short, and X and Y are often large and fixed, the only safe course is to begin migrating immediately. The inequality is why the world cannot simply wait and watch, and why the defense — new mathematical codes deployable today on ordinary computers — had to begin years before the threat itself could possibly materialize. Forging that new armor became one of the most consequential collaborations in the history of cryptography.

Chapter 52: The New Armor — Post-Quantum Cryptography

The defense against the quantum threat to cryptography is, at first glance, surprisingly low-tech. It does not require quantum computers, quantum physics, or exotic hardware of any kind. It requires only *new mathematics* — new one-way functions, based on problems that are hard for classical *and* quantum computers alike, deployed as software on the ordinary computers we already have. This defense is called **post-quantum cryptography**, or PQC, and its central insight is elegant: Shor's algorithm breaks specific mathematical problems (factoring and discrete logarithms), but it does *not* break every hard problem. There are other mathematical problems, of a quite different character, for which no efficient quantum algorithm is known — problems believed to be hard even for a large quantum computer. Build your cryptography on *those* problems instead, and it will resist quantum attack while running perfectly well on today's classical machines. Post-quantum cryptography is the project of finding such problems and turning them into practical, standardized cryptographic systems, and it is the world's principal line of defense against Q-Day.

The most important family of post-quantum systems is based on **lattices**, and while the underlying mathematics is intricate, the flavor can be conveyed. A lattice is a regular grid of points stretching through a high-dimensional space — imagine the intersections of an infinite, multidimensional graph paper, but in hundreds of dimensions. Certain problems about lattices — such as finding the lattice point closest to a given location, or finding the shortest vector in the lattice — become extraordinarily hard as the number of dimensions grows, hard in a way that, crucially, no known quantum algorithm cracks. Shor's algorithm exploits the special periodic structure of factoring and discrete logarithms; lattice problems have no such structure for it to exploit, and so they stand firm against it. Cryptographers have learned to build encryption and digital-signature schemes whose security rests on the hardness of these lattice problems, and these lattice-based systems have emerged as the leading candidates for the post-quantum age — efficient enough for practical use, and resistant, as far as anyone knows, to both classical and quantum attack. Other families of post-quantum systems exist too, based on different hard problems — the mathematics of error-correcting codes, of multivariate equations, of hash functions, of isogenies between elliptic curves — providing diversity in case one family should fall.

The decisive milestone in post-quantum cryptography came from an unglamorous but supremely important source: a standards body. Beginning in 2016, the U.S. **National Institute of Standards and Technology** (NIST) — the same agency that standardized earlier cryptographic systems used worldwide — ran a multi-year, open, international competition to select the best post-quantum algorithms, inviting cryptographers everywhere to submit candidate schemes and then subjecting them to years of intense public scrutiny and attempted attacks, winnowing dozens of submissions down to a handful of winners. This open, adversarial process — where the world's best cryptanalysts try to break every candidate, and only the survivors are trusted — is how cryptographic standards earn confidence, and it did claim casualties: some promising candidates were dramatically broken during the process, a reminder of how hard it is to be sure a system is secure. In August 2024, after eight years of effort, NIST published the first finalized post-quantum cryptography standards — a genuine landmark, the moment the world's new cryptographic armor was officially forged and made ready for deployment.

The standards NIST finalized in 2024 deserve to be named, because they are now being deployed across the world's digital infrastructure and will, over the coming years, quietly replace the vulnerable systems in every secure connection. The primary standard for **key encapsulation** — the mechanism for establishing shared secret keys, the successor to Diffie–Hellman and the RSA key exchange — is **FIPS 203**, called **ML-KEM**, derived from a lattice-based scheme originally known as **CRYSTALS-Kyber**. For **digital signatures** — the mechanism that authenticates identities and certifies software, the successor to RSA and elliptic-curve signatures — NIST standardized **FIPS 204**, called **ML-DSA**, derived from the lattice-based **CRYSTALS-Dilithium**, alongside **FIPS 205**, called **SLH-DSA**, a hash-based signature scheme derived from **SPHINCS+** that offers a mathematically conservative alternative resting on different assumptions. These names will mean little to most readers, but they are the new locks of the internet, the systems that will, over the next decade, be woven invisibly into browsers, servers, phones, and networks, replacing the quantum-vulnerable cryptography that currently protects the digital world. The migration is already underway; major technology companies have begun deploying post-quantum key exchange in web browsers and messaging apps, hedging against the harvest-now-decrypt- later threat by protecting today's traffic with tomorrow's cryptography.

The primary defense against the quantum threat is not, as one might expect, some exotic quantum technology of our own, but something far more practical and immediately deployable: new mathematical codes, runnable on the ordinary classical computers already sitting on every desk, designed from the ground up to resist attack even by a large quantum computer. This field is called post-quantum cryptography, or sometimes quantum-resistant cryptography, and the effort to develop, vet, and standardize it has been one of the most important and successful large-scale scientific collaborations of the past decade — a rare case of the world seeing a threat clearly and mobilizing against it before disaster struck.

The core idea is elegant in its simplicity. Recall that RSA and elliptic-curve cryptography are vulnerable because they rest on specific mathematical problems — factoring large numbers and computing discrete logarithms — that Shor's algorithm happens to solve efficiently. But Shor's algorithm is a specialist: it cracks those particular problems by exploiting their hidden periodic structure, and it does NOT provide a general-purpose way to break all hard problems. So the strategy of post-quantum cryptography is to build new encryption schemes on DIFFERENT mathematical problems — problems believed to be hard even for a quantum computer, problems that have no known efficient quantum algorithm and no periodic structure for Shor's method to exploit. Find a lock that quantum machines cannot pick, and build the new internet on that. Crucially, these new schemes run on ordinary computers; they do not require anyone to own a quantum machine. Everyone can deploy them today, using existing hardware, to protect against the quantum computers of tomorrow.

The most important family of these quantum-resistant problems comes from the mathematics of lattices — regular, repeating grids of points extending through many-dimensional space. Certain problems about lattices — such as finding the shortest vector in a high-dimensional lattice, or the "learning with errors" problem, which involves solving noisy systems of equations — are believed to be extraordinarily hard for both classical AND quantum computers, with no efficient algorithm known for either despite decades of study. Lattice-based cryptography builds its public and private keys out of these problems, so that breaking the encryption would require solving a hard lattice problem that even Shor's algorithm cannot touch. Other families exist too — schemes based on error-correcting codes, on multivariate equations, on the mathematics of hash functions, and on the properties of mathematical objects called isogenies — each resting on a different hard problem, providing valuable diversity so that if one family is unexpectedly broken, others remain. This diversity is a deliberate hedge, and a wise one.

The vetting of these schemes was organized as a remarkable public competition, and it is a model of how such things should be done. Beginning in 2016, the U.S. National Institute of Standards and Technology (NIST) — the same body that had standardized earlier encryption — ran an open, international, multi-year contest, inviting cryptographers worldwide to submit candidate post-quantum algorithms and then subjecting them to years of relentless public scrutiny, in which teams from around the globe tried their utmost to break one another's proposals. This adversarial, transparent process is exactly how cryptographic trust is properly built: a scheme earns confidence not because its designers vouch for it but because the world's best code-breakers have attacked it for years and failed. The competition was dramatic in its own quiet way — several promising candidates were spectacularly broken during the process, some by ordinary classical computers in a matter of hours or days, a humbling reminder that designing cryptography that resists attack is fiendishly hard and that untested schemes are dangerous. One prominent candidate, an isogeny-based scheme called SIKE that had survived to a late round, was shattered in 2022 by a clever classical attack, underscoring how essential the years of scrutiny were.

In 2024, after nearly a decade of work, NIST published the first finalized post-quantum cryptography standards, a milestone of genuine historic importance: the world now had official, vetted, quantum-resistant encryption algorithms ready for deployment. The chosen standards included a lattice-based scheme for key establishment — the crucial task of two strangers agreeing on a shared secret — now named ML-KEM (derived from an algorithm called Kyber), and lattice-based and hash-based schemes for digital signatures, named ML-DSA (from Dilithium) and SLH-DSA (from SPHINCS+). These are not research proposals but finished standards, and the global migration to deploy them across the internet's infrastructure — web browsers, servers, operating systems, messaging apps, financial networks — has begun in earnest, a vast, years-long undertaking to replace the vulnerable foundations of digital security before Q-Day arrives.

The elegance of the post-quantum strategy is worth restating, because it dispels a common assumption that the only defense against a quantum computer must itself be quantum. It need not be. Shor's algorithm is a *specialist*: it breaks RSA and elliptic-curve cryptography by exploiting the specific hidden periodic structure of the factoring and discrete-logarithm problems, and it provides no general-purpose way to break all hard problems. So the strategy of post-quantum cryptography is simply to build new encryption on *different* mathematical problems — ones believed hard even for a quantum computer, with no known efficient quantum algorithm and no periodic structure for Shor's method to exploit. Find a lock that quantum machines cannot pick, and rebuild the internet on it. Crucially, these new schemes run on ordinary classical computers; they require no quantum hardware whatsoever. Everyone can deploy them today, using the machines already on every desk, to protect against the quantum computers of tomorrow. The primary defense against the quantum threat is not exotic technology but mathematics, downloadable as a software update.

The public competition through which these schemes were vetted deserves admiration as a model of how cryptographic trust should be built, because trust in a code cannot be asserted — it must be earned through relentless attack. Beginning in 2016, the U.S. National Institute of Standards and Technology ran an open, international, multi-year contest, inviting cryptographers worldwide to submit candidate post-quantum algorithms and then subjecting them to years of public scrutiny, in which teams around the globe tried their hardest to break one another's proposals. This adversarial, transparent process is exactly right: a scheme earns confidence not because its designers vouch for it but because the world's best code-breakers have attacked it for years and failed. And the competition was dramatic in its quiet way — several promising candidates were spectacularly broken during the process, some by ordinary classical computers in hours or days. One prominent late-round candidate, an isogeny-based scheme called SIKE, was shattered in 2022 by a clever classical attack, a humbling reminder of how essential the years of scrutiny were and how dangerous an untested scheme can be.

The outcome, and its limits, are worth stating plainly. In 2024, after nearly a decade of work, NIST published the first finalized post-quantum cryptography standards — a milestone of genuine historic importance, giving the world official, vetted, quantum-resistant encryption ready for deployment. The chosen standards, drawn largely from the mathematics of lattices, include a scheme for key establishment now named ML-KEM (from the algorithm Kyber) and signature schemes named ML-DSA (from Dilithium) and SLH-DSA (from SPHINCS+), and the vast, years-long global migration to deploy them across browsers, servers, operating systems, and financial networks has begun in earnest. But honesty requires naming the caveats: these schemes rest on the *belief* that certain problems are hard for quantum computers, a belief well supported but not proven, which is why cryptographers hedge with diverse families and continue to scrutinize; the new algorithms tend to require larger keys and more computation than those they replace; and the migration itself is enormous and slow, touching virtually every piece of software that uses encryption. The new armor is being forged and fitted — humanity's main line of defense against Q-Day — even as a second, radically different approach roots security not in mathematics but in the unbreakable laws of physics themselves.

Chapter 53: Quantum Keys — Security From Physics Itself

There are two fundamentally different ways to defend communication against a quantum computer, and they could hardly be more different in spirit. The first, post-quantum cryptography, fights quantum computers with *harder mathematics* — new problems that even quantum machines cannot solve — and it runs on ordinary classical hardware. The second fights quantum computers with *quantum physics itself,* building communication whose security rests not on the difficulty of any mathematical problem but on the fundamental laws of nature, which forbid an eavesdropper from copying or observing quantum information without leaving a trace. This second approach is called **quantum cryptography**, and its flagship application, **quantum key distribution** or QKD, offers a form of security that is, in principle, unbreakable by any computer, quantum or classical, now or ever — a remarkable promise, though one hedged with serious practical limitations.

The foundational insight of quantum cryptography, which we met glancingly in Part I, is that the very laws that make quantum information fragile can be turned into a security guarantee. Recall the **no-cloning theorem**: an unknown quantum state cannot be copied. And recall that *measuring* a quantum system disturbs it, collapsing superpositions and leaving evidence of the intrusion. Together these two facts mean that quantum information cannot be secretly intercepted. An eavesdropper who tries to tap a quantum communication cannot make a perfect copy to examine at leisure (no-cloning forbids it), and cannot measure the quantum signals in transit without disturbing them in a way the legitimate parties can detect (measurement collapses the state). Any attempt to spy on quantum-encoded information necessarily leaves fingerprints. This is the physical basis of quantum key distribution: it lets two parties establish a shared secret key with the guarantee that if anyone eavesdropped, they will know, because the eavesdropping will have introduced detectable errors into their quantum exchange.

The first and most famous QKD protocol is **BB84**, devised by Charles Bennett and Gilles Brassard in 1984 — a foundational contribution that predates even Shor's algorithm and shows how early the seeds of quantum information were sown. In BB84, one party, Alice, sends the other, Bob, a stream of single photons, each prepared in one of several possible quantum states (for instance, polarizations of light). Through a clever protocol involving random choices and a public comparison of some of those choices afterward, Alice and Bob can distill a shared secret key from the photons that arrived — and, crucially, can estimate whether anyone tampered with the transmission by checking the error rate. If the error rate is low, no one was listening, and the key is safe; if it is high, an eavesdropper was present, and they discard the key and try again. A second protocol, **E91**, proposed by Artur Ekert in 1991, achieves a similar end using entangled photon pairs and the violation of Bell's inequality (Part I) as the guarantee of security — a beautiful closing of the loop, in which the very phenomenon Einstein invoked to attack quantum mechanics becomes the certification of unbreakable communication. Either way, the security rests on physics, not on any assumption about the eavesdropper's computational power. A quantum computer, however large, cannot break a properly implemented

QKD system, because the security does not depend on the hardness of any calculation.

Quantum key distribution has moved from theory to deployment, and the achievements are real and impressive. QKD systems are commercially available and have been used to secure banking transactions, elections, and government communications; fiber-optic QKD networks have been built in several countries; and China, through the physicist Jian-Wei Pan's group, achieved a spectacular milestone in 2016 with the **Micius satellite**, which distributed entangled photons and quantum keys between ground stations separated by more than a thousand kilometers, demonstrating space-based quantum communication and pointing toward a future global quantum-secured network. China has invested heavily in QKD infrastructure, including a long fiber "quantum backbone" between Beijing and Shanghai, and regards quantum communication as a strategic technology. These are genuine engineering triumphs, and they show that security-from-physics is not merely a theoretical curiosity but a working technology.

Yet quantum key distribution faces serious practical limitations that have kept it from displacing ordinary cryptography, and honesty requires naming them. QKD requires special hardware — single-photon sources and detectors, carefully controlled optical channels — and cannot simply run as software on existing computers the way post-quantum cryptography can. It is limited in *range*: photons are lost as they travel through fiber, so terrestrial QKD is confined to distances of a few hundred kilometers without trusted intermediate relay stations, which themselves become points of vulnerability, and truly long-distance QKD awaits the development of **quantum repeaters** (still largely experimental) or satellite links. It secures only the *key exchange*, not the bulk communication, and it addresses only confidentiality, not the authentication that digital signatures provide — and it even requires a separately authenticated classical channel to work at all. And it is expensive. For these reasons, the mainstream security community, including agencies such as the U.S. National Security Agency, has generally favored *post-quantum cryptography* over QKD as the primary defense against the quantum threat, precisely because PQC is cheap, software-based, and deployable everywhere at once, while QKD is costly, hardware-bound, and range-limited. QKD remains a valuable niche technology, unbreakable in principle and impressive in practice, but it is a complement to, not a replacement for, the post-quantum cryptography that will do the bulk of the world's quantum-era defending.

There is a second and utterly different approach to securing communications in the quantum age, and where post-quantum cryptography builds better mathematical locks, this approach abandons mathematics as the source of security altogether and roots secrecy in the laws of physics themselves. It is called quantum key distribution, or QKD, and it offers something no mathematical code can: security guaranteed not by the assumed difficulty of some problem, but by the fundamental properties of quantum mechanics — a security that would hold even against an adversary with unlimited computing power, quantum or otherwise. It is one of the most elegant ideas in the whole field, and unlike quantum computing itself, it is a real, deployed, commercially available technology today.

The purpose of QKD is narrow but crucial: to let two parties share a secret key — a random string of bits — with the guarantee that no eavesdropper could have obtained it without being detected. Once they share such a key, they can use it with ordinary, unbreakable symmetric encryption (the one-time pad, or a strong cipher like AES) to communicate in perfect secrecy. The entire security problem thus reduces to sharing the key safely, and this is exactly what QKD accomplishes using quantum physics. The foundational protocol, called BB84, was devised in 1984 by Charles Bennett — the same Bennett of reversible computing and quantum teleportation — and Gilles Brassard, and it is a small masterpiece of turning quantum weirdness into practical advantage.

The genius of QKD rests on two quantum principles we have already met, now turned into a security guarantee. The first is that measuring a quantum system unavoidably disturbs it: you cannot look at a qubit without, in general, changing it, because measurement collapses superpositions. The second is the no-cloning theorem: an unknown quantum state cannot be copied. Put these together and you get a wiretap detector woven into the laws of nature. To share a key, one party sends the other a stream of individual photons, each prepared in a quantum state chosen at random from a set of possibilities (for instance, polarized at one of several angles). An eavesdropper who tries to intercept the photons to learn the key faces an impossible bind: she cannot copy them (no-cloning forbids it), so she must measure them to read them — but measuring them, without knowing how each was prepared, unavoidably disturbs their states, introducing detectable errors. When the two legitimate parties afterward compare a sample of their results over a public channel, they will find those tell-tale errors and know that someone was listening, and they simply discard the compromised key and try again. An eavesdropper cannot learn the key without leaving fingerprints. Secrecy is enforced not by a hard math problem but by the fact that eavesdropping physically corrupts the message.

This is a genuinely different and, in a sense, stronger kind of security than any code can offer. Mathematical cryptography, including post-quantum cryptography, rests on the assumption that some problem is hard to solve; if that assumption fails — if someone finds a clever algorithm — the security collapses, potentially retroactively. QKD rests instead on the laws of physics, which, so far as we know, cannot be circumvented by any amount of cleverness or computing power. A QKD-shared key is secure against an adversary with a quantum computer, a classical supercomputer, or any conceivable future machine, because breaking it would require violating quantum mechanics itself. This is why QKD is sometimes called "unconditionally secure" or "information-theoretically secure" — its security does not depend on any unproven assumption about computational difficulty.

QKD is not a laboratory curiosity but a working, deployed technology, and its real-world reach is impressive. Commercial QKD systems have been sold for years, banks and governments have used QKD-secured links to protect high-value communications, and China in particular has invested heavily, building a QKD network stretching over two thousand kilometers between Beijing and Shanghai and, most spectacularly, using the Micius satellite — launched in 2016 by Jian-Wei Pan's team, the same group behind the Jiuzhang photonic computer — to distribute quantum keys between ground stations thousands of kilometers apart, demonstrating intercontinental quantum-secured communication. The technology is real, it works, and it is being built into critical infrastructure.

It is worth appreciating the sheer conceptual audacity of quantum key distribution: it roots the secrecy of a message not in the difficulty of a mathematical puzzle but in the fundamental laws of physics, offering a kind of security that no advance in computing — quantum or classical — could ever break. Every mathematical cipher, including the post-quantum schemes of the previous chapter, rests on the *assumption* that some problem is hard to solve; if that assumption fails — if someone finds a clever algorithm — the security collapses, potentially retroactively exposing everything ever encrypted with it. Quantum key distribution rests on no such assumption. Its security follows from the fact that measuring a quantum system unavoidably disturbs it and that unknown quantum states cannot be copied — properties as fundamental as any in physics. A key shared by QKD is secure against an adversary with a quantum computer, a classical supercomputer, or any conceivable future machine, because breaking it would require violating quantum mechanics itself. This is why QKD is called "unconditionally" or "information-theoretically" secure — its safety does not hinge on any unproven belief about what is hard to compute.

The mechanism by which QKD turns quantum weirdness into a wiretap detector deserves a second, appreciative pass, because it is so elegant. To share a key, one party sends the other a stream of individual photons, each prepared in a quantum state chosen at random from a set of possibilities. An eavesdropper who wants to learn the key faces an impossible bind. She cannot copy the photons, because the no-cloning theorem forbids duplicating an unknown quantum state. So to read them she must measure them — but measuring them, without knowing how each was prepared, unavoidably disturbs their states, introducing errors. When the two legitimate parties afterward compare a sample of their results over a public channel, they find those tell-tale errors and know someone was listening; they simply discard the compromised key and try again. The eavesdropper cannot learn the key without leaving fingerprints. Secrecy is enforced not by making the code hard to crack but by making eavesdropping physically self-announcing — the act of spying corrupts the very message it seeks to steal.

Yet honesty requires a clear account of QKD's limitations, because they are the reason it complements rather than replaces mathematical cryptography. QKD needs special hardware — single-photon sources and detectors, and either dedicated optical fiber or line-of-sight to a satellite — so it cannot simply run over the existing internet the way software-based post-quantum cryptography can. Photons are absorbed as they travel through fiber, limiting the range of direct QKD to a few hundred kilometers before the signal fades, and extending it requires either trusted relay stations (which reintroduce points of vulnerability) or quantum repeaters that do not yet practically exist. It secures only point-to-point links between parties with the requisite equipment, not the sprawling any-to-any connectivity of the global internet, and it addresses only key *sharing*, not the many other tasks cryptography performs, such as digital signatures and authentication. For these reasons most experts see QKD as a valuable armor for particular high-security links — a bank's inter-office connection, a government's most sensitive channel — rather than a wholesale replacement for cryptography, with post-quantum mathematical schemes doing the broad work of securing the everyday internet. Yet QKD points toward something larger: if quantum information can be sent securely between distant points, then quantum machines and sensors might be linked across the globe into an entirely new kind of network — a quantum internet, whose grand possibilities and profound challenges are the subject of the next chapter.

Chapter 54: The Quantum Internet

Running alongside the story of quantum *computing* — of machines that calculate with qubits — is a parallel story of quantum *communication*, of networks that transmit quantum information across distance, and the two are converging toward a grand vision: a **quantum internet**, a global network that distributes entanglement and quantum states as readily as today's internet distributes bits. This vision is more speculative and further from realization than the near-term applications of quantum computing, but it is pursued seriously by physicists and funded seriously by governments, and it represents the ultimate destination of the communication branch of quantum information science. This chapter surveys what a quantum internet would be, what it could do, and how far off it remains.

The core idea is to build a network whose fundamental service is the distribution of **entanglement** — the delivery of entangled qubit pairs to distant parties, on demand, reliably. Recall from Part IV that entanglement is a resource: a shared entangled pair enables quantum teleportation (transferring a quantum state from one place to another), enables the most secure forms of key distribution, and can link separate quantum processors so they act as one. A quantum internet would be, at bottom, an infrastructure for generating entanglement between any two nodes that want it, however far apart, and then letting them spend that entanglement on whatever quantum task they wish. Where today's internet moves classical bits, the quantum internet would move — or rather, establish — quantum correlations, weaving distant quantum systems into entangled wholes.

What would such a network be good for? Several applications, of varying maturity, motivate the effort. The most immediate is **secure communication**: a quantum internet would enable quantum key distribution (the previous chapter) at global scale, over any distance, providing physics-guaranteed security everywhere. A second is **distributed quantum computing**: linking multiple quantum processors by entanglement could let them cooperate as a single larger machine, potentially a path to scaling up quantum computing power by networking many modest processors rather than building one enormous one — an approach some hardware efforts (notably networked trapped-ion and photonic architectures) already pursue internally. A third is **enhanced quantum sensing**: entangled sensors distributed across a network could measure with a precision beyond what any isolated sensor achieves, enabling more sensitive telescopes, more accurate clocks, and better detection of faint signals — quantum sensing being, as the next part notes, one of the areas where quantum technology already delivers real value. A fourth is a set of cryptographic and coordination tasks — secure distributed computation, clock synchronization, consensus protocols — that entanglement enables in ways classical networks cannot. The quantum internet is thus not a single application but an enabling infrastructure, whose full uses, like those of the classical internet in its early days, are only partly foreseeable.

The central technical obstacle to a quantum internet is the same fragility that haunts quantum computing, now manifested over distance: quantum states, and entanglement in particular, degrade as they travel. Photons carrying quantum information are absorbed and scattered in optical fiber, so that the probability of a photon surviving falls exponentially with distance, limiting direct quantum transmission to a few hundred kilometers. In classical communication this problem is solved trivially by **repeaters** — amplifiers that boost the signal along the way — but the no-cloning theorem forbids simply copying and amplifying a quantum signal, so classical repeaters are useless. The quantum internet therefore requires **quantum repeaters**, devices that extend entanglement over long distances without cloning, by cleverly stitching together entanglement across shorter links using a technique called entanglement swapping and storing quantum states in quantum memories along the way. Quantum repeaters are the key enabling technology for a long-distance quantum internet, and they remain largely experimental — a subject of intense research, with laboratory demonstrations of the necessary ingredients but no mature, deployable device yet. Until practical quantum repeaters exist, long-distance quantum communication depends on either trusted relay stations (which compromise the end-to-end security guarantee) or satellite links like China's Micius (which sidestep fiber loss by sending photons through the near-vacuum of space).

The quantum internet, then, is a vision in progress rather than a reality — further from fruition than the quantum computers this book chiefly concerns, its full realization dependent on quantum repeaters and quantum memories that are still being invented, its timeline as uncertain as that of large-scale quantum computing itself. Prototype quantum networks have been built — small entanglement-distribution networks in the Netherlands, China, the United States, and elsewhere, linking a few nodes over modest distances — and they are growing, node by node, much as the early classical internet grew from a handful of connected computers. Whether they will one day knit together into a true global quantum internet, and what that internet will ultimately be used for, are questions for the coming decades. What is clear is that the communication branch of quantum information science — quantum cryptography, quantum networks, the quantum internet — is a rich and consequential field in its own right, intertwined with but distinct from quantum computing, and that the two together constitute the broader quantum technological revolution now underway.

Beyond individual quantum computers and secure quantum keys lies a grander and more speculative vision: the quantum internet, a network that would connect quantum devices — computers, sensors, and communication nodes — across cities, nations, and eventually the globe, transmitting not ordinary bits but quantum information, entanglement, and qubits themselves. It is one of the most ambitious long-term goals in the field, and while it remains largely a work in progress, understanding what it would be, why it is so hard to build, and what it might enable illuminates both the promise and the profound technical challenges of the quantum future.

It is essential first to be clear about what a quantum internet is NOT, because the name invites serious misunderstanding. A quantum internet would not replace the ordinary internet for everyday use; it would not make your web browsing faster, your video calls sharper, or your downloads quicker. There would be no point in transmitting cat videos or emails as quantum information — ordinary bits do that job perfectly well, and quantum communication is far more delicate and expensive. Rather, a quantum internet is a SPECIALIZED network for specific quantum tasks that are impossible or inferior over classical channels: distributing entanglement between distant parties, linking quantum computers together, connecting quantum sensors into coordinated arrays, and enabling forms of secure communication and distributed computation that have no classical equivalent. It is a complement to the classical internet for quantum purposes, not a successor to it — a parallel network for a different kind of information.

The core function of a quantum internet is to distribute ENTANGLEMENT between distant locations, and from this single capability a surprising wealth of applications flows. If two distant parties share entangled pairs of qubits, they can perform quantum teleportation to transmit quantum states between them — essential for connecting far-flung quantum computers into a larger distributed machine. They can perform quantum key distribution for provably secure communication, as we saw in the previous chapter. They can link quantum sensors — atomic clocks, telescopes, gravitational detectors — into entangled networks that achieve precision beyond any single instrument, enabling, for instance, arrays of telescopes to act as one vastly larger virtual telescope, or networks of clocks to test fundamental physics. And they could enable "blind quantum computing," in which a user sends an encrypted computation to a remote quantum computer that runs it without ever learning what the computation actually is — allowing secure use of cloud quantum computers that cannot spy on their users. Entanglement distributed across distance is the master resource, and the quantum internet exists to spread it.

The central obstacle to building a quantum internet is the same fragility that plagues quantum computing, now compounded by distance, and it is severe. Quantum information cannot be amplified or copied — the no-cloning theorem forbids it — so the trick that makes the classical internet work over long distances, in which signals are repeatedly boosted by repeaters along the route, is flatly unavailable. And photons, the natural carriers of quantum information, are absorbed as they travel through optical fiber, so that after a few hundred kilometers the probability of a photon surviving the journey becomes vanishingly small, and you cannot simply amplify the fading signal without destroying its quantum character. This is the fundamental barrier: quantum signals fade with distance and cannot be boosted. Overcoming it requires a device of extraordinary subtlety called a quantum repeater, which would extend the range of entanglement not by copying or amplifying the signal but by cleverly stitching together entanglement across successive short segments — using entanglement swapping and quantum memories to link a chain of manageable hops into one long-distance entangled link, all without ever measuring or cloning the precious quantum state. Practical quantum repeaters that work over useful distances do not yet exist; they remain one of the great engineering challenges standing between us and a true global quantum internet.

Despite these formidable obstacles, real progress is being made, and the vision is not idle dreaming. Researchers have demonstrated entanglement distribution over metropolitan-scale fiber networks, built small multi-node quantum networks linking a handful of quantum processors in the laboratory (notably a pioneering three-node network in the Netherlands, at Delft), and, as we have seen, used satellites to distribute entanglement across continental and intercontinental distances, sidestepping fiber losses by sending photons through the near-vacuum of space. Prototype quantum memories that can store a qubit and release it on demand — an essential component of a repeater — are steadily improving. National and international efforts to build quantum-network testbeds are underway in Europe, China, and the United States. The building blocks are being assembled, piece by piece, even as the full architecture remains distant.

It is worth stating plainly, once more, what a quantum internet is *not*, because the name invites a serious and common misunderstanding that inflates expectations absurdly. A quantum internet would not replace the ordinary internet for everyday use; it would not make your web browsing faster, your video calls sharper, or your downloads quicker. There would be no point whatever in transmitting emails or videos as quantum information — ordinary bits do that job perfectly well, and quantum communication is far more delicate, expensive, and limited. A quantum internet is a *specialized* network for specific quantum tasks that are impossible or inferior over classical channels: distributing entanglement between distant parties, linking quantum computers together, connecting quantum sensors into coordinated arrays, and enabling forms of secure communication and distributed computation with no classical equivalent. It is a complement to the classical internet for quantum purposes, a parallel network for a different kind of information — not a faster successor to the web but a wholly different thing serving wholly different ends.

The wealth of applications that flow from the single capability of distributing entanglement is worth cataloguing, because it explains why the vision commands such effort despite its difficulty. If two distant parties share entangled qubits, they can teleport quantum states between them — essential for connecting far-flung quantum computers into a larger distributed machine, pooling their power. They can perform quantum key distribution for provably secure communication. They can link quantum sensors — atomic clocks, telescopes, gravitational detectors — into entangled networks achieving precision beyond any single instrument, letting arrays of telescopes act as one vastly larger virtual telescope or networks of clocks test fundamental physics. And they could enable "blind quantum computing," in which a user sends an encrypted computation to a remote quantum computer that runs it without ever learning what the computation is — allowing secure use of cloud quantum machines that cannot spy on their users. Entanglement distributed across distance is the master resource, and from it these diverse capabilities all spring.

The central obstacle — that quantum signals fade with distance and, uniquely, cannot be boosted — deserves emphasis, because it is what makes the quantum internet so much harder to build than its classical counterpart and why it remains largely a vision. The classical internet works over long distances because signals can be repeatedly amplified by repeaters along the route. This trick is flatly unavailable to quantum information: the no-cloning theorem forbids copying or amplifying an unknown quantum state, and photons are absorbed as they travel through fiber, so that after a few hundred kilometers the chance of a photon surviving becomes vanishingly small, and you cannot simply boost the fading signal without destroying its quantum character. Overcoming this requires a device of extraordinary subtlety — a quantum repeater — that would extend entanglement not by copying or amplifying but by cleverly stitching together entanglement across successive short segments, using entanglement swapping and quantum memories to link a chain of manageable hops into one long-distance entangled link. Practical quantum repeaters over useful distances do not yet exist; they are among the great unsolved engineering challenges of the field. Yet real progress is being made — entanglement distributed across metropolitan fiber networks, small multi-node networks linking laboratory processors, satellites distributing entanglement across continents, and steadily improving quantum memories — so the building blocks are being assembled even as the full architecture remains distant. The quantum internet is a long-horizon vision whose earliest fruit, quantum-secured communication, is already partly realized, while its grander promise of linked machines and global sensor networks awaits the repeater. And the deepest justification for building quantum machines at all was never cryptography or communication but Feynman's founding dream of simulating nature — to which, having surveyed the threats and defenses, we now return.

Part XI — What It's Good For

Chapter 55: Simulating Molecules

If quantum computers ever change the world for the better — as opposed to merely threatening its cryptography — the change will most likely begin with molecules. Of all the constructive applications proposed for quantum computing, the simulation of quantum chemistry and materials is the one that experts most consistently rank as both the most valuable and the most likely to arrive first, and for a reason we have met repeatedly since Part II: it is the application that plays directly to the machine's deepest strength, the one Feynman identified at the very birth of the idea. A quantum computer is, at bottom, a controllable quantum system, and the natural thing for a controllable quantum system to do is to mimic *other* quantum systems — which is exactly what molecules and materials are. This chapter examines that promise, and, in keeping with this book's insistence on honesty, its present limits.

Recall the simulation problem from Part II. The behavior of molecules and materials is governed by the quantum mechanics of their electrons, and the exact description of an entangled system of electrons explodes exponentially with their number, defeating classical computers for all but the smallest cases. Chemists and materials scientists are therefore forced to approximate, with a toolbox of methods that work well for many systems and fail for the most interesting and difficult ones — the strongly correlated materials, the transition-metal catalysts, the complex reactions where the approximations break down precisely where accurate prediction would be most valuable. A quantum computer, as Lloyd proved in 1996 (Part V), could in principle simulate such systems directly, electron for qubit, entanglement for entanglement, without the exponential classical blowup — computing the properties of molecules and materials that classical methods cannot reach. This is not code-breaking; it is a constructive, creative capability, the power to predict and eventually design matter at the quantum level.

The potential applications read like a wish list for twenty-first-century science and industry. In **chemistry and catalysis**, quantum simulation could enable the design of better catalysts — substances that speed chemical reactions — with enormous practical stakes. The showcase example, cited so often it has become a cliché of quantum- computing promotion (and one we should treat with appropriate caution), is the **Haber–Bosch process** for fixing nitrogen from the air into ammonia for fertilizer, which consumes on the order of one to two percent of the world's total energy because the industrial catalyst is crude and requires high temperature and pressure; certain bacteria, by contrast, fix nitrogen at ordinary conditions using an enzyme called nitrogenase, whose active site involves a cluster of metal atoms whose quantum behavior classical computers cannot accurately simulate. A quantum computer that could simulate nitrogenase might reveal how to design a better industrial catalyst, saving a meaningful fraction of global energy use. In **medicine and drug discovery**, quantum simulation could predict how candidate drug molecules bind to their protein targets, accelerating the design of new medicines. In **materials science**, it could aid the design of better batteries, more efficient solar cells, high-temperature superconductors, and novel materials with tailored properties. In each case the promise is the same: to replace slow, expensive trial- and-error in the laboratory with accurate quantum prediction, letting scientists design molecules and materials on a computer before ever making them.

Now the honest caveats, which the promotional literature tends to bury. First, these applications are, for the most part, *not yet realized*. Quantum simulations performed on the NISQ machines of the mid-2020s have been small demonstrations — simulating tiny molecules like hydrogen, or modest instances of model systems — impressive as proofs of principle but far short of solving problems beyond classical reach or delivering practical value. The molecules and materials of real industrial interest require more qubits, higher fidelity, and greater circuit depth than current machines provide, and in many cases they require *fault-tolerant* machines with error-corrected logical qubits, which do not yet exist at the needed scale. Second, the classical competition is not standing still: classical computational chemistry is a mature, powerful, ingeniously optimized field, and for a quantum computer to prove *advantage* it must beat not a naive classical method but the best classical methods, which are formidable — and it has not yet clearly done so on a problem of real value. Third, even where quantum computers will eventually help, they will not simply replace classical methods but complement them, handling the specific hard cores of problems that defeat classical approaches while classical computers do the rest. The realistic picture is of quantum computers as specialized accelerators for the hardest quantum-chemical problems, not universal oracles of molecular design.

Of all the potential applications of quantum computers, the one that most experts believe will prove genuinely transformative — the killer application that could justify the entire enterprise — is the simulation of molecules and chemical reactions. This was Feynman's original dream, the very reason quantum computing was first imagined, and after the long detour through cryptography and supremacy, the field is circling back to it with growing conviction that here, in the quiet domain of chemistry, lies the quantum computer's most important work. To understand why, we must appreciate both the staggering difficulty of the problem and the reason a quantum machine is uniquely suited to solve it.

Chemistry is, at its most fundamental level, quantum mechanics in action. The properties of every molecule — how it is shaped, how strongly its atoms bind, how it reacts with other molecules, what color it is, how it conducts or resists electricity — are determined by the behavior of its electrons, and electrons are quantum particles that superpose and, crucially, ENTANGLE with one another in ways that grow explosively complex as the number of electrons increases. Recall the exponential wall: to describe the entangled quantum state of many electrons, a classical computer must track a number of amplitudes that doubles with each electron, so that even a modestly sized molecule of interest — the active site of an enzyme, a novel catalyst, a candidate drug — presents a quantum state far too large for any classical computer to represent exactly. This is not a temporary limitation to be overcome by faster classical machines; it is a fundamental barrier rooted in the exponential structure of quantum entanglement. The most powerful supercomputer ever built cannot exactly simulate the electrons of a molecule with more than a few dozen strongly correlated electrons.

Faced with this wall, chemists have spent a century developing ingenious approximation methods — clever mathematical shortcuts that capture enough of the quantum behavior to be useful while remaining classically computable. These approximations, with names like density functional theory and coupled cluster, are genuine triumphs of science and underpin much of modern chemistry, materials science, and drug discovery; they run on classical computers and deliver enormously valuable predictions. But they are approximations, and for the hardest, most strongly entangled molecules — precisely the ones involved in some of the most important and mysterious chemistry — they break down, sometimes badly, giving unreliable or simply wrong answers, and there is no way to fix them by throwing more classical computing power at the problem. The wall does not move. These "strongly correlated" systems, where electrons are entangled too intricately for any classical shortcut to capture, are exactly where our understanding fails and where a quantum computer could see clearly.

The reason a quantum computer can succeed where classical machines fail is the beautiful inversion at the heart of Feynman's insight: a quantum computer is ITSELF a controllable quantum system, so it can naturally hold and manipulate the exponentially large quantum states that describe molecules, not by laboriously storing amplitudes in memory but by physically embodying them in its own qubits. To simulate a molecule, you map the molecule's electrons onto the computer's qubits, prepare the qubits to represent the molecule's quantum state, evolve them according to the same physical laws the electrons obey, and measure the qubits to extract the quantities you want — the molecule's energy, its structure, its reactivity. The quantum computer speaks the molecule's own language; it does not translate quantum reality into an exponentially bloated classical description and choke, but mirrors it directly, qubit for electron, entanglement for entanglement. Seth Lloyd's proof, which we met earlier, established that this can be done efficiently and generally, and specialized algorithms — such as quantum phase estimation for finding molecular energies precisely, and hybrid quantum-classical methods like the variational quantum eigensolver designed for near-term noisy machines — have been developed to carry it out.

It is important to be precise about what molecular simulation on a quantum computer does and does not promise, because the hype here runs especially hot. A quantum computer will not produce a magical movie of a chemical reaction for you to watch and instantly understand; it computes specific quantities — ground-state energies, reaction rates, spectroscopic properties — that reveal how a molecule behaves. Extracting useful numbers efficiently, and preparing the right starting quantum state for the molecule of interest, are themselves genuine challenges that active research is still working to solve; simulation is a powerful instrument, not an oracle. And, decisively, useful molecular simulation of the hard, classically-intractable cases will require large, error-corrected quantum computers with many high-quality logical qubits — machines that do not yet exist. The near-term, noisy machines of today can simulate only small molecules that classical computers can already handle, so they have not yet delivered practical chemical advantage. The prize is real but it awaits the fault-tolerant machines of the future.

It is worth grasping, concretely, why chemistry is the natural home of the quantum computer, because the argument is airtight in a way that the arguments for most other applications are not. Chemistry is, at its foundation, quantum mechanics in action: every property of every molecule — its shape, the strength of its bonds, how it reacts, what it does inside a living cell — is dictated by the behavior of its electrons, and electrons are quantum particles that superpose and, crucially, *entangle* in ways that grow explosively complex as their number increases. To describe the entangled quantum state of many electrons, a classical computer must track a number of amplitudes that doubles with each electron, so even a modestly sized molecule of real interest presents a state too large for any classical machine to represent exactly. This is not a temporary limitation awaiting a faster classical computer; it is the exponential wall, rooted in the structure of entanglement itself. A quantum computer, being itself a controllable quantum system, holds that same exponential wealth of amplitudes not by storing them but by *being* them — which is precisely why the match between quantum machines and quantum chemistry is not a marketing analogy but a mathematical inevitability.

The distinction between exact simulation and the classical

approximations chemists rely on deserves emphasis, because it locates exactly where a quantum computer would earn its keep. Faced with the exponential wall, chemists have spent a century developing ingenious approximation methods — density functional theory, coupled cluster, and their kin — clever shortcuts that capture enough quantum behavior to be useful while remaining classically computable. These are genuine triumphs that underpin much of modern chemistry, materials science, and drug discovery. But they are approximations, and for the hardest, most strongly entangled molecules — precisely the ones involved in the most important and mysterious chemistry, from enzyme active sites to exotic magnetic materials — they break down, sometimes badly, and there is no way to fix them by buying a bigger classical computer. The wall does not move. These "strongly correlated" systems, where electrons are entangled too intricately for any classical shortcut to capture, are exactly where our understanding fails and where a quantum computer could see clearly, computing what no approximation can.

Honesty about the limits and the timeline is essential here, because molecular simulation is where quantum hype runs especially hot and disappointment is easiest. A quantum computer does not produce a little movie of atoms bouncing around for you to watch and understand; it computes specific quantities — ground-state energies, reaction rates, spectroscopic properties — that reveal how a molecule behaves. Extracting useful numbers efficiently, and preparing the right starting quantum state for the molecule of interest, are themselves genuine unsolved challenges; simulation is a powerful instrument, not an oracle. And decisively, useful simulation of the hard, classically-intractable cases will require large, error-corrected machines with many high-quality logical qubits — machines that do not yet exist. Today's noisy devices can simulate only small molecules that classical computers already handle, so no practical chemical advantage has yet been demonstrated. The prize is real, and most experts believe it is the application most likely to justify the entire enterprise — touching nitrogen fixation, superconductivity, batteries, catalysts, and drug design — but it awaits the fault-tolerant machines of the future. It is Feynman's founding dream, matured into a concrete agenda and still, for now, over the horizon.

Chapter 56: Materials and Medicine — the Promise and the Hype

The promise that quantum computers will revolutionize the design of materials and medicines is among the most frequently invoked justifications for the billions being poured into the field, and it appears in nearly every corporate presentation and government white paper on quantum technology. It is a promise with a real foundation, as the previous chapter established — quantum simulation genuinely fits the machine's nature — but it is also a promise peculiarly prone to exaggeration, and this chapter examines both the substance and the hype, because separating them is essential to an honest reckoning with quantum computing's constructive potential.

Start with the substance, which is real. Many of the most important problems in materials science and drug discovery are, at their core, quantum problems that classical computers struggle to solve. The behavior of electrons in a novel material — whether it will conduct, insulate, or superconduct; how it will respond to light or magnetism; whether it will be strong or brittle — is governed by quantum mechanics that becomes intractable to simulate classically for the most interesting, strongly correlated materials. The action of a drug — how a molecule docks into the pocket of a target protein, how tightly it binds, how it reacts — likewise depends on quantum chemistry that classical methods approximate imperfectly. In principle, a sufficiently capable quantum computer could simulate these systems accurately, letting scientists screen candidate materials and drugs *in silico*, on a computer, with quantum precision, before synthesizing anything in a laboratory — compressing years of costly trial-and-error into computational search. The potential value is genuinely enormous: better batteries for electric vehicles and grid storage, more efficient catalysts to reduce industrial energy use, new superconductors, lighter and stronger materials, and faster, cheaper development of life-saving drugs. This is not fantasy; it is a plausible extrapolation of the machine's demonstrated nature, and it is why serious pharmaceutical and chemical companies have established quantum-computing research programs.

Now the hype, which is equally real and must be named. The gap between "quantum computers could in principle simulate these systems" and "quantum computers are about to revolutionize drug discovery" is vast, and it is routinely elided in promotional material. Several hard truths temper the promise. The molecules and materials of real commercial interest are large and complex, requiring far more qubits and far higher fidelity than NISQ machines possess, and in most cases requiring fault-tolerant machines still years from existing. Drug discovery, moreover, is not bottlenecked solely by quantum-chemical calculation; it is a long, multifarious process — biology, toxicology, clinical trials, regulation — in which the quantum-simulable step is only one part, so that even a perfect quantum chemistry engine would accelerate only a portion of the pipeline, not the whole. The classical competition, again, is formidable: classical computational chemistry and the newer machine-learning-based methods for predicting molecular properties are advancing rapidly, raising the bar a quantum computer must clear to prove advantage. And the field has a history of over-promising specific breakthroughs "within a few years" that then fail to materialize on schedule, feeding a cycle of inflated expectation and quiet disappointment that risks a backlash.

A particularly instructive example of the hype dynamic is the ubiquitous invocation of the nitrogenase enzyme and the Haber–Bosch process, mentioned in the previous chapter. It is a genuinely apt illustration of the *kind* of problem quantum computers could eventually help with — a real, valuable, classically intractable quantum-chemical puzzle. But it has been repeated so often, with so little acknowledgment of how far current machines are from actually solving it, that it has become almost a shibboleth of quantum marketing, cited to lend concreteness to a capability that does not yet exist. Careful resource estimates suggest that accurately simulating the nitrogenase active site would require a large fault-tolerant quantum computer — millions of physical qubits, error-corrected — of the kind that is, on any honest reckoning, years or decades away. The example is real; the implication that it is imminent is hype. Learning to hear the difference — to recognize when a genuine long-term possibility is being dressed up as a near-term certainty — is one of the most useful skills a reader of quantum-computing news can cultivate.

If molecular simulation is the quantum computer's most promising application, then materials science and medicine are the two domains where that promise could most profoundly reshape human life — the places where cracking the exponential wall of quantum chemistry might translate into cleaner energy, better technology, and cures for disease. These are the concrete, world-changing prizes that motivate the enormous investment in quantum computing, and while they lie beyond the reach of today's noisy machines, they are worth examining in detail, because they show what "useful" quantum computing might actually mean for ordinary people.

Consider first the grand challenge of nitrogen fixation, perhaps the single most cited example of a problem a quantum computer might crack, and a genuinely astonishing one. Nearly half of humanity is fed thanks to synthetic fertilizer, and synthetic fertilizer depends on converting inert nitrogen gas from the air into a form plants can use — a process performed industrially by the Haber-Bosch method, which demands enormous heat and pressure and consumes on the order of one to two percent of the entire world's energy supply, much of it from natural gas. Yet certain humble bacteria, using an enzyme called nitrogenase, perform the very same nitrogen-fixing chemistry at ordinary temperature and pressure, effortlessly, in the soil. If we understood exactly how nitrogenase works, we might design catalysts to fix nitrogen far more efficiently, saving a substantial fraction of global energy and transforming agriculture. But we do NOT understand it, and the reason is precisely the exponential wall: the enzyme's active site, a cluster of iron and molybdenum atoms, is a knot of strongly entangled electrons that no classical computer can simulate accurately. It is a problem seemingly tailor-made for a future quantum computer, which could model that entangled cluster directly and reveal nature's secret. A single successful simulation could reverberate through the world's food and energy systems.

Materials science offers a parallel wealth of prizes, all resting on the same quantum-mechanical foundation. High-temperature superconductivity — the ability of certain materials to conduct electricity with zero resistance at temperatures far above absolute zero — remains, decades after its discovery, only partially understood, because it arises from the collective, entangled behavior of vast numbers of electrons, again beyond exact classical simulation. A quantum computer that could illuminate the mechanism might guide the design of superconductors that work at everyday temperatures, which would revolutionize power transmission (eliminating the enormous losses in the electrical grid), transportation (frictionless magnetic levitation), and computing itself. Beyond superconductors, quantum simulation could accelerate the design of better batteries — crucial for electric vehicles and renewable-energy storage — by modeling the complex electrochemistry of novel electrode materials; more efficient solar cells; lighter and stronger alloys; and catalysts for cleaner industrial chemistry and carbon capture. In each case, the bottleneck is the classical intractability of simulating strongly correlated electrons, and in each case a quantum computer offers a way through the wall.

In medicine, the most discussed application is drug discovery, and understanding both its promise and its limits guards against the considerable hype. Designing a drug involves finding a molecule that binds precisely to a target — usually a protein in the body — to produce a therapeutic effect, and predicting how strongly and specifically a candidate molecule will bind requires understanding the quantum-mechanical interactions between the drug and its target. Current computational drug design relies on classical approximations that are useful but imperfect, and much of pharmaceutical research remains slow, expensive trial and error, with candidate drugs taking many years and enormous sums to develop. A quantum computer capable of accurately simulating the relevant molecular interactions could, in principle, predict binding and reactivity with far greater accuracy, allowing researchers to screen and design candidates computationally with a precision impossible today, potentially compressing years of laboratory work and slashing costs. The vision is of drugs designed rationally, from a true quantum understanding of the chemistry, rather than found by laborious searching.

Honesty requires tempering these dazzling prospects with hard caveats, because the gap between promise and delivery is where much quantum hype curdles into disappointment. Every one of these applications — nitrogen fixation, superconductivity, batteries, drug design — requires large, fault-tolerant, error-corrected quantum computers with many high-quality logical qubits, machines that are years or likely decades away. Today's noisy machines cannot simulate any molecule large enough to matter that a classical computer cannot already handle, so no practical materials or medical advantage has yet been demonstrated. Moreover, quantum computers will not replace classical methods wholesale but complement them, handling the specific hard cores of problems where classical methods fail while classical computers do the surrounding work; drug discovery in particular involves vast challenges beyond quantum chemistry — biology, toxicity, clinical trials — that no computer of any kind can shortcut. The quantum computer is a powerful specialized tool for the quantum-mechanical heart of these problems, not a magic wand for the whole of medicine or materials.

The nitrogen-fixation example deserves to be lingered over, because it is the single most compelling illustration of what "useful quantum computing" might actually mean for ordinary human life. Nearly half of humanity is fed thanks to synthetic fertilizer, and synthetic fertilizer depends on converting inert nitrogen gas from the air into a usable form — a process performed industrially by the Haber-Bosch method, which demands enormous heat and pressure and consumes on the order of one to two percent of the entire world's energy supply, much of it from natural gas. Yet humble soil bacteria, using an enzyme called nitrogenase, perform the very same chemistry at ordinary temperature and pressure, effortlessly. If we understood exactly how nitrogenase does it, we might design catalysts to fix nitrogen far more efficiently, saving a substantial fraction of global energy and transforming agriculture. But we do *not* understand it, precisely because the enzyme's active site — a cluster of iron and molybdenum atoms — is a knot of strongly entangled electrons beyond exact classical simulation. It is a problem seemingly tailor-made for a future quantum computer, and cracking it could reverberate through the world's food and energy systems. That a single successful simulation might help feed the planet more cheaply captures, better than any abstraction, why the effort matters.

The parallel prizes in materials science all rest on the same quantum-mechanical foundation, and it is worth seeing them as a family rather than a list. High-temperature superconductivity — the ability of certain materials to carry electricity with zero resistance well above absolute zero — remains, decades after its discovery, only partially understood, because it arises from the collective, entangled behavior of vast numbers of electrons, again beyond exact classical simulation; a quantum computer that illuminated the mechanism might guide the design of superconductors working at everyday temperatures, remaking power grids (eliminating the enormous losses in transmission), transport (frictionless magnetic levitation), and computing itself. Beyond superconductors lie better batteries for electric vehicles and renewable-energy storage, more efficient solar cells, stronger and lighter alloys, and catalysts for cleaner industrial chemistry and carbon capture. In every case the bottleneck is identical — the classical intractability of simulating strongly correlated electrons — and in every case a quantum computer offers a way through the wall. These are not separate speculative bets but a single bet, placed on the quantum computer's unique ability to model entangled matter.

The caveats that temper these dazzling prospects are worth stating firmly, because the gap between promise and delivery is where much quantum hype curdles into disillusionment. Every one of these applications requires large, fault-tolerant, error-corrected machines with many high-quality logical qubits — machines years or likely decades away. Today's noisy devices cannot simulate any molecule or material large enough to matter that a classical computer cannot already handle, so no practical materials or medical advantage has yet been demonstrated. Quantum computers will not replace classical methods wholesale but complement them, handling the specific hard cores of problems where classical methods fail while classical machines do the surrounding work — and in drug discovery especially, the vast challenges of biology, toxicity, and clinical trials lie far beyond what any computer of any kind can shortcut. The quantum computer is a powerful specialized tool for the quantum-mechanical heart of these problems, not a magic wand for medicine or materials. Still, the balance of expert opinion holds these simulation applications to be the likeliest to deliver transformative value, and the ones that could justify the whole enterprise. Beyond them lies a broader, more commercially hyped, and far shakier category — optimization — that demands especially careful, skeptical examination.

Chapter 57: Optimization and Finance — QAOA, Monte Carlo, and the Fine Print

Two application areas have generated more commercial excitement — and, arguably, more inflated claims — than any others in quantum computing: **optimization** and **finance**. The pitch is intuitive and seductive. Optimization problems — finding the best arrangement among astronomically many possibilities — are everywhere in industry: routing delivery trucks, scheduling airlines, managing supply chains, allocating investments, designing networks. Finance is a domain built on computation: pricing complex derivatives, managing risk, optimizing portfolios, detecting fraud. If quantum computers could speed up these tasks, the commercial value would be immense, and so a whole industry of quantum-optimization and quantum-finance ventures has sprung up, promising transformation. This chapter examines those promises with particular care, because they are, on the whole, the shakiest in the field — the applications where the gap between marketing and demonstrated reality is widest, and where a critical reader must be most on guard.

Consider optimization first. The hope rests on quantum algorithms designed for the NISQ era, chief among them the **Quantum Approximate Optimization Algorithm**, or **QAOA**, a hybrid quantum-classical algorithm that uses a quantum computer to explore the landscape of possible solutions to a combinatorial optimization problem, guided by a classical computer that tunes the quantum circuit's parameters. QAOA is elegant and much studied, and it is designed to run on the imperfect, non-error-corrected machines available today, which is part of its appeal. But the honest assessment of QAOA and related quantum-optimization approaches, after years of intensive research, is sobering: it has proven *very difficult to demonstrate that they provide any advantage* over the best classical optimization algorithms, which are themselves extraordinarily sophisticated and effective. Classical optimization is a mature field with decades of ingenious heuristics — simulated annealing, branch-and-bound, sophisticated solvers — that solve enormous real-world problems routinely and well, and beating them is a high bar. Numerous studies have found that QAOA, on the noisy machines of the NISQ era and often even in idealized analysis, fails to clearly outperform classical methods on practically relevant problems, and some theoretical results suggest the speedups, where they exist, may be modest. The quantum-annealing approach of D-Wave, discussed in Part VIII, faces the same difficulty: after more than a decade of commercial machines, a clear, decisive quantum advantage on a practical optimization problem remains elusive.

Finance presents a similar pattern of real theoretical promise obscured by heavy hype. The most credible quantum- finance application rests on **amplitude estimation**, a quantum technique related to Grover's algorithm that can, in principle, provide a *quadratic* speedup for **Monte Carlo simulation** — the workhorse method by which banks price complex derivatives and estimate financial risk by simulating many random scenarios. A quadratic speedup on Monte Carlo would be genuinely valuable, since these simulations consume vast computational resources on Wall Street. But the same caveats apply, and they are decisive for the near term. The quadratic speedup requires fault-tolerant quantum computers with error correction — not the NISQ machines of today — because it demands deep, precise circuits that current noisy hardware cannot execute; realizing it awaits hardware years away. And a quadratic speedup, while real, is modest compared with the exponential magic of Shor's algorithm: it helps, but it does not transform, and whether it will ultimately justify the cost and complexity of quantum hardware for financial computation, once the substantial overheads of error correction are accounted for, is genuinely uncertain. The many banks and financial firms that have established quantum-computing research teams are, for the most part, prudently *exploring* and *positioning* for a possible future advantage, not deploying quantum computers that outperform their classical systems today — a distinction their press releases do not always make clear.

The fine print, then, on optimization and finance is this: these are the application areas where the promise is *least* established and the hype *most* pronounced. Unlike quantum simulation of chemistry, which fits the machine's nature and enjoys a strong theoretical foundation, quantum optimization and much of quantum finance rest on hoped-for speedups that have proven stubbornly hard to demonstrate against formidable classical competition, and that in many cases require fault-tolerant hardware not yet available. This does not mean they are worthless — amplitude estimation offers a real if modest speedup, and future hardware may yet find optimization niches — but it does mean that claims of imminent quantum revolution in logistics, supply chains, or trading floors should be treated with strong skepticism. Much of the commercial quantum-optimization and quantum-finance activity of the mid-2020s is better understood as *positioning* — companies acquiring expertise and hedging against a possible future — than as the deployment of machines that beat classical computers today, which, on any honest accounting, they do not.

Beyond the simulation of nature lies a second broad category of proposed quantum applications, one that has generated enormous commercial excitement and, it must be said, a great deal of inflated and premature hype: optimization. Optimization problems — finding the best solution among a vast number of possibilities — are ubiquitous in business, finance, logistics, and industry, which is exactly why companies have been so eager to believe that quantum computers will solve them, and why a sober, skeptical examination is so necessary. This chapter tries to separate the genuine possibility from the marketing.

Optimization is everywhere in the practical world. A delivery company wants the shortest route visiting many cities; an airline wants to schedule crews and aircraft at minimum cost; a factory wants to arrange its production for maximum efficiency; a financial firm wants to construct an investment portfolio balancing return against risk; a telecommunications network wants to route traffic optimally. All of these are optimization problems: find, among an astronomically large number of possible configurations, the one that minimizes cost or maximizes value. Many such problems are computationally hard — the number of possible configurations explodes exponentially with the size of the problem, so that checking them all is impossible, and even finding a provably optimal solution can be intractable for classical computers. Because these problems are so valuable and so hard, the prospect of a quantum computer that could solve them faster has been irresistible to industry, and enormous commercial hope — and hype — has been invested in it.

The trouble is that the case for a large, general quantum advantage in optimization is far weaker and more uncertain than the case for quantum simulation, and this is the crucial point that the marketing tends to obscure. Several quantum approaches to optimization exist. There is Grover's algorithm, which offers a quadratic speedup for searching unstructured possibilities — real, but modest, merely a square-root improvement, and often insufficient to overcome the enormous overhead of running a quantum computer. There is quantum annealing, the approach embodied by D-Wave's machines, which we examined earlier and which has struggled for years to demonstrate any clear, general advantage over the best classical algorithms. And there are hybrid quantum-classical algorithms designed for near-term noisy machines, such as the quantum approximate optimization algorithm (QAOA), which have generated much research interest but for which convincing evidence of practical advantage over good classical methods remains elusive. In no case is there a proven, general, exponential quantum speedup for the optimization problems that industry cares about, of the kind Shor's algorithm provides for factoring. The speedups on offer are either modest (quadratic at best) or unproven or both.

The reason for this disappointing state of affairs goes back to the deep lesson of the whole book: quantum computers provide dramatic speedups only for problems with special hidden mathematical structure that interference can exploit, and most real-world optimization problems do not obviously possess such structure. Grover's quadratic speedup, provably optimal for unstructured search, is the ceiling for problems with no exploitable pattern — and a quadratic speedup, while genuine, is often not enough to matter in practice, because the quantum computer's per-operation overhead (the slow gates, the error correction, the sheer expense) can easily swamp a mere square-root gain, so that a classical computer running a well-tuned heuristic finishes first anyway. Meanwhile, classical optimization algorithms are extraordinarily sophisticated and have been refined over many decades; they are a formidable, fast-moving competitor, and the bar for a quantum machine to beat them on a real problem is very high. The history of quantum optimization claims is littered with results that looked promising until a cleverer classical algorithm matched them.

None of this means quantum computers will be useless for optimization — the honest verdict is one of genuine uncertainty rather than dismissal. It remains possible that specific, structured optimization problems will be found where quantum machines offer a real and useful advantage; that quadratic speedups will matter for certain high-value problems where even a modest acceleration is worth the cost; or that clever future algorithms, not yet invented, will unlock advantages we cannot currently foresee. Finance in particular has attracted intense quantum interest — for portfolio optimization, risk analysis, derivative pricing (via quantum versions of the Monte Carlo sampling methods used to price complex financial instruments, which do offer a proven quadratic speedup), and fraud detection — and major banks have established quantum research teams to be ready if and when advantage materializes. The commercial world is hedging its bets, and reasonably so. But the crucial point for the reader is skepticism: claims of quantum advantage in optimization and finance should be scrutinized carefully, distinguished sharply from the better-founded case for quantum simulation, and treated as unproven possibilities rather than established facts.

The gap between the *commercial excitement* around quantum optimization and the *scientific evidence* for it is worth confronting directly, because nowhere else in the field is the mismatch so large or so consequential for investors and executives. Optimization problems — finding the best configuration among an astronomical number of possibilities — are ubiquitous in business, finance, logistics, and industry, which is exactly why companies have been so eager to believe quantum computers will solve them, and why sober skepticism is so necessary. The delivery company wants the shortest route; the airline wants the cheapest crew schedule; the financial firm wants the optimal portfolio; the factory wants the most efficient production plan. All are optimization problems, many are computationally hard, and all are enormously valuable — so the prospect of a quantum machine that could solve them faster has been irresistible, and enormous hope, and hype, has been invested in it. The trouble is that the scientific case for a large, general quantum advantage in optimization is far weaker and more uncertain than the case for quantum simulation, and the marketing tends to obscure exactly this.

The reasons for the disappointing evidence go straight back to the deepest lesson of the whole book, and they are worth spelling out because they explain why optimization is not the easy win it appears. Quantum computers provide dramatic speedups only for problems with special hidden mathematical structure that interference can exploit — and most real-world optimization problems do not obviously possess such structure. The available quantum approaches all run into this wall. Grover's algorithm offers a speedup for searching unstructured possibilities, but only a *quadratic* one — a square-root improvement, provably the best possible for unstructured problems, and often insufficient to overcome the enormous per-operation overhead of running a quantum computer, so that a classical machine with a well-tuned heuristic finishes first anyway. Quantum annealing, the D-Wave approach, has struggled for years to demonstrate any clear, general advantage. And near-term hybrid methods like the quantum approximate optimization algorithm have generated much research interest but little convincing evidence of practical advantage. In no case is there a proven, general, exponential quantum speedup for the optimization problems industry cares about, of the kind Shor's algorithm provides for factoring. Meanwhile classical optimization algorithms are extraordinarily sophisticated, refined over decades, and a formidable, fast-moving competitor.

None of this means quantum computers will be useless for optimization — the honest verdict is genuine uncertainty rather than dismissal, and finance illustrates both the hope and the discipline required. It remains possible that specific structured optimization problems will be found where quantum machines offer real advantage, that quadratic speedups will matter for certain high-value problems where even a modest acceleration is worth the cost, or that clever future algorithms will unlock advantages we cannot now foresee. Finance in particular has drawn intense quantum interest — for portfolio optimization, risk analysis, derivative pricing, and fraud detection — and one genuinely well-founded case exists there: quantum versions of the Monte Carlo sampling methods used to price complex financial instruments do offer a proven *quadratic* speedup, and major banks have established quantum teams to be ready if and when advantage materializes. But the crucial posture for the reader is skepticism: claims of quantum advantage in optimization and finance should be scrutinized carefully, distinguished sharply from the better-founded case for quantum simulation, and treated as unproven possibilities rather than established facts. Learning to tell the solid from the speculative is the single most valuable skill in reading quantum news — and nowhere is it more necessary than in the domain where the two most overhyped technologies of the age, quantum computing and artificial intelligence, are joined together.

Chapter 58: Quantum Machine Learning — and the Dequantization Backlash

In the mid-2010s, as the artificial-intelligence revolution gathered force and machine learning transformed one industry after another, it was perhaps inevitable that someone would propose marrying the two hottest technologies in computing: quantum machine learning, or QML. The pitch was irresistible to investors and headline writers alike — quantum computers, with their exponential state spaces, would supercharge machine learning, which is fundamentally about finding patterns in high-dimensional data, delivering AI systems of unprecedented power. A wave of excitement followed, and a series of quantum machine-learning algorithms were proposed that promised dramatic, even exponential, speedups over classical machine learning for tasks like recommendation systems, principal component analysis, and solving the large systems of linear equations that underlie much of machine learning. For a few years, QML was among the most hyped corners of an already-hyped field. Then a teenager took it apart, and the episode became one of the most instructive cautionary tales in the recent history of quantum computing.

The central QML algorithms relied on a quantum subroutine for handling large amounts of data efficiently, often building on the **HHL algorithm** (named for Harrow, Hassidim, and Lloyd) for solving systems of linear equations, which promised an exponential speedup under certain conditions. Several prominent quantum machine-learning algorithms — most famously a quantum recommendation-system algorithm published in 2016 — claimed exponential advantages over any classical method for their tasks. These claims were taken seriously and widely publicized as evidence that quantum computers would revolutionize AI. But there was a subtlety that the excitement glossed over, one that would prove the undoing of the strong claims: the quantum algorithms achieved their speedups only under special assumptions about how the data was provided and what kind of answer was required, and no one had carefully checked whether *classical* algorithms, given comparable assumptions, might do just as well.

In 2018, an eighteen-year-old undergraduate named **Ewin Tang**, working with the computer scientist Scott Aaronson at the University of Texas, checked. Aaronson had given Tang the quantum recommendation-system algorithm as a research problem, expecting her to prove that no fast classical algorithm could match it — to establish the quantum advantage rigorously. Instead, Tang did the opposite: she discovered a *classical* algorithm that solved the same problem essentially as fast as the quantum one, stripping away the claimed exponential advantage entirely. Her classical algorithm was "inspired" by the quantum one — it borrowed the mathematical techniques and the same data-access assumptions the quantum algorithm had used — and it showed that once you granted a classical computer the same kind of access to the data that the quantum algorithm assumed, the classical computer could keep up. The quantum speedup had been an illusion, an artifact of comparing the quantum algorithm against a *naive* classical baseline rather than the best possible one. Tang's result, and the wave of similar "**dequantization**" results that followed as researchers applied her techniques to other QML algorithms, showed that a whole class of celebrated quantum machine-learning speedups evaporated under careful scrutiny.

The dequantization episode carries a lesson that echoes the supremacy wars of Part IX and deserves to be stated as a general principle, because it applies far beyond machine learning: **a quantum speedup is a claim about the quantum algorithm relative to the best classical alternative, and establishing it requires ruling out clever classical approaches — which is hard, and often not done carefully.** Many claimed quantum advantages, in machine learning and elsewhere, have turned out to rest on comparisons against classical straw men, and when the classical competition is taken seriously — as Tang took it seriously, as the classical simulators of Sycamore took it seriously — the quantum advantage shrinks or vanishes. This does not mean quantum computers offer no advantages; Shor's exponential speedup for factoring is real and has survived decades of scrutiny, and Grover's quadratic speedup is proven. But it means that claims of quantum advantage must be regarded with skepticism until the classical competition has been genuinely and rigorously bested, and that the history of the field is littered with premature claims that did not survive that test. The teenager who dequantized the recommendation algorithm did the field a service, puncturing hype and forcing a more honest reckoning.

Where does quantum machine learning stand after the dequantization backlash? Chastened, but not dead. Researchers continue to explore genuinely quantum approaches to machine learning — using quantum computers to process intrinsically quantum data (such as the states produced by quantum sensors or simulations, where there is no classical data to dequantize), or seeking quantum-learning advantages that do not fall to Tang-style attacks — and some of these remain live possibilities. But the grandiose early claims that quantum computers would broadly and exponentially accelerate classical machine learning have been substantially deflated, and the honest current view is that quantum machine learning is a fascinating research area with uncertain practical payoff, not the imminent revolution once advertised. It is, in miniature, the story of quantum computing's applications generally: real scientific interest, genuine open possibilities, and a great deal of hype that does not survive careful, skeptical examination.

When two of the most hyped technologies of the age — quantum computing and artificial intelligence — are combined in a single phrase, the result is a marketing department's dream and a skeptic's nightmare. "Quantum machine learning" promises to marry the exponential power of quantum computers to the transformative capabilities of AI, and the claims made on its behalf have been among the boldest and least substantiated in the entire field. This chapter approaches the subject with the sharp skepticism it demands, while acknowledging the genuine research questions that lie beneath the froth.

The intuitive appeal is easy to see and easy to oversell. Machine learning, the technology behind modern AI, involves finding patterns in enormous quantities of data and performing vast numbers of mathematical operations, particularly operations on the large arrays of numbers called matrices and vectors that represent data and the parameters of AI models. Quantum computers, the pitch goes, can represent and manipulate exponentially large mathematical objects using relatively few qubits, and certain quantum algorithms perform linear-algebra operations — exactly the kind of operations at the heart of machine learning — with apparent exponential speedups. Put these together and you seem to have a recipe for revolutionizing AI: quantum computers that train machine-learning models exponentially faster, or that recognize patterns no classical AI could find. The syllogism is seductive, and it has launched countless startups and press releases.

The reality is far more sobering, and it rests on a set of technical obstacles that the marketing conveniently omits. The most celebrated quantum machine-learning algorithm, a method for solving systems of linear equations known as the HHL algorithm (after Harrow, Hassidim, and Lloyd, 2009), does indeed offer an apparent exponential speedup for a certain linear-algebra task — but it comes wrapped in a thicket of fine-print caveats that severely limit its practical usefulness, and understanding these caveats is a masterclass in reading quantum claims critically. First, there is the INPUT problem: to process classical data on a quantum computer, you must first load that data into quantum states, and for large datasets this loading can itself take so long that it erases the quantum speedup entirely — getting the data in is as slow as the classical computation you hoped to avoid. Second, there is the OUTPUT problem: a quantum algorithm's result is a quantum state, and measuring it yields only limited classical information (a few numbers, or a sample), not the full detailed answer — so even when the quantum computer has done its work, extracting a useful classical result can be impossible or slow. Third, the apparent speedups often depend on restrictive technical assumptions about the data that may not hold in practice. When these caveats are properly accounted for, many of the claimed exponential advantages shrink dramatically or vanish.

The skepticism has been sharpened by a remarkable series of results that turned the tables in the most instructive way possible. Beginning around 2018, a young researcher named Ewin Tang — then an undergraduate advised by Scott Aaronson — produced a sequence of "dequantization" results: she and others showed that several celebrated quantum machine-learning algorithms, thought to offer exponential speedups, could be MATCHED by clever CLASSICAL algorithms inspired by the quantum ones. In other words, the supposed quantum advantage was illusory: once someone looked closely, a classical computer could do essentially the same task about as fast, using techniques the quantum algorithms had inadvertently pointed the way toward. Tang's work, accomplished while she was still a teenager, was a stunning and clarifying blow, demonstrating that many quantum-machine-learning speedup claims did not survive rigorous scrutiny, and that the field had been too quick to assume advantage where none existed. It stands as one of the great cautionary tales of quantum hype, and a testament to the value of hard-nosed skepticism.

This does not mean quantum machine learning is entirely without merit — the honest position, once again, is genuine uncertainty rather than blanket dismissal, and there remain real and interesting research directions. There may be advantages for machine-learning tasks involving intrinsically QUANTUM data — for instance, learning about quantum systems, quantum states, or the outputs of quantum experiments, where the data is naturally quantum and the input problem dissolves. There is active research into quantum neural networks and quantum-enhanced models whose ultimate value is still being explored. And it remains possible that future insights will uncover genuine quantum advantages for some learning problems that we cannot currently identify. The field is young and the questions are open. But the sweeping claims that quantum computers will imminently and generally revolutionize artificial intelligence are, on present evidence, unsupported and should be treated with deep skepticism.

The combination of two hyped technologies in a single phrase should, by itself, set off alarms, and it is worth naming why quantum machine learning warrants the very sharpest skepticism this book has to offer. Quantum computing and artificial intelligence are, individually, among the most oversold technologies of the age; joined together as "quantum machine learning," their hype multiplies rather than their substance, and the claims made on the field's behalf have been among the boldest and least supported anywhere in quantum computing. The intuitive pitch is seductive: machine learning involves vast numbers of operations on the large arrays of numbers — matrices and vectors — that represent data, and quantum computers can represent and manipulate exponentially large mathematical objects with relatively few qubits, and certain quantum algorithms perform linear-algebra operations with apparent exponential speedups. Put these together and you seem to have a recipe for revolutionizing AI. The syllogism has launched countless startups and press releases — and it dissolves, on close inspection, into a thicket of caveats that the marketing conveniently omits.

The caveats surrounding the most celebrated quantum machine-learning algorithm — the HHL algorithm for solving linear systems — are worth spelling out, because understanding them is a masterclass in reading quantum claims critically. HHL offers an apparent exponential speedup for a certain linear-algebra task, but it comes wrapped in fine print that severely limits its practical usefulness. First, the *input* problem: to process classical data on a quantum computer you must first load it into quantum states, and for large datasets this loading can take so long that it erases the quantum speedup entirely — getting the data in is as slow as the classical computation you hoped to avoid. Second, the *output* problem: a quantum algorithm's result is a quantum state, and measuring it yields only limited classical information — a few numbers or a sample, not the full detailed answer — so even when the quantum computer has done its work, extracting a useful result can be impossible or slow. Third, the speedups often depend on restrictive assumptions about the data that may not hold in practice. When these caveats are properly accounted for, many of the claimed exponential advantages shrink dramatically or vanish outright. The lesson generalizes: for any quantum machine-learning claim, ask how the data gets in, how the answer comes out, and what assumptions are buried in between.

The "dequantization" results deserve pride of place in any honest account, because they turned the tables on the field in the most instructive way imaginable. Beginning around 2018, a young researcher named Ewin Tang — then an undergraduate advised by Scott Aaronson — produced a sequence of results showing that several celebrated quantum machine-learning algorithms, thought to offer exponential speedups, could be *matched* by clever classical algorithms inspired by the quantum ones. The supposed quantum advantage was illusory: once someone looked closely, a classical computer could do essentially the same task about as fast, using techniques the quantum algorithms had inadvertently pointed the way toward. That this was accomplished by a teenager, puncturing claims that had generated enormous excitement, made it one of the great cautionary tales of quantum hype and a testament to the value of hard-nosed skepticism. None of this means quantum machine learning is entirely without merit — there may be genuine advantages for tasks involving intrinsically *quantum* data, and research into quantum-enhanced models continues — but the sweeping claims that quantum computers will imminently and generally revolutionize artificial intelligence are, on present evidence, unsupported. The reader armed with the essential questions — how does the data get in, how does the answer come out, has anyone checked whether a classical algorithm can do the same — is well equipped to see through the froth, and to appreciate the one application area, quieter and less glamorous than AI, that is already delivering real value today: the exquisite sensitivity of quantum sensors.

Chapter 59: Sensing and Metrology — the Quiet Quantum Revolution That Already Works

Amid the noisy promises and contested claims of quantum computing, there is a corner of quantum technology that makes no headlines proportionate to its success, requires no error-corrected logical qubits, awaits no future breakthrough, and yet delivers real, deployed, world-beating performance *today*. It is **quantum sensing and metrology** —

the use of quantum systems to measure physical quantities with a precision beyond anything classical methods can achieve — and it deserves a prominent place in any honest accounting of the quantum revolution, precisely because it is the part that has quietly, undeniably worked. If quantum computing is the field's dazzling, unfulfilled promise, quantum sensing is its unglamorous, delivered reality.

The principle behind quantum sensing is, in a sense, the flip side of the fragility that so bedevils quantum computing. The quantum states used in computing are maddeningly sensitive to their environment — that is the whole problem of decoherence. But sensitivity to the environment is exactly what you *want* in a sensor. A quantum system that responds to the faintest magnetic field, the tiniest change in temperature, the slightest acceleration, the subtlest variation in time, makes an exquisite detector of precisely those quantities. Quantum sensing turns the enemy of quantum computing into an asset: it uses the delicacy of quantum states, and quantum effects like superposition and entanglement, to measure the world with a precision that classical sensors cannot match. Where the quantum computer must fight to isolate its qubits from every disturbance, the quantum sensor deliberately exposes its quantum system to the quantity it wishes to measure, and reads the world from the quantum system's exquisitely sensitive response.

The applications are already numerous and consequential, and they span science and industry. The most familiar are **atomic clocks**, the most precise timekeepers ever built, which use the quantum transitions of atoms as pendulums of unearthly regularity — so precise that the best optical atomic clocks would not lose or gain a second over the age of the universe. Atomic clocks are not a laboratory curiosity; they are the invisible foundation of the Global Positioning System, which works by precisely timing signals from satellites, and of the synchronization of telecommunications networks and financial markets. Every time your phone's map locates you, quantum metrology is at work. Beyond clocks, quantum sensors include **magnetometers** of extraordinary sensitivity — some using the nitrogen-vacancy centers in diamond mentioned in Part VIII — capable of detecting the faint magnetic fields produced by neural activity in the brain or by distant geological formations, with applications in medical imaging, mineral exploration, and navigation. Quantum **gravimeters** measure tiny variations in gravity, useful for detecting underground structures, monitoring volcanoes and aquifers, and navigation without satellites. And quantum-enhanced measurement has already transformed fundamental science: the **LIGO** gravitational-wave detectors, which in 2015 first detected the ripples in spacetime from colliding black holes — one of the great scientific achievements of the century — use "squeezed" quantum states of light to push their sensitivity below what classical optics would allow, a direct and spectacular payoff of quantum metrology.

Why does quantum sensing succeed where quantum computing still struggles? The answer is illuminating and worth dwelling on, because it clarifies the whole landscape of quantum technology. Quantum sensing is *far less demanding* than quantum computing. A quantum sensor typically uses a small number of quantum systems, does not need to entangle thousands of qubits into a fragile collective state, does not need to perform millions of precise gate operations in sequence, and does not need error correction. It needs only to prepare a quantum system, expose it to the world, and read its response — a far simpler task than orchestrating a long quantum computation. The extreme difficulty of quantum computing lies in *scale* and *coherence over long computations*, requirements that quantum sensing largely sidesteps. This is why quantum sensing has delivered real technology while quantum computing is still building toward its first useful advantage: sensing asks much less of the quantum world, and the quantum world has obliged.

Amid all the excitement and hype surrounding quantum COMPUTERS, it is easy to overlook a quieter branch of quantum technology that is already delivering real, practical value today — not in some speculative future but right now, in laboratories, hospitals, and navigation systems. This is quantum sensing and metrology: the use of quantum effects to measure physical quantities with a precision unattainable by any classical means. It is arguably the most mature and immediately useful of all quantum technologies, and it deserves recognition precisely because it lacks the glamour that draws attention to computing.

The foundational insight of quantum sensing is that the very fragility which makes quantum states such a nuisance for computing makes them SUPERB for sensing. Recall that a quantum superposition is exquisitely sensitive to its environment — the slightest disturbance, a whisper of a magnetic field, a flicker of temperature, a trace of a stray force, will alter it. For a quantum computer this sensitivity is the enemy, the source of decoherence that destroys computations. But for a sensor, sensitivity is exactly the point: if a quantum state changes measurably in response to a tiny external influence, then by carefully measuring that change you can detect the influence with extraordinary precision. The quantum system becomes a probe of unmatched delicacy, registering perturbations far too small for any classical instrument to notice. What is a bug for the computer is a feature for the sensor.

The applications are already remarkably diverse and consequential. The most precise clocks ever built — atomic clocks, and their successors, optical-lattice clocks — rely on quantum transitions in atoms to keep time so accurately that the best would not gain or lose a second over the entire age of the universe. These clocks are not laboratory curiosities: they are the beating heart of the satellite navigation systems, like GPS, that guide ships, aircraft, and the phone in your pocket, and they underpin the synchronization of telecommunications networks and financial markets. Quantum magnetometers — sensors of magnetic fields — can detect fields so faint that they are being developed to map the tiny magnetic signals produced by the electrical activity of the human brain and heart, offering new windows into medicine, and to detect submarines, mineral deposits, and archaeological remains. Quantum gravimeters, which sense minute variations in gravity by dropping clouds of ultra-cold atoms and watching their quantum behavior, can detect underground structures — tunnels, pipes, voids, aquifers, oil reservoirs — from the surface, and could one day map the interior of the Earth or aid navigation where GPS fails.

A particularly elegant example, and one worth dwelling on, is the nitrogen-vacancy center in diamond: a tiny defect where a nitrogen atom sits beside a gap in the diamond's carbon lattice, trapping electrons whose quantum spin states are exquisitely sensitive to magnetic and electric fields, temperature, and strain — and, remarkably, can be read out and controlled with ordinary light at room temperature. These diamond sensors can measure magnetic fields with nanoscale spatial resolution, enabling scientists to probe the magnetic properties of individual molecules and materials, image electric currents in microchips, and pursue applications from geology to biology. That a flaw in a diamond can serve as one of the world's most sensitive and versatile quantum sensors, working at room temperature without exotic refrigeration, captures the surprising practicality of the field.

Quantum sensing also exploits entanglement to push precision past what even independent quantum probes can achieve — the domain of quantum metrology proper. There is a fundamental limit to how precisely a given number of independent probes can measure a quantity, set by the statistics of random noise; but by ENTANGLING the probes, one can, in principle, surpass this classical limit and approach a more fundamental bound set by quantum mechanics itself, squeezing more precision from the same resources. This idea has already borne spectacular fruit in an unexpected place: the LIGO gravitational-wave detectors, which listen for the almost unimaginably faint ripples in spacetime produced by colliding black holes billions of light-years away, now use "squeezed light" — a quantum-engineered state of light — to reduce quantum noise and sharpen their sensitivity, allowing them to detect more and fainter cosmic collisions. Quantum metrology is helping humanity hear the universe.

The reversal at the heart of quantum sensing — that the very fragility which dooms quantum computers is precisely what makes quantum sensors superb — is worth savoring, because it turns the field's greatest curse into a blessing. A quantum superposition is exquisitely sensitive to its environment: the slightest magnetic field, temperature shift, or stray force will alter it. For a quantum computer this sensitivity is the enemy, the source of the decoherence that destroys computations and demands the whole elaborate apparatus of error correction. For a sensor, sensitivity is the entire point. If a quantum state changes measurably in response to a tiny external influence, then by carefully measuring that change you can detect the influence with extraordinary precision. The quantum system becomes a probe of unmatched delicacy, registering perturbations far too faint for any classical instrument to notice. What is a fatal bug for the computer is the defining feature of the sensor — the same physics, read with opposite intent.

The practical reach of quantum sensing is already remarkably broad and consequential, and it is worth cataloguing because it corrects the impression that all quantum technology lies in an uncertain future. Atomic clocks and their successors, optical-lattice clocks, keep time so accurately that the best would not gain or lose a second over the age of the universe — and they are not curiosities but the beating heart of the satellite navigation systems that guide ships, aircraft, and the phone in your pocket, and the synchronization of telecommunications and financial markets. Quantum magnetometers detect fields so faint they are being developed to map the tiny magnetic signals of the brain and heart, and to find submarines, minerals, and buried archaeology. Quantum gravimeters, sensing minute variations in gravity by watching clouds of ultra-cold atoms fall, can detect underground tunnels, pipes, voids, and aquifers from the surface. And the nitrogen-vacancy center in diamond — a defect whose electron spin is exquisitely sensitive to fields and temperature yet readable with ordinary light at room temperature — serves as a versatile nanoscale sensor probing individual molecules and microchips. That a flaw in a diamond, at room temperature, is one of the world's most sensitive quantum instruments captures the surprising, unglamorous practicality of the whole field.

The deepest triumph of quantum metrology — surpassing the classical limits of precision by entangling the probes themselves — deserves emphasis, because it has already helped humanity hear the universe. There is a fundamental limit to how precisely a given number of *independent* probes can measure a quantity, set by the statistics of random noise; but by *entangling* the probes, one can in principle surpass this classical limit and approach a more fundamental bound set by quantum mechanics itself, squeezing more precision from the same resources. This idea has borne spectacular fruit in the LIGO gravitational-wave detectors, which listen for almost unimaginably faint ripples in spacetime from colliding black holes billions of light-years away, and which now use quantum-engineered "squeezed light" to reduce quantum noise and sharpen their sensitivity, detecting more and fainter cosmic collisions. Quantum sensing, then, is the mature, humble, already-useful cousin of the glamorous quantum computer — sharing the same deep mastery of preparing, controlling, and measuring delicate quantum states, serving as a proving ground for the control techniques computing also needs, and reminding us that quantum technology is not merely a distant promise but a present reality, delivering real value today. With this survey of applications complete, we can return to the central drama of the computer itself, and to the state of the art, where after decades of struggle the field has begun at last to cross the threshold that separates noisy demonstration from genuine, error-corrected computation.

Part XII — The Error-Correction Revolution

Chapter 60: Utility Before Advantage — IBM's Error-Mitigation Era

By the early 2020s, quantum computing found itself in an awkward middle passage. The supremacy demonstrations of 2019 and 2020 had proved that quantum hardware could reach a regime beyond easy classical simulation, but on useless tasks. Full fault tolerance — error-corrected logical qubits running long, reliable computations — remained years away. The great question was whether the noisy, intermediate-scale machines of the NISQ era could do anything *useful* in the meantime, or whether the field would simply have to wait, perhaps a decade, for fault tolerance before any value appeared. Into this uncertainty IBM injected a reframing that would characterize the middle years of the decade: the pursuit not of "quantum advantage" in the strong sense, but of what it called quantum **utility** — the demonstration that a NISQ machine could produce accurate, useful results on a genuinely hard scientific problem, even without error correction, by cleverly managing rather than eliminating the noise. This chapter examines that development and the technique at its heart, **error mitigation**, which bridged the NISQ present and the fault-tolerant future.

The core problem, as always, was noise. NISQ machines have no error correction, so as a computation proceeds, errors accumulate and the results drift away from the truth, until for a sufficiently deep circuit the output is meaningless. Full error *correction* — the surface codes and threshold theorem of Part VI — would fix this but requires far more qubits and higher fidelity than NISQ machines possess. **Error mitigation** is a cleverer, more modest middle path: instead of correcting errors as they happen (which requires the full apparatus of logical qubits), it runs the noisy computation many times, deliberately varying the noise, and then uses classical post- processing and statistical extrapolation to *estimate what the answer would have been with no noise at all.* It does not produce a noise-free computation; it produces a statistical estimate of the noise-free result, extracted from many noisy runs. Techniques with names like "zero-noise extrapolation" and "probabilistic error cancellation" run the circuit at several artificially amplified noise levels and extrapolate back to the zero-noise limit, or characterize the noise precisely enough to subtract its average effect. Error mitigation is cheaper than error correction — it needs no extra qubits, only extra runs and classical computation — but it is also more limited: its cost grows steeply with circuit size, and it cannot enable arbitrarily long computations the way true error correction can. It is a bridge technology, a way to squeeze useful results from noisy hardware in the interim before fault tolerance arrives.

IBM's landmark demonstration came in June 2023, published in *Nature*, and it was carefully framed to claim *utility*, not supremacy or advantage. Using a 127-qubit Eagle processor and sophisticated error-mitigation techniques, IBM's team simulated the behavior of a model magnetic material — a system of interacting quantum spins, a genuine physics problem — at a scale and accuracy that pushed against the limits of classical simulation. The claim was pointed and careful: not that the quantum computer had done something *impossible* for classical computers, but that it had produced *accurate results* on a hard problem in a regime where the standard classical approximation methods become unreliable, demonstrating that a NISQ machine with error mitigation could be a useful scientific instrument. IBM called this "quantum utility" — evidence that quantum computers could deliver value *before* the arrival of fault tolerance, in the here and now of the NISQ era.

The reaction was instructive and, again, an example of the healthy skeptical dynamic that governs the field. Within weeks and months, classical-computing researchers responded to IBM's utility claim exactly as they had responded to Google's supremacy claim: by demonstrating improved classical methods that could reproduce IBM's results. Several groups showed that clever classical techniques — tensor-network methods and other advanced simulation approaches — could in fact match or approximate the specific quantum simulation IBM had performed, suggesting that the particular problem was not truly beyond classical reach after all. IBM's demonstration, like Google's, had reached an impressive frontier but had not decisively crossed it; the classical competition, once motivated, could keep pace. The episode reinforced the recurring lesson: claims of quantum advantage or utility must survive the assault of the best classical methods, and the boundary between quantum and classical capability is a contested, moving frontier, not a fixed line.

As the 2020s unfolded, the language and the ambitions of the quantum computing field underwent a subtle but important shift, one that reflected both hard-won maturity and a certain retreat from the grand promises of the supremacy era. The new watchwords became "quantum utility" and "utility before advantage," and understanding this reframing is essential to grasping where the field honestly stands today — past the useless benchmarks of supremacy, not yet at the transformative applications of a fault-tolerant future, but somewhere meaningful in between.

Recall the milestones the field had passed. Quantum SUPREMACY, achieved by Sycamore and Jiuzhang, meant doing something — anything, however useless — that classical computers could not. It was a scientific proof of principle, but the tasks were deliberately pointless, and the achievements were perpetually harried by improving classical algorithms. Quantum ADVANTAGE, the more meaningful goal, would mean solving a USEFUL problem faster or better than any classical computer — a genuine practical payoff. But quantum advantage on a real, valuable problem had not been achieved and remained, by most honest assessments, years or decades away, awaiting the large error-corrected machines that do not yet exist. The field faced an uncomfortable gap: supremacy was real but useless; advantage was useful but distant. What could be claimed in between?

The answer, championed notably by IBM around 2023, was "quantum utility": the idea that today's noisy, intermediate-scale quantum computers — machines of a few hundred imperfect qubits, without full error correction — could serve as useful SCIENTIFIC TOOLS for exploring quantum systems and running computations at a scale where they become difficult to simulate classically, even if they had not achieved a clear, proven speedup over the very best classical methods. In a prominent 2023 demonstration, IBM used a 127-qubit processor to simulate the behavior of a complex quantum magnetic system and produce accurate results in a regime where straightforward classical simulation struggled — arguing that this represented "utility," a genuine use of a quantum computer to compute something scientifically meaningful and hard, a step beyond mere supremacy benchmarks toward real usefulness. It was framed explicitly as "utility before advantage": the machine was useful as a scientific instrument even before it could provably out-compute all classical rivals.

The reception of the quantum-utility framing was revealing, and it followed the now-familiar pattern that should by this point feel like an old friend. IBM's 2023 claim was quickly met by classical-computing researchers who devised clever classical methods — using tensor-network techniques and other approaches — that could reproduce IBM's results on ordinary computers, arguing that the specific computation was not, after all, beyond classical reach, and that "utility" had been overstated. And IBM and others responded, as ever, by pushing to larger and harder instances. The eternal arms race between quantum claims and classical rebuttals continued, exactly as it had since Sycamore, a perpetual contest that both disciplines and advances the field. The lesson, once again, was that the boundary between quantum and classical capability is contested, mobile, and mapped only through fierce competition — and that any claim of quantum usefulness must be scrutinized against the ever-improving classical state of the art.

Yet beneath the marketing skirmishes, the utility framing captured something genuinely true and important about the era, and it is worth stating fairly. The noisy machines of the present, for all their limitations, ARE useful in real ways: as research instruments for studying quantum systems and quantum error correction; as testbeds for developing and refining quantum algorithms; as platforms for training a generation of quantum programmers and engineers; and as engines for exploring the fascinating physics of many-qubit entangled systems, which is scientifically valuable in its own right regardless of practical payoff. The current machines are not useless, even if they are not yet transformative — they are the essential proving ground on which the fault-tolerant future is being built. And they are available, through cloud services, to researchers and companies worldwide, seeding a broad ecosystem of experimentation. "Utility before advantage" is, in this reading, an honest acknowledgment of a genuine intermediate stage: the machines can do real work, even as the world-changing applications remain over the horizon.

The three-way distinction the "utility" framing tries to navigate — among supremacy, utility, and advantage — is worth laying out cleanly, because these terms are constantly confused and the confusion matters. Quantum *supremacy*, achieved by Sycamore and Jiuzhang, means doing something useless that classical computers cannot; it is a scientific proof of principle, perpetually harried by improving classical algorithms. Quantum *advantage*, the meaningful goal, means solving a genuinely *useful* problem faster or better than any classical computer — a real practical payoff, not yet achieved and by most honest assessments years or decades away. Between the useless-but-real achievement of supremacy and the useful-but-distant goal of advantage yawned an uncomfortable gap, and "quantum utility" was the field's attempt to name something meaningful within it: the idea that today's noisy machines could serve as useful *scientific tools*, running computations at a scale where they become hard to simulate classically, even without a proven speedup over the very best classical methods. It was, in essence, an honest acknowledgment that the machines can do real work even before they can provably out-compute all classical rivals — "utility before advantage."

The now-familiar cycle that greeted IBM's 2023 utility claim is worth recounting once more, because its very predictability has become one of the most reliable features of the field. IBM used a 127-qubit processor to simulate a complex quantum magnetic system and produce accurate results in a regime where straightforward classical simulation struggled, framing it as a demonstration of utility. Almost immediately, classical-computing researchers devised clever methods — using tensor-network techniques and other approaches — that reproduced IBM's results on ordinary computers, arguing that the specific computation was not, after all, beyond classical reach and that "utility" had been overstated. IBM and others responded, as ever, by pushing to larger and harder instances. The eternal arms race between quantum claims and classical rebuttals continued exactly as it had since Sycamore — claim, classical counterattack, larger quantum machine, repeat — a perpetual contest that both disciplines and advances the field, and that should by now feel to the reader like an old and trustworthy friend. The lesson, once again: the boundary between quantum and classical capability is contested, mobile, and mapped only through fierce competition.

Yet beneath the marketing skirmishes, the utility framing captured something genuinely true about the era, and it deserves a fair hearing. The noisy machines of the present, for all their limitations, *are* useful in real ways — as research instruments for studying quantum systems and quantum error correction, as testbeds for developing and refining quantum algorithms, as platforms for training a generation of quantum programmers and engineers, and as engines for exploring the fascinating physics of many-qubit entangled systems, which is scientifically valuable in its own right regardless of practical payoff. Available worldwide through cloud services, they are seeding a broad ecosystem of experimentation. "Utility before advantage" is, in this reading, an honest recognition of a genuine intermediate stage: the machines can do real work, even as the world-changing applications remain over the horizon, and the path to those applications runs through a long stretch of useful-but-not-revolutionary labor. But the utility framing was, in the end, a way of making the best of the noisy present while the field waited for the achievement everything ultimately depends upon — real, working, scalable quantum error correction, the crossing of the threshold that would finally tame the fragility that had haunted the field from the start. And in December 2024, in the most important experimental result since Sycamore, Google announced it had crossed that threshold at last.

Chapter 61: Below the Threshold — Google's Willow and the Turning Point

In December 2024, Google announced a result that, if it holds up as its significance suggests, may be remembered as one of the genuine turning points in the history of quantum computing — more important, in the long run, than the supremacy claim that made bigger headlines five years earlier. The announcement concerned a new superconducting processor named **Willow**, and its central claim was not about speed or qubit count but about something subtler and deeper: Willow had demonstrated quantum error correction that *improves as the system grows* — that crosses, at last, **below the threshold** that the threshold theorem of Part VI had defined as the boundary between possible and impossible fault-tolerant computing. To understand why this matters so much, and why it excited the field more than another supremacy demonstration would have, we must return to what the threshold actually means.

Recall the central promise and peril of the threshold theorem from Part VI. Quantum error correction protects a logical qubit by encoding it across many physical qubits and continuously detecting and fixing errors. But the correction machinery is itself made of noisy components, and there was always the nightmare that adding more physical qubits to a logical qubit might introduce errors faster than it removed them — that scaling up the code would make things *worse*, not better, drowning the computation in the noise of its own repair. The threshold theorem said that this nightmare is avoided *if and only if* the physical error rate is below a critical threshold: below threshold, making the code bigger *reduces* the logical error rate, exponentially; above threshold, making it bigger increases the error rate, and fault tolerance is impossible. For thirty years, crossing below the threshold on real hardware — actually demonstrating that a larger error-correcting code gives a *lower* logical error rate — had been the holy grail of experimental quantum computing, the proof that error correction genuinely *works* and that scalable, fault- tolerant quantum computing is achievable in practice and not merely in theory.

Willow, Google claimed, crossed that line. The demonstration used the **surface code** (Kitaev's code from Part VI), and the crucial experiment was to build surface-code logical qubits at several different sizes — encoding a logical qubit in a small patch of physical qubits, then a larger patch, then a larger one still — and to measure the logical error rate at each size. The long-sought result was that *each increase in the code size reduced the logical error rate*, and reduced it by a consistent factor — the exponential suppression the threshold theorem predicts for a system operating below threshold. Google reported that going from a smaller code to a larger one cut the logical error rate roughly in half at each step, the signature of below-threshold operation, and that the logical qubit could be kept alive — its encoded information preserved by continuous error correction — for longer than any of its individual physical qubits would survive. That last point is worth pausing on: the error-corrected logical qubit outlived its physical components, meaning the correction was genuinely *adding* protection, genuinely winning the race against noise. This is precisely what "below threshold" means in practice, and it is the milestone the field had chased for three decades.

The significance of this result, if it stands and scales as hoped, is hard to overstate, and it is different in kind from the supremacy demonstrations. Supremacy proved that quantum hardware could do *some* task beyond classical reach — a milestone about raw capability on a contrived problem. Below-threshold error correction proves something about *scalability and reliability* — that the path to arbitrarily long, arbitrarily accurate quantum computation is open, that error correction genuinely works and improves with scale, that the fundamental obstacle which many feared might forever prevent useful quantum computing has, in principle, been overcome. It is the difference between showing that a machine can sprint and showing that it can be built to run any distance without falling apart. Willow did not run a useful computation — it demonstrated the *mechanism* that will make useful computation possible, the error correction that fault-tolerant machines will need. In the arc of this book, it is the moment the fragility problem of Part VI — the objection that nearly killed the field in the 1990s, the decoherence that seemed insurmountable — is answered not merely in theory (as Shor and the threshold theorem answered it decades ago) but in demonstrated, scalable hardware. The uncertain machine took a decisive step from noisy adolescence toward fault-tolerant maturity.

In December 2024, Google announced a result that many in the field regard as more important than the quantum-supremacy claim that had made headlines five years earlier — a result that struck at the very heart of the fragility problem and offered the first convincing experimental proof that the long-promised solution actually works. The announcement centered on a new quantum chip named Willow, and its central achievement was to demonstrate, for the first time, error correction operating BELOW the threshold — the crossing of the exact line the threshold theorem had defined nearly thirty years before. To understand why this mattered so profoundly, we must recall the theory and see it, finally, made real.

The threshold theorem, proven in the late 1990s, had promised that IF the physical error rate of a quantum computer's components could be pushed below a certain critical value, THEN adding more qubits to an error-correcting code would REDUCE the logical error rate rather than increase it — and by scaling up the code, the errors could be suppressed as low as desired, enabling arbitrarily long reliable computation. This was the theoretical salvation of quantum computing, the answer to the pessimists who had thought decoherence insurmountable. But for decades it remained just that — a theorem, a promise on paper. In practice, hardware error rates had hovered near or above the threshold, and crucially, no one had yet demonstrated the theorem's signature prediction: that making the error-correcting code BIGGER actually makes the logical qubit BETTER. Until you could show that scaling up the code suppressed errors rather than amplifying them, error correction remained an unproven hope, and the whole edifice of fault-tolerant quantum computing rested on faith.

Willow provided exactly that demonstration, and it did so with beautiful directness. Google's team took their surface-code error correction and ran it at increasing sizes — encoding a logical qubit first in a small grid of physical qubits (a "distance-3" code, a 3-by-3 arrangement), then a larger one (distance-5, a 5-by-5 grid), then larger still (distance-7, a 7-by-7 grid) — and measured the logical error rate at each size. The result was the one the theorem had promised and the field had waited three decades to see: each time they made the code bigger, the logical error rate dropped, roughly halving with each increase in code size. This is the unmistakable signature of operating below threshold — the crossover point where scaling up the error correction wins, where more physical qubits mean a better logical qubit rather than a worse one. Google had crossed the line. The logical qubit, protected by the surface code, became more reliable as it grew larger, exactly as Shor, Kitaev, Preskill, and the other architects of error-correction theory had foretold. The promise on paper had become a fact in the laboratory.

The significance of this can hardly be overstated, and it is worth being precise about why it eclipses even the supremacy milestone in the eyes of many experts. Quantum supremacy proved that a quantum machine could enter a regime beyond classical simulation — impressive, but with noisy qubits whose errors would eventually swamp any long computation. Willow's below-threshold demonstration proved something more fundamental and more necessary: that the errors themselves can be systematically SUPPRESSED, that the path to arbitrarily reliable, arbitrarily long quantum computation is genuinely open. Without error correction working below threshold, quantum computing has no future beyond short, noisy demonstrations; with it, the road to the large fault-tolerant machines that could run Shor's algorithm or simulate complex molecules is, at last, demonstrably passable. Willow did not build that large machine — it protected only a single logical qubit — but it proved the principle on which every future machine depends. It converted the threshold theorem from a hopeful abstraction into an engineering reality.

The Willow chip carried a second, more familiar headline as well, and it is worth noting mainly to set it in proper perspective. The 105-qubit processor also performed a random-circuit-sampling benchmark — the same kind of supremacy task as Sycamore — that Google estimated would take a leading classical supercomputer an almost incomprehensible ten septillion years (a one followed by twenty-five zeros), a figure so vast it exceeds the age of the universe many times over. This grabbed attention, but it was, in truth, the less important of Willow's two results: another entry in the ongoing, contested supremacy arms race, subject as always to classical rebuttal and to the caveat that the task is useless. The below-threshold error-correction demonstration was the real breakthrough — quieter, less flashy, but far more consequential, because it addressed the fundamental obstacle rather than a benchmark. Willow's true achievement was not doing something useless very fast; it was proving that quantum errors can be beaten.

To appreciate why the Willow result struck so many experts as *more* important than the 2019 supremacy claim, it helps to remember what supremacy had and had not settled. Supremacy proved that a quantum machine could enter a regime beyond easy classical simulation — impressive, but with noisy qubits whose accumulating errors would eventually swamp any long computation. It said nothing about whether those errors could be *beaten*. And beating the errors was the true precondition for everything: without error correction working below threshold, quantum computing has no future beyond short, noisy demonstrations, no matter how many qubits are piled up. Willow addressed that deeper, more necessary question. It did not merely enter a hard-to-simulate regime; it proved that the errors themselves can be *systematically suppressed*, that the path to arbitrarily reliable, arbitrarily long computation is genuinely open. Supremacy showed the machine could sprint; Willow showed it could be taught not to stumble — and only the second lesson leads anywhere useful.

The directness of the demonstration is worth restating, because its very simplicity is what makes it so convincing. For nearly thirty years the threshold theorem had *promised* that below-threshold hardware would let a bigger error-correcting code yield a *better* logical qubit, but no one had shown the theorem's signature prediction in the flesh. Google's team simply ran their surface-code error correction at increasing sizes — a distance-3 code on a 3-by-3 grid of physical qubits, then a distance-5 code on a 5-by-5 grid, then a distance-7 code on a 7-by-7 grid — and measured the logical error rate at each size. The result was the one the theorem had foretold and the field had waited three decades to see: each time the code grew, the logical error rate *dropped*, roughly halving with each step up in size. That is the unmistakable signature of operating below threshold — the crossover where scaling up the error correction wins, where more physical qubits mean a better logical qubit rather than a worse one. The promise on paper had become a fact in the laboratory, and the architects of error-correction theory — Shor, Kitaev, Preskill, and the rest — were vindicated in the most concrete possible way.

It is important, amid the excitement, to keep Willow's two results in proper proportion, because the flashier one is the less important. The 105-qubit chip also performed a random-circuit-sampling benchmark — the same kind of supremacy task as Sycamore — that Google estimated would take a leading supercomputer some ten septillion years (a one followed by twenty-five zeros), a figure so vast it exceeds the age of the universe many times over. This grabbed headlines, but it was in truth the *less* important of Willow's two results: another entry in the ongoing, contested supremacy arms race, subject as always to classical rebuttal and to the caveat that the task is useless. The below-threshold error-correction demonstration was the real breakthrough — quieter, less flashy, but far more consequential, because it addressed the fundamental obstacle rather than a benchmark. Willow's true achievement was not doing something useless very fast; it was proving that quantum errors can be beaten. And with that proof, the entire industry reoriented around the race it inaugurated: the race to scale from one logical qubit to many, from a single protected qubit to the thousands that useful, error-corrected computation will require. That race, and the concept of the logical qubit at its center, is where the story of the present truly lives.

Chapter 62: Logical Qubits — the New Scoreboard

For most of quantum computing's public history, the scoreboard was simple and misleading: count the qubits. Each new processor was announced with its qubit number as the headline — 53, 127, 433, 1,121 — as though the race were a straightforward climb to ever-higher counts. This book has argued repeatedly that raw physical-qubit count is a poor measure of a machine's power, and by the middle years of the 2020s the field itself had come to agree, decisively shifting its attention to a new and far more meaningful scoreboard: the count of **logical qubits** — error- corrected, reliable qubits, each built from many physical ones — and the quality of the error correction protecting them. This shift, from physical to logical qubits as the measure of progress, marks the maturing of quantum computing from its noisy adolescence toward fault-tolerant adulthood, and understanding it is essential to reading the field's news in the fault-tolerant era now dawning.

Recall the distinction, so central it bears restating one final time. A **physical qubit** is an actual piece of hardware — a superconducting circuit, a trapped ion, a neutral atom — noisy and error-prone, surviving only microseconds to milliseconds before decoherence corrupts it. A **logical qubit** is a reliable, error-corrected qubit that many physical qubits are marshaled together to emulate, protected by continuous error correction so that it can, in principle, store quantum information indefinitely and support computations of arbitrary length. The transformative machines — the ones that will run Shor's algorithm, simulate complex molecules, and deliver on quantum computing's promises — will be measured not in physical qubits but in logical ones, because it is logical qubits that can actually perform long, reliable computations. A machine with a million physical qubits but no error correction is still a NISQ machine, powerful for narrow tasks but unable to run the great algorithms; a machine with even a few hundred good *logical* qubits would be a different order of being, capable of things no NISQ machine can touch. The passage from counting physical qubits to counting logical qubits is the passage across the great divide of the field.

The mid-2020s saw this new scoreboard come alive, with multiple platforms demonstrating genuine logical qubits — error-corrected qubits that outperform their physical constituents — for the first time. Google's Willow, as the previous chapter described, demonstrated a single logical qubit whose error rate fell as its code grew, crossing below threshold. But Google was not alone, and the competition on logical qubits became one of the most exciting races in the field. **Quantinuum**, with its high-fidelity trapped ions, demonstrated small numbers of logical qubits with impressively low error rates, exploiting the superior physical fidelity of ions to make error correction cheaper. And most strikingly, the **neutral-atom** platform surged to the forefront of the logical-qubit race: a collaboration involving Harvard, MIT, and the company QuEra demonstrated, in a widely noted 2023 result, dozens of logical qubits assembled from hundreds of physical neutral atoms, exploiting the reconfigurability of atom arrays to implement error-correcting codes with a flexibility the fixed architectures lacked. By late 2025, neutral-atom and trapped-ion systems were reporting logical-qubit counts in the dozens, marshaling hundreds of physical qubits into tens of error-corrected logical ones — the first stirrings of genuinely error-corrected quantum computing at meaningful scale. The scoreboard had changed, and the numbers on it — dozens of logical qubits, not thousands of physical ones — told a truer story of progress.

It is important to keep these logical-qubit milestones in honest perspective, neither dismissing nor inflating them, for they occupy a genuinely intermediate position. Demonstrating dozens of logical qubits is a real and important achievement, proof that error correction works across multiple platforms and that the logical-qubit era has begun. But *dozens* of logical qubits is still far short of the *thousands* that the great algorithms require: Shor's algorithm on a cryptographically relevant key needs several thousand logical qubits, each built from perhaps a thousand physical ones, implying millions of physical qubits — orders of magnitude beyond the hundreds that today's best logical-qubit demonstrations employ. The logical qubits demonstrated so far are also, in many cases, still of limited quality and limited number, and performing complex *computations* on them (as opposed to merely demonstrating their existence and error suppression) remains an emerging capability. The field has crossed a threshold — logical qubits exist, error correction works, the scoreboard has shifted — but it stands near the beginning of the fault-tolerant era, not its fulfillment. The distance from dozens of logical qubits to the thousands needed for the killer applications is vast, and crossing it will require both more physical qubits and, crucially, *more efficient* error correction that demands fewer physical qubits per logical one.

With error correction demonstrated below threshold, the field's central metric — the number that will ultimately determine when quantum computers become genuinely useful — shifted decisively from the PHYSICAL qubit counts that had dominated headlines to a far more meaningful figure: the number of LOGICAL qubits. Understanding the distinction between physical and logical qubits, and why the logical count is the one that truly matters, is perhaps the single most important thing for making sense of quantum computing's present and future, and for cutting through the qubit-counting hype that has misled the public for years.

A PHYSICAL qubit is an actual, individual piece of hardware — a single superconducting circuit, a single trapped ion, a single atom in a tweezer. These are the qubits that companies count in their headline announcements: "127 qubits," "1,000 qubits." But physical qubits, as we have seen throughout this book, are noisy and fragile, subject to decoherence and gate errors, unable on their own to sustain the long computations that useful algorithms require. A LOGICAL qubit, by contrast, is an error-CORRECTED qubit — a single, reliable, high-quality qubit constructed out of many physical qubits working together through an error-correcting code like the surface code, so that the logical qubit's information is protected and its effective error rate is driven far below that of any individual physical component. The logical qubit is the qubit that actually matters for computation, because it is the one reliable enough to run algorithms on. Physical qubits are the raw, noisy material; logical qubits are the refined, usable product.

The catch — the brutal, defining catch of the entire fault-tolerant era — is the overhead: it takes MANY physical qubits to make ONE logical qubit. With the surface code, depending on how good the physical qubits are and how low the logical error rate needs to be, a single logical qubit may require anywhere from dozens to over a thousand physical qubits. This is the "overhead" that shapes every roadmap in the industry, and it explains the enormous gap between the machines of today and the useful machines of the future. To run Shor's algorithm against a real 2,048-bit encryption key requires perhaps a few thousand logical qubits executing billions of error-corrected operations; multiply those few thousand logical qubits by the hundreds or thousands of physical qubits each requires, and you arrive at the sobering estimate that breaking RSA might demand a machine with a MILLION or more physical qubits — against the few hundred that today's largest machines possess. The gap is not a factor of two or ten; it is a factor of thousands, and it must be closed in every dimension at once: more qubits, better qubits, and the error correction to weld them into reliable logical units.

This reframing utterly transforms how one should read quantum computing news, and it is worth stating the corrected perspective bluntly. A machine with a thousand PHYSICAL qubits might, once error correction is switched on, yield only a handful of LOGICAL qubits — or, if its physical qubits are too noisy to be below threshold, none at all. So the headline physical-qubit counts that companies trumpet are, by themselves, nearly meaningless as measures of real capability; a machine with fewer but better physical qubits, capable of forming reliable logical qubits, is far more powerful than a machine with more but noisier ones. The sophisticated observer now asks, of any quantum computer: not how many physical qubits does it have, but how many LOGICAL qubits can it form, and at what error rate? That is the number that determines what the machine can actually do, and it is the number the industry has, belatedly and rightly, begun to emphasize.

The state of the art in logical qubits, as of the mid-2020s, is early but genuinely encouraging — the field has moved from zero logical qubits to a small but growing handful, a transition of real historic weight. Google's Willow demonstrated a single logical qubit protected below threshold, its error rate falling as the code grew. Quantinuum, using its high-quality trapped ions, has created multiple logical qubits with logical error rates well below their physical error rates, and has achieved among the best physical-to-logical encoding ratios of any platform, using efficient codes to make dozens of logical qubits from around a hundred physical ones. Neutral-atom teams, notably the Harvard-QuEra collaboration, have demonstrated dozens of logical qubits and entangling operations between them, exploiting the atoms' mobility to implement the necessary code structures. The numbers are still tiny — a handful to a few dozen logical qubits, against the thousands needed for the great applications — but the crucial transition has begun: the field has genuinely entered the logical-qubit era, moving from merely having noisy physical qubits to building reliable logical ones. The foundation of fault-tolerant computing is being laid, one logical qubit at a time.

The reason the field's central metric shifted, after Willow, from physical to logical qubits deserves to be made vivid, because it marks a genuine turning point in how progress is measured. For years, companies had trumpeted physical-qubit counts — 127, 433, over a thousand — and the public had learned, wrongly, to read these as measures of power. But a physical qubit is merely raw material: noisy, fragile, unable on its own to sustain the long computations useful algorithms require. What matters for computation is the *logical* qubit — an error-corrected qubit built from many physical ones, its information protected and its effective error rate driven far below that of any individual component. Once error correction was demonstrated to work below threshold, the meaningful question was no longer "how many physical qubits?" but "how many reliable logical qubits can you build, and at what error rate?" The physical count is the raw ore; the logical count is the refined metal, and only the refined metal can be forged into a working machine.

The brutal arithmetic of overhead is worth confronting head-on, because it explains the enormous gap between today's machines and the useful machines of the future. It takes *many* physical qubits to make *one* logical qubit — with the surface code, anywhere from dozens to over a thousand, depending on the hardware quality and the error rate required. This overhead shapes every roadmap in the industry. To run Shor's algorithm against a real 2,048-bit encryption key requires perhaps a few thousand logical qubits executing billions of error-corrected operations; multiply those thousands of logical qubits by the hundreds or thousands of physical qubits each demands, and you arrive at the sobering estimate that breaking RSA might need a machine with a *million or more* physical qubits — against the few hundred that today's largest machines possess. The gap is not a factor of two or ten but a factor of thousands, and it must be closed in every dimension at once: more qubits, better qubits, and the error correction to weld them into reliable logical units. This is why a machine that can factor real encryption keys might one day be advertised as having a million qubits while possessing only a few thousand logical ones.

The state of the art, as of the mid-2020s, is early but genuinely historic, and it is worth naming the milestones because the transition they mark — from zero logical qubits to a growing handful — is the true beginning of the fault-tolerant era. Google's Willow demonstrated a single logical qubit protected below threshold, its error rate falling as the code grew. Quantinuum, using its pristine trapped ions, has created multiple logical qubits with error rates well below their physical rates and achieved among the best physical-to-logical encoding ratios of any platform, making dozens of logical qubits from around a hundred physical ones using efficient codes. Neutral-atom teams, notably the Harvard-QuEra collaboration, have demonstrated dozens of logical qubits and entangling operations between them, exploiting the atoms' mobility to implement the necessary code structures. The numbers remain tiny — a handful to a few dozen logical qubits against the thousands the great applications require — but the crucial transition has begun: the field has genuinely entered the logical-qubit era, laying the foundation of fault-tolerant computing one logical qubit at a time. The challenge now is to scale that handful into hundreds and thousands while keeping the error rates low — the defining engineering problem of the coming years, and the one where the surface code's brutal overhead bites hardest, driving the search for the more efficient codes that could bring the whole fault-tolerant future dramatically closer.

Chapter 63: qLDPC and the Low-Overhead Future

The single greatest obstacle standing between the quantum computers of the mid-2020s and the useful, fault-tolerant machines the field aspires to is not qubit count as such, but *overhead* — the enormous number of physical qubits that error correction demands for each reliable logical qubit. The surface code, the workhorse of the error- correction era, is voracious: at realistic error rates it consumes a thousand or more physical qubits to protect a single logical qubit robustly. Multiply that by the thousands of logical qubits the great algorithms require, and you arrive at the daunting figure of millions of physical qubits — a target so distant that it makes useful fault- tolerant computing seem decades away. Much of the field's most important recent research attacks this overhead directly, seeking error-correcting codes that protect logical qubits with far fewer physical ones, and the leading candidates go by an unlovely acronym — **qLDPC**, for quantum low-density parity-check codes — that may nonetheless name the key to the fault-tolerant future.

To appreciate why overhead is so central, consider the arithmetic once more, now as an engineering constraint rather than an abstraction. If a useful application needs, say, two thousand logical qubits, and each logical qubit under the surface code needs a thousand physical qubits, the machine needs two million physical qubits — an almost unimaginable number given that the best machines of the mid-2020s have hundreds to about a thousand. But if a better code could protect a logical qubit with only, say, a hundred physical qubits instead of a thousand — a tenfold reduction in overhead — then the same application would need only two hundred thousand physical qubits, still enormous but ten times closer, potentially pulling the fault-tolerant future years or even decades nearer. Overhead is the multiplier that turns thousands of logical qubits into millions of physical ones, and shrinking it is the most powerful lever available for accelerating the arrival of useful quantum computing. A breakthrough in code efficiency is worth as much as a breakthrough in hardware.

The **qLDPC codes** are the most promising route to lower overhead, and they represent a genuine theoretical advance of the 2020s. The surface code, recall, has a great practical virtue — it requires only local, nearest-neighbor interactions on a two-dimensional grid — but this locality is precisely what makes it expensive, forcing a large patch of qubits for each logical qubit. The qLDPC codes achieve dramatically better efficiency by relaxing the strict two-dimensional locality, allowing each check to involve qubits that are not all immediate neighbors, which permits codes that pack far more logical qubits into a given number of physical qubits while maintaining strong error protection. Theoretical work in the early 2020s proved the existence of "good" qLDPC codes with asymptotically optimal properties — a landmark in coding theory — and showed that they could, in principle, reduce the overhead of fault tolerance by an order of magnitude or more compared with the surface code. The catch is that their non-local checks are harder to implement in hardware than the surface code's tidy neighbor-only pattern, requiring qubits that are not physically adjacent to interact, which is challenging for solid-state chips but more natural for architectures with flexible connectivity, such as trapped ions, neutral atoms, or superconducting chips with long-range couplers.

IBM, in particular, made qLDPC codes central to its

roadmap for fault-tolerant computing, betting that they are the key to reaching useful machines with manageable qubit counts. In the mid-2020s IBM announced specific qLDPC codes (sometimes called "bicycle" codes) that it claimed could reduce the overhead of error correction roughly tenfold relative to the surface code, and unveiled processors — with names like Loon — designed to demonstrate the long- range connectivity that qLDPC codes require, including the fast, real-time decoding of error syndromes that fault tolerance demands. Other groups pursued qLDPC codes on neutral-atom and trapped-ion hardware, whose flexible connectivity suits the codes' non-local structure. By late 2025, the demonstration of qLDPC-based error correction, and of the real-time decoding needed to make it work, had become one of the hottest frontiers in the field — the race to prove that low-overhead codes could work in practice, not just on paper, and so bring the millions-of-qubits target down toward something achievable. IBM framed its roadmap around delivering a large-scale fault-tolerant machine — it named a target system, "Starling," for the end of the decade — built on qLDPC error correction, an ambitious bet that the overhead problem can be tamed.

The single greatest obstacle standing between today's small quantum machines and the large fault-tolerant computers that could transform the world is the crushing overhead of error correction — the hundreds or thousands of physical qubits required to build each reliable logical qubit under the surface code. If that overhead could be dramatically reduced, the entire timeline to useful quantum computing would shorten, and the million-physical-qubit machines that seem so distant might shrink to something far more attainable. In recent years, a new generation of error-correcting codes has emerged that promises exactly this reduction, and it has generated enormous excitement as perhaps the most important theoretical advance in the field since Kitaev's surface code itself. These are the quantum low-density parity-check codes, or qLDPC codes.

To appreciate why qLDPC codes matter, recall the surface code's central trade-off. Its great virtue, which made it the workhorse of the industry, is that it requires only NEAREST-NEIGHBOR interactions — each physical qubit need only talk to its immediate neighbors on a two-dimensional grid, which is exactly what solid-state chips can naturally provide. But this locality comes at a steep price: the surface code is profligate with qubits, needing a large two-dimensional patch of physical qubits to protect just one logical qubit, and the number grows rapidly as you demand lower error rates. The surface code buys practicality with extravagance, spending qubits lavishly to keep the wiring simple. For a machine needing thousands of logical qubits, this extravagance balloons into the millions of physical qubits that make useful quantum computing seem so remote.

The qLDPC codes offer a fundamentally better bargain: they can protect the same number of logical qubits using DRAMATICALLY fewer physical qubits — potentially ten times fewer, or even better — a reduction in overhead that could transform the practical prospects of the entire field. The "low-density parity-check" name, borrowed from a celebrated family of classical error-correcting codes used in everything from Wi-Fi to deep-space communication, refers to the sparse, efficient structure of the error-checking measurements. Theorists proved that certain qLDPC codes achieve near-optimal efficiency, packing far more protected logical information into a given number of physical qubits than the surface code can. On paper, qLDPC codes shatter the overhead barrier that has made fault-tolerant quantum computing look so forbiddingly expensive. If the surface code's overhead put useful machines decades away, qLDPC codes might pull that future substantially closer.

There is, inevitably, a catch, and it goes to the heart of why the surface code was chosen in the first place. The efficiency of qLDPC codes comes from allowing error-checking measurements that involve qubits which are NOT nearest neighbors — qubits scattered across the chip must be connected and made to interact, requiring long-range connectivity that flat, two-dimensional, nearest-neighbor hardware cannot easily provide. This is precisely the connectivity the surface code was designed to avoid, because building long-range connections between distant qubits is hard. So qLDPC codes trade the surface code's qubit extravagance for a demand for richer connectivity, and whether that trade is worthwhile depends entirely on whether the hardware can deliver the required long-range links. Different platforms face this challenge differently: neutral atoms, which can physically MOVE qubits around to bring distant ones together, and trapped ions, with their natural all-to-all connectivity, may be better suited to qLDPC codes than fixed superconducting chips — though superconducting architectures are exploring clever ways, such as long-range couplers and modular designs, to provide the needed connections. The choice of code and the choice of hardware are deeply intertwined.

The excitement around qLDPC codes has been intense precisely because the potential payoff is so large, and major players have embraced them. IBM, notably, restructured its long-term roadmap around qLDPC codes, announcing plans for a fault-tolerant machine — dubbed Starling and targeted for around 2029 — that would use these efficient codes to build roughly 200 logical qubits capable of executing on the order of a hundred million error-corrected operations, a scale that would begin to enable genuinely useful fault-tolerant computation, with a path beyond to still larger modular machines. That a company would rebuild its entire roadmap around qLDPC codes signals how transformative the reduced overhead is believed to be. Other groups across academia and industry are racing to develop qLDPC codes tailored to their hardware and to demonstrate them experimentally, and early experimental steps have begun to appear.

The strategic importance of reducing error-correction overhead is worth stating starkly, because it is the lever that could most dramatically shorten the timeline to useful quantum computing. The single greatest obstacle between today's small machines and the large fault-tolerant computers that could transform the world is the crushing overhead of the surface code — the hundreds or thousands of physical qubits required for each reliable logical qubit, which balloons into the millions of physical qubits a useful machine seems to demand. If that overhead could be slashed — if logical qubits could be built from *far fewer* physical qubits — the entire enterprise would come closer at a stroke, the million-qubit machines shrinking toward something far more attainable. This is precisely what the new quantum low-density parity-check codes, the qLDPC codes, promise: to protect the same number of logical qubits using perhaps ten times fewer physical qubits, or better — a reduction so large it could, in effect, do the work of a decade of hardware improvement through mathematics alone.

The trade-off at the heart of qLDPC codes is worth understanding, because it explains both their power and why they were not simply adopted long ago. The surface code's great virtue is that it requires only *nearest-neighbor* interactions — each physical qubit need only talk to its immediate neighbors on a two-dimensional grid, exactly what solid-state chips naturally provide. But this locality is bought with extravagance: the surface code is profligate with qubits. The qLDPC codes achieve their efficiency by allowing error-checking measurements that involve qubits which are *not* nearest neighbors — qubits scattered across the chip must be connected and made to interact, requiring the very long-range connectivity that flat, two-dimensional hardware struggles to provide. So qLDPC codes trade the surface code's qubit extravagance for a demand for richer connectivity, and whether that trade is worthwhile depends entirely on whether the hardware can deliver the required long-range links. Different platforms face this differently: neutral atoms, which can physically move qubits to bring distant ones together, and trapped ions, with their natural all-to-all connectivity, may suit qLDPC codes better than fixed superconducting chips — though superconducting architectures are exploring long-range couplers and modular designs to provide the needed connections. Code and hardware are, as ever, deeply intertwined.

The seriousness with which the industry has embraced qLDPC codes is the clearest signal of their transformative potential, and IBM's reaction is the most striking example. IBM restructured its entire long-term roadmap around these efficient codes, announcing plans for a fault-tolerant machine — dubbed Starling and targeted for around 2029 — that would use qLDPC codes to build roughly 200 logical qubits capable of executing on the order of a hundred million error-corrected operations, a scale that would begin to enable genuinely useful fault-tolerant computation, with a path beyond to still larger modular machines. That a company would rebuild its whole roadmap around a new class of code signals how transformative the reduced overhead is believed to be, and other groups across academia and industry are racing to develop qLDPC codes tailored to their hardware and to demonstrate them experimentally. The advent of these codes is a reminder that the road to useful quantum computing runs not only through better hardware but through better *theory* — that a clever new code can do the work of a tenfold hardware improvement — and it exemplifies the deep, productive interplay, present throughout this history, between the theorists who devise the codes and the experimentalists who build the machines. Whether qLDPC codes fully deliver on their promise remains to be seen, but they have injected fresh optimism into the timeline and reshaped the roadmaps of the industry's leaders, who are now locked in a race whose stakes reach far beyond science.

Chapter 64: The Race to Advantage — Nighthawk, Neutral Atoms, and 2026

As this book is written, in the middle years of the 2020s, quantum computing stands at a moment of unusual intensity and unusual promise. The below-threshold error-correction demonstration of late 2024 proved that the field's deepest obstacle could be overcome; the shift to counting logical qubits marked its maturing; the qLDPC codes promised a path to manageable overhead. And into this moment the major players poured their most advanced machines and their boldest predictions, racing toward two intertwined goals: the first demonstration of genuine *quantum advantage on a useful problem* — the strong-sense milestone that, as Part IX stressed, has never yet been clearly achieved — and, beyond it, the first *fault-tolerant* machines capable of running the great algorithms. This chapter surveys that race at its current frontier, with the strong caveat that in a field moving this fast, any snapshot is provisional and the specific claims dated.

The superconducting giants pressed forward with new processors aimed squarely at demonstrated advantage. IBM, continuing its detailed public roadmap, unveiled in late 2025 a processor it called **Nighthawk** — reported to have around 120 qubits connected by a new generation of tunable couplers designed to reduce the crosstalk errors that plague dense superconducting chips, and positioned by IBM as the machine most likely to demonstrate a verifiable quantum advantage on a useful problem within 2026. IBM framed the near-term goal explicitly as *verified quantum advantage* — not another contrived supremacy demonstration, but a genuine, checkable computational advantage on a problem of real interest — and set 2026 as a target year for achieving it, while pointing beyond, via its qLDPC-based error-correction roadmap, toward a large-scale fault-tolerant system (the "Starling" machine) later in the decade. IBM also reported processors, such as **Loon**, built to validate the long-range connectivity and fast real-time decoding that its fault-tolerant architecture requires — claiming, for instance, real-time decoding of error syndromes in well under a microsecond, a crucial capability for error correction that must keep pace with errors as they occur. Google, for its part, continued to build on Willow, pursuing both larger error-corrected systems and, in 2025, a new claim it called **Quantum Echoes** — an algorithm run on Willow that Google described as the first *verifiable* quantum advantage, computing a physical quantity (an out-of-time-order correlator, a probe of how quantum information scrambles) reportedly some thirteen thousand times faster than a leading classical supercomputer, and, crucially, verifiable in a way earlier supremacy demonstrations were not.

The neutral-atom and trapped-ion camps, meanwhile, pressed their own advantages in the logical-qubit race. Neutral- atom systems from QuEra, Pasqal, and Atom Computing, exploiting the reconfigurability and scalability of atom arrays, reported some of the largest logical-qubit counts — QuEra, for instance, announced results marshaling hundreds of physical atoms into dozens of logical qubits, closing in on the numbers where error-corrected computation becomes genuinely useful. Quantinuum and IonQ, with their high-fidelity trapped ions, continued to post leading fidelities and strong logical-qubit results, arguing that quality would carry them to useful fault tolerance efficiently. The result, by late 2025, was a genuine multi-platform race, with superconducting, neutral-atom, and trapped-ion systems each claiming leadership on different metrics — raw scale, logical-qubit count, fidelity, verifiable advantage — and each with a credible path forward. No single platform had pulled decisively ahead; the winnowing that usually marks a maturing technology had not yet come.

A necessary word of caution frames all of these claims, and this book has earned the right to insist on it. Every one of the milestones just described is, at the time of writing, a *recent, fast-moving, often vendor-originated claim*, and the field's history — the eroded supremacy figures, the dequantized machine-learning speedups, the contested utility demonstrations, the retracted Majorana papers — counsels strongly against taking any such claim at face value before it has weathered independent scrutiny and the assault of the best classical methods. "Verifiable quantum advantage" is a meaningful phrase precisely because so many earlier claims were *not* verifiable and did not survive; whether Google's Quantum Echoes, IBM's Nighthawk ambitions, and the various logical-qubit records hold up as their promoters suggest will be determined not by press releases but by the slow, adversarial process of scientific confirmation that this book has repeatedly shown to be the field's true arbiter. The reader encountering quantum-computing news in the years after this book is written should apply the same discipline: ask whether the task is useful or contrived, whether the classical competition has genuinely been bested, whether the qubits are physical or logical, and whether the claim has survived independent replication. These questions separate real progress from hype, and they are the enduring tools this book has tried to place in the reader's hands.

Quantum computing in the mid-2020s is no longer a quiet academic pursuit but a global race involving some of the world's largest corporations, best-funded startups, and most powerful governments, all pouring extraordinary resources into the effort to build the first genuinely useful quantum computer. Understanding the shape of this race — who the players are, what they are betting on, and what "winning" would even mean — is essential to grasping the present moment and to reading the flood of quantum news with a clear eye. It is a race with many runners, no clear leader, and a finish line that keeps being redefined.

The corporate landscape is a study in contrasting bets. The technology giants lead in resources and visibility: Google, with its Sycamore and Willow superconducting chips and its landmark supremacy and below-threshold demonstrations; IBM, with its aggressive public roadmap, its bird-named processors, and its pivot to qLDPC codes and a fault-tolerant target around 2029; Microsoft, with its long, patient, high-risk wager on topological qubits and its role as a cloud provider of others' quantum hardware; and Amazon, which offers quantum machines through its cloud and pursues its own hardware research. Alongside the giants stand focused, well-funded startups, each embodying a different technological bet: IonQ and Quantinuum on trapped ions, prizing qubit quality; QuEra, Pasqal, and Atom Computing on neutral atoms, riding that platform's recent surge; PsiQuantum and Xanadu on photonics, wagering on manufacturability; and a host of others pursuing silicon spins, novel architectures, and specialized approaches. No single technology has won; the race remains genuinely open, and the diversity of bets is itself a sign of a field still finding its way.

What would it mean to WIN this race? Here the honest answer requires the distinctions this book has labored to draw, because "winning" has several possible meanings, achieved at very different times. Quantum SUPREMACY — doing something useless that classical computers cannot — has already been achieved, repeatedly, and is behind us. Quantum ADVANTAGE on a USEFUL problem — solving some valuable real-world task better or faster than any classical computer — has NOT been achieved, and is the next great milestone, the one that would prove quantum computers earn their keep. Beyond that lies FAULT-TOLERANT quantum computing at scale — machines with thousands or millions of logical qubits able to run Shor's algorithm and simulate complex molecules — the ultimate prize, and the one still likely a decade or more away. When a company claims to be "leading" or "winning," it is essential to ask which of these very different finish lines they mean, because conflating them is the source of much confusion and hype.

The claims of near-term useful advantage deserve especially sharp scrutiny, because the commercial pressure to declare victory is intense and the temptation to overstate is correspondingly great. Companies have made various claims of quantum advantage on specific problems — a widely noted example being a collaboration in which a quantum machine was applied to a medical-device or materials simulation and reported to outperform a particular classical approach by some margin. Such claims should be examined with the same skepticism the whole history of the field teaches: Was the comparison against the VERY BEST classical algorithm, or merely a convenient one? Was the problem genuinely useful, or contrived to favor the quantum machine? Has the result been independently reproduced and survived classical rebuttal? More often than not, early advantage claims have been narrowed or overturned by improved classical methods, following the pattern established by Sycamore and repeated ever since. Genuine, durable, useful quantum advantage — the kind that survives the full assault of classical competition — remains, as of the mid-2020s, unproven, though the field inches toward it.

The investment pouring into this race is staggering and worth registering, because it shapes the field's trajectory and its incentives. Governments have committed enormous sums — the United States through its National Quantum Initiative, the European Union through its Quantum Flagship program, and China through investments widely reported to exceed those of any other nation, building vast quantum research facilities. Private venture capital has flowed into quantum startups by the billions, and the technology giants spend heavily from their own vast resources. This flood of money has accelerated progress and built a broad ecosystem of talent, tools, and infrastructure — but it also creates powerful incentives to overstate progress, to claim milestones prematurely, and to sustain the hype that keeps funding flowing, which is precisely why the informed skepticism this book cultivates is so valuable. Money accelerates science, but it also inflates claims.

The most important thing to understand about the race to build a useful quantum computer is that "winning" means different things achieved at very different times — and conflating them is the single greatest source of hype and confusion, so it is worth pinning down. Quantum *supremacy* — doing something useless no classical computer can match — has already been achieved, repeatedly, and is behind us. Quantum *advantage* on a *useful* problem — solving some valuable real-world task better or faster than any classical computer — has *not* been achieved, and is the next great milestone, the one that would prove quantum computers earn their keep. Beyond that lies *fault-tolerant* quantum computing at scale — machines with thousands or millions of logical qubits able to run Shor's algorithm and simulate complex molecules — the ultimate prize, still likely a decade or more away. When a company claims to be "leading" or "winning," always ask which of these very different finish lines it means, because the word "winning" is doing an enormous amount of quiet, misleading work.

The claims of near-term useful advantage deserve especially sharp scrutiny, because the commercial pressure to declare victory is intense and the temptation to overstate is correspondingly great. Companies have made various claims of quantum advantage on specific problems — a widely noted example being a collaboration in which a quantum machine was applied to a materials or medical-device simulation and reported to outperform a particular classical approach by some margin. Such claims should be examined with the skepticism this whole history teaches: Was the comparison against the *very best* classical algorithm, or merely a convenient one? Was the problem genuinely useful, or contrived to favor the quantum machine? Has the result been independently reproduced and survived classical rebuttal? More often than not, early advantage claims have been narrowed or overturned by improved classical methods, following the pattern established by Sycamore and repeated ever since. Genuine, durable, useful quantum advantage — the kind that survives the full assault of classical competition — remains, as of the mid-2020s, unproven, even as the field inches toward it.

The flood of investment shaping this race is worth registering, because it both accelerates progress and distorts incentives. Governments have committed enormous sums — the United States through its National Quantum Initiative, the European Union through its Quantum Flagship, and China through investments widely reported to exceed those of any other nation. Private venture capital has flowed into quantum startups by the billions, and the technology giants spend heavily from their own vast resources. This money has accelerated progress and built a broad ecosystem of talent, tools, and infrastructure — but it also creates powerful incentives to overstate progress, to claim milestones prematurely, and to sustain the hype that keeps funding flowing, which is precisely why the informed skepticism this book cultivates is so valuable. The race to advantage is real, intense, global, and consequential, but it is a marathon, not a sprint, with the most important finish lines still ahead and frequently misrepresented. Holding both truths at once — remarkable genuine progress and vast genuine distance remaining — is the mark of a clear-eyed understanding. And because this race is not merely commercial but strategic, entangled with national security and great-power competition, it has drawn in the governments of the world's most powerful nations, transforming the quest for the quantum computer into an arena of geopolitics.

Part XIII — Stakes and Horizons

Chapter 65: The Geopolitics of Qubits

Quantum computing is not only a scientific enterprise and a commercial one; it is, increasingly, a geopolitical one — a technology that governments regard as strategically vital, worthy of national investment, export controls, and the kind of great-power competition once reserved for nuclear weapons and space rockets. The reasons are not hard to find, and they run through everything this book has described: a nation that first builds a cryptographically relevant quantum computer gains the ability to break its rivals' encryption while (with post-quantum defenses) protecting its own; a nation that leads in quantum simulation gains advantages in designing materials, drugs, and weapons; and the prestige and economic spillovers of leading a foundational technology are considerable. Quantum computing has thus become a field where laboratories and companies operate against a backdrop of national strategy, and no account of its stakes is complete without reckoning with the race among nations.

The cryptographic dimension is the sharpest edge of the geopolitical competition, and it explains much of the secrecy and urgency surrounding the field. As Part X established, a large fault-tolerant quantum computer running Shor's algorithm could break the public-key cryptography protecting the world's communications, and the harvest-now-decrypt-later strategy means that intelligence agencies are believed to be already recording encrypted data against the day such a machine exists. This creates a powerful strategic incentive for nations to be *first* — to reach the code-breaking capability before rivals do, gaining a potentially decisive and possibly temporary intelligence advantage — and simultaneously to defend, by migrating their own systems to post-quantum cryptography before rivals' quantum computers come online. The quantum race is, in significant part, a cryptographic arms race, conducted partly in the open among companies and partly in the classified programs of intelligence and defense agencies, and it lends the whole field an undertone of strategic competition that pure science would not carry.

The principal competitors are the United States and China, with important efforts also in Europe, the United Kingdom, Canada, Australia, Japan, India, Israel, and elsewhere. The **United States** hosts the leading corporate efforts — Google, IBM, Microsoft, and a vibrant ecosystem of startups (IonQ, Rigetti, Quantinuum's US operations, QuEra, Atom Computing, PsiQuantum) — backed by substantial federal investment channeled through the **National Quantum Initiative**, enacted in 2018 to coordinate and fund quantum research across agencies and national laboratories. **China** has mounted an effort of comparable ambition and, in some areas, clear leadership: it has invested enormous state resources in quantum technology, achieved landmark results in quantum communication (the Micius satellite, the world's first space-based quantum key distribution, and the longest terrestrial QKD networks), and demonstrated quantum supremacy by both photonic (Jiuzhang) and superconducting routes under the leadership of Jian-Wei Pan and others. The scale of China's investment and its progress, particularly in quantum communication, prompted alarm in Washington and helped drive the West's own initiatives, in a dynamic reminiscent of earlier technological competitions. **Europe** launched its billion-euro **Quantum Flagship** program in 2018 and hosts strong efforts in the Netherlands (QuTech, spin qubits), France (Pasqal, neutral atoms), Germany, Finland (IQM), and elsewhere; the United Kingdom, Canada (a longtime quantum-research power, home to D-Wave, Xanadu, and major academic centers), Australia (silicon spin qubits), Japan, India, and Israel all sustain significant national programs.

The competition has hardened into the familiar instruments of great-power technological rivalry. Nations have declared quantum technology a strategic priority in official policy; substantial public funding flows to research and to national quantum initiatives; and, increasingly, **export controls** and investment restrictions have been imposed on quantum-related technologies, as governments seek to protect their leads and deny rivals access to key hardware and expertise — the United States, for instance, has moved to restrict exports of certain quantum technologies, echoing controls on semiconductors and other strategic goods. The 2025 designation of an International Year of Quantum Science and Technology, marking the centenary of quantum mechanics, celebrated the field's scientific universality even as the underlying technology grew more entangled with national competition. There is a tension here worth noting: quantum science is a global, collaborative, open enterprise, built over a century by scientists of every nation sharing results freely, and yet quantum *technology* is increasingly treated as a strategic asset to be guarded and raced for — the same tension that has long attended other dual-use sciences.

Quantum computing has become a matter of national security and great-power competition, drawing in the governments of the world's most powerful nations and transforming what was once an obscure scientific pursuit into an arena of geopolitical rivalry with stakes measured in the security of nations. Understanding this dimension is essential to grasping why quantum computing attracts such enormous government investment and attention, and why the quest for the quantum computer is increasingly framed in the language of strategic competition, national advantage, and even a "quantum arms race."

The security stakes flow directly from Shor's algorithm and the specter of Q-Day. A nation that built a large, fault-tolerant quantum computer capable of running Shor's algorithm before its rivals could, in principle, break the encryption protecting its adversaries' military communications, intelligence, financial systems, and state secrets — while its own secrets, if protected by post-quantum cryptography, remained secure. This asymmetry would confer an intelligence advantage of extraordinary magnitude, a capability some have compared in strategic significance to the breaking of the Enigma cipher in the Second World War, or even to the development of nuclear weapons. And because of the "harvest now, decrypt later" threat, the mere PROSPECT of such a machine already shapes national security policy today: intelligence agencies harvest encrypted data in anticipation, and governments race to migrate their systems to quantum-resistant cryptography before an adversary's machine can crack what has been stored. The quantum computer is, in the eyes of national-security planners, a potential master key to the secrets of the world.

The competition is dominated, like so much of twenty-first-century technological rivalry, by the United States and China, and the contest between them has been explicit and intensifying. China has made quantum technology a national priority, reportedly investing sums exceeding those of any other nation, constructing enormous dedicated research facilities, and achieving world-class results — Jian-Wei Pan's group producing the Jiuzhang photonic supremacy demonstration, the Micius quantum-communication satellite, and the world's most extensive quantum-communication networks. These achievements demonstrated that China was not merely catching up but leading in important areas, particularly quantum communication, and they galvanized concern in Washington. The United States responded with its National Quantum Initiative, substantial funding, and — significantly — export controls and restrictions aimed at limiting China's access to the advanced technology and expertise needed to build quantum computers, treating quantum capability as a strategic asset to be guarded much like advanced semiconductors. The two superpowers now eye each other's quantum progress warily, each determined not to fall behind in a technology that could reshape the balance of intelligence and security.

Other nations and blocs are far from absent, and the global map of quantum ambition is broad. The European Union launched its Quantum Flagship, a major coordinated research program, and individual European nations — Germany, France, the Netherlands (home to pioneering work at Delft), and others — host significant quantum efforts and companies. The United Kingdom established a National Quantum Technologies Programme early and maintains strong academic and commercial activity. Canada punches well above its weight, home to important research and to companies like D-Wave and Xanadu. Japan, Australia (a leader in silicon spin qubits), India, and others have national quantum initiatives. Quantum technology has become a marker of national technological prowess, a field in which serious nations feel they must participate or risk strategic and economic marginalization. The result is a genuinely global effort, though one increasingly bifurcated by the geopolitical fault line between the United States and China and their respective partners.

It is worth injecting a note of perspective, however, lest the "quantum arms race" framing be taken too breathlessly, for it carries its own distortions. The analogy to a nuclear arms race is imperfect and can mislead: a useful, cryptographically relevant quantum computer does not yet exist and may be a decade or more away, if it arrives on the anticipated timeline at all; the primary defense against the quantum threat — post-quantum cryptography — is a mathematical solution deployable now on ordinary computers, not itself a quantum technology, so the "race" to defend is not the same as the race to build; and much of quantum computing research is open, international, and collaborative, conducted in the published scientific literature rather than in secret weapons laboratories. The competition is real and consequential, but it is not a simple sprint toward a doomsday device, and the more lurid framings serve hype and budget requests as much as they reflect reality. Sober strategists focus less on winning a race to a superweapon than on the prudent, unglamorous work of securing systems against an uncertain future threat.

The precise nature of the strategic stakes is worth spelling out, because it explains why governments treat quantum computing not merely as a promising technology but as a matter of national security. The stakes flow directly from Shor's algorithm. A nation that built a large, fault-tolerant quantum computer before its rivals could, in principle, break the encryption protecting its adversaries' military communications, intelligence, financial systems, and state secrets — while its own secrets, if protected by post-quantum cryptography, remained secure. This asymmetry would confer an intelligence advantage of extraordinary magnitude, comparable in strategic significance, some have argued, to the breaking of the Enigma cipher in the Second World War. And because of the harvest-now-decrypt-later threat, the mere *prospect* of such a machine already shapes national-security policy today: intelligence agencies harvest encrypted data in anticipation, and governments race to migrate their systems to quantum-resistant cryptography before an adversary's machine can crack what has been stored. The quantum computer is, to national-security planners, a potential master key to the secrets of the world — and that is reason enough for states to pour resources into building it and defending against it.

The bipolar structure of the competition, dominated by the

United States and China, deserves a clear-eyed description, because it increasingly frames how the whole field is discussed. China has made quantum technology a national priority, reportedly investing sums exceeding those of any other nation, building enormous dedicated research facilities, and achieving world-class results — Jian-Wei Pan's group producing the Jiuzhang photonic supremacy demonstration, the Micius quantum-communication satellite, and the world's most extensive quantum-secured networks, particularly in the domain of quantum communication where China leads. The United States responded with its National Quantum Initiative, substantial funding, and — significantly — export controls aimed at limiting China's access to the advanced technology and expertise needed to build quantum computers, treating quantum capability as a strategic asset to be guarded much like advanced semiconductors. Europe (through the Quantum Flagship and strong national efforts in Germany, France, the Netherlands, and the United Kingdom), Canada, Japan, Australia, and India all maintain significant programs, so the effort is genuinely global — though increasingly bifurcated along the geopolitical fault line between the United States and China and their respective partners.

A note of perspective is essential here, lest the "quantum arms race" framing be swallowed too breathlessly, for the analogy carries its own distortions. The comparison to a nuclear arms race is imperfect and can mislead in several ways. A useful, cryptographically relevant quantum computer does not yet exist and may be a decade or more away, if it arrives on the anticipated timeline at all. The primary defense against the quantum threat — post-quantum cryptography — is a mathematical solution deployable now on ordinary computers, not itself a quantum technology, so the "race" to defend is not the same as the race to build. And much of quantum computing research is open, international, and collaborative, conducted in the published scientific literature rather than in secret weapons laboratories. The competition is real and consequential, but it is not a simple sprint toward a doomsday device, and the more lurid framings serve hype and budget requests as much as they reflect reality. Sober strategists focus less on winning a race to a superweapon than on the prudent, unglamorous work of securing systems against an uncertain future threat. The geopolitics of qubits adds a final dimension to why quantum computing matters and commands such resources — a domain of national power and strategic rivalry entangled with the great questions of twenty-first-century geopolitics — but it should be understood soberly, not sensationally. Having surveyed the science, the machines, the applications, the threats, the defenses, the players, and the politics, we are ready at last to step back and take the long view.

Chapter 66: Hype, Winter, and the Long View

We come, at last, to the question that hangs over everything this book has described: what is quantum computing actually going to mean for the world, and when? It is the question the reader has surely been asking throughout, and it is the question the field's promoters answer with confident visions of transformation and its detractors answer with weary dismissal. The honest answer is more difficult and more interesting than either, and giving it fairly — holding wonder and skepticism in balance, as this book has tried to do from its first page — is the proper way to close. Quantum computing is real, profound, and advancing; it is also unfinished, oversold, and genuinely uncertain in its destination. Both halves of that judgment are true, and wisdom about the field lies in refusing to collapse it into either the hype or the cynicism.

Consider first the case for genuine optimism, which is strong and grounded. Quantum computing rests on sound physics and rigorous mathematics; the theory is not in doubt. The machines have advanced enormously — from a two-qubit factoring of fifteen to processors of over a thousand qubits, from decoherence times of nanoseconds to nearly milliseconds, from noisy chaos to below-threshold error correction demonstrated in hardware. The deepest obstacle, the fragility that nearly killed the field in the 1990s, has been overcome in principle by error correction and, in late 2024, in demonstrated practice. Logical qubits are multiplying; low-overhead codes promise to tame the overhead; multiple well-funded platforms race forward; and the applications, especially quantum simulation of chemistry and materials, rest on the machine's deepest strength and could deliver real value. A reasonable optimist, surveying the trajectory, may well believe that fault-tolerant machines capable of useful work will arrive within a decade or two, and that when they do, they will transform parts of chemistry, materials science, and cryptography, and open possibilities we cannot yet foresee — much as the classical computer, once built, transformed the world far beyond what its inventors imagined. This is not fantasy; it is a sober extrapolation of real progress.

Now the case for caution, which is equally grounded and equally necessary. Quantum computing has, as of the mid-2020s, *not yet done anything useful that a classical computer cannot do* — not one practically valuable computation, not one deployed application beyond quantum sensing (which is a different technology). The supremacy demonstrations were on useless tasks and were partly eroded by classical methods; the useful-advantage milestone remains unclaimed; many hyped applications (optimization, finance, machine learning) have seen their promised speedups shrink or vanish under scrutiny; and the fault-tolerant machines that could run the great algorithms require millions of physical qubits that do not exist and may be a decade or more away. The field has a documented history of over-promising — of specific breakthroughs predicted "within five years" that did not arrive, of claims that did not survive independent scrutiny — and it rides a wave of investment and expectation that could, if useful applications are slow to materialize, break into disappointment. The history of technology knows the phenomenon of the "winter" — the artificial-intelligence field endured two of them, decades of dashed expectations and withdrawn funding after periods of hype — and a "quantum winter" is a real possibility if the gap between promise and delivery stretches too long. A reasonable skeptic, surveying the same trajectory as the optimist, may well believe that useful quantum computing is further off and less transformative than the enthusiasm suggests, and that much current investment will disappoint.

The wise view, I think, holds these two assessments together rather than choosing between them, and it takes the long view that the history in this book supports. Quantum computing is a genuine and profound scientific achievement, advancing faster than the deep pessimists expected and slower than the enthusiasts hoped, with a plausible but uncertain path to transformative applications on a timeline no one can honestly predict. It will probably not deliver the imminent, sweeping revolution of the press releases; it will probably not prove the elaborate boondoggle of the debunkers; it will most likely follow the pattern of many foundational technologies — slower and less magical in the near term than the hype promised, then, over decades, more consequential than the skeptics allowed, arriving not with a single dramatic breakthrough but through a long accumulation of engineering, each advance modest, the sum transformative. The classical computer took decades to move from room-sized curiosities factoring artillery tables to the devices that reshaped civilization; the quantum computer, if it fulfills its promise, will likely follow a similarly long arc, and we are, at best, in its early middle chapters.

We come, at the end of this long journey through the strange and consequential world of quantum computing, to the most important question of all — the one that should guide how you, the reader, interpret every breathless headline and every skeptical dismissal you will encounter in the years ahead: What should you actually believe about quantum computing? Cutting through both the extravagant hype and the reflexive cynicism to a clear-eyed, honest assessment is the final and most valuable thing this book can offer, and it is worth stating plainly, without either the salesmanship or the sneering that so often distort the subject.

Begin with the hype, because it is the more seductive distortion and the more likely to mislead. Quantum computing has been subjected to some of the most extravagant overselling of any technology in recent memory. Headlines proclaim that quantum computers will imminently break all encryption, revolutionize artificial intelligence, cure every disease, solve every optimization problem, and render classical computers obsolete. Companies and startups, competing for investment and attention, make sweeping claims of near-term advantage that repeatedly fail to survive scrutiny. Much of this hype rests on the fundamental misconception this book has fought from its opening chapters — the false image of a quantum computer as a magic machine that "tries all answers at once" and accelerates everything. The reality, as we have seen at length, is that quantum computers offer dramatic advantages only for a narrow class of problems with special mathematical structure, modest advantages for some others, and no advantage at all for the overwhelming majority of computing tasks. Anyone promising a universal quantum acceleration of everything is either confused or selling something.

But cynicism is equally a distortion, and the reflexive dismissal of quantum computing as pure hype is as wrong as the breathless overselling. The achievements are genuine and profound. Quantum supremacy was really demonstrated — programmable machines really did enter a regime beyond classical simulation. Error correction has really crossed below the threshold — Google's Willow really did show that scaling up a code suppresses errors, vindicating a thirty-year-old theorem and proving the path to reliable computation is open. Logical qubits are really being built, their numbers really growing. The algorithms of Shor and Grover are real mathematical facts, and Shor's threat to cryptography is real enough that the world is genuinely, urgently rebuilding its cryptographic foundations. Quantum sensing already delivers practical value today. The physics is sound, the progress is real, and the long-term potential — especially for simulating nature, Feynman's founding dream — is genuinely transformative. To dismiss all this as vaporware is to be as badly wrong as the hypesters, merely in the opposite direction.

The field has, in fact, weathered cycles of inflated expectation and subsequent disappointment — what observers, borrowing a term from the history of artificial intelligence, sometimes call the risk of a "quantum winter," a period of disillusionment and withdrawn funding following a peak of overpromising. Artificial intelligence endured several such winters before its eventual triumphs, and quantum computing may yet face its own, if the gap between the hype and the deliverable near-term reality provokes a backlash among disappointed investors and a public told for years that a revolution was imminent. The honest truth — that useful, transformative quantum computers are probably a decade or more away, and that the intervening years will bring incremental progress rather than sudden revolution — is not what fuels investment bubbles, and a correction of inflated expectations would not be surprising. But a quantum winter, if it comes, would be a matter of sentiment and funding cycles, not of the underlying science, which continues its steady advance regardless of the mood. The physics does not care about the hype cycle.

So what is the long view — the sober, balanced assessment with which to leave the reader? It is this. Quantum computing is a real and important technology, grounded in sound physics, that has achieved genuine milestones and holds genuine transformative promise, particularly for simulating quantum systems in chemistry and materials science, and genuine disruptive threat to current cryptography. It is also a technology whose difficulties are immense, whose useful large-scale realization remains years or decades away and is not absolutely guaranteed to arrive on any particular timeline, and whose promise has been persistently and badly oversold. The mature attitude holds both truths at once, without collapsing into either the salesman's enthusiasm or the cynic's dismissal: profound long-term potential, hard and uncertain near-term reality; real achievements, real remaining distance; genuine revolution possible, genuine patience required. The reader who holds this balance — who greets each new quantum headline by asking whether it concerns supremacy or advantage, physical qubits or logical ones, a proven result or a marketing claim, a problem with real structure or a contrived benchmark — possesses the single most valuable tool for navigating the quantum future: not credulity, not cynicism, but informed, patient, curious skepticism.

It is worth naming, as a final discipline, the specific mental habit that separates those who truly understand quantum computing from those who merely react to its headlines — because that habit, more than any single fact, is what this book has tried to instill. It is the habit of holding two truths at once, refusing to collapse into either the salesman's enthusiasm or the cynic's dismissal. The genuine achievements are real: supremacy demonstrated, error correction crossed below threshold, logical qubits multiplying, cryptography genuinely threatened, sensing already useful, and the long-term potential — especially for simulating nature — genuinely transformative. And the genuine difficulties are equally real: useful large-scale machines remain years or decades away and are not guaranteed on any timeline; the overhead of error correction is brutal; the engineering of a million-qubit machine is forbidding; and the promises have been persistently, badly oversold. Neither truth cancels the other. The mature attitude is to feel the wonder and the skepticism together, and to greet each new development by asking not "is quantum computing real or hype?" — a false binary — but "what, precisely, has been shown, and what remains to be done?"

The specific danger of a "quantum winter" deserves a sober word, because the pattern is familiar and the risk is real. Artificial intelligence endured several "winters" — periods of disillusionment and withdrawn funding following peaks of overpromising — before its eventual triumphs, and quantum computing may yet face its own. The honest truth, that useful transformative machines are probably a decade or more away and that the intervening years will bring incremental progress rather than sudden revolution, is not what fuels investment bubbles; and a public told for years that a revolution was imminent may, when it fails to arrive on schedule, swing to disappointment and dismissal. A correction of inflated expectations would not be surprising. But it is crucial to see that a quantum winter, if it comes, would be a matter of sentiment and funding cycles, not of the underlying science, which continues its steady advance regardless of the mood. The physics does not care about the hype cycle. The qubits grow more numerous and more reliable, the codes more efficient, the error rates lower, whether or not the headlines are enthusiastic — and a field can make profound progress in the quiet of a winter, as artificial intelligence did, emerging stronger when the thaw comes.

And there is, finally, a reason to care about quantum computing that transcends any practical payoff, and it is the note on which to end — for it returns us to where this book began, to the strangeness itself. Quantum computing is, at bottom, a profound exploration of the nature of reality: of superposition, entanglement, and the deep structure of information and physics. In building these machines, humanity is not merely creating a new tool but probing the foundations of the universe, learning to speak nature's own quantum language, testing how far the strange laws that govern the very small can be harnessed by the deliberate hand of engineering. In a single long arc, we took the most disturbing puzzles of twentieth-century physics — Einstein's spooky action, Schrodinger's suspended cat, Heisenberg's irreducible uncertainty — and, rather than merely arguing about what they meant, learned to *use* them: entanglement became a fuel, superposition a register, interference an algorithm, and the very fragility of quantum states became, in the sensor, a gift of unmatched precision. Whether or not the great machines arrive on the timeline the optimists hope, that transformation — from paradox to engineering — is already one of the deepest intellectual achievements of our age. The uncertain machine is, in the end, a mirror held up to an uncertain universe, and in learning to build it we are learning, more deeply than ever before, what kind of universe we live in. That is reason enough to watch what comes next with patience, with rigor, and with awe — and the story, which began with a glowing oven in Berlin in the last December of the nineteenth century, is very far from over.

Epilogue: The Uncertain Machine

I began this book with an image: a golden chandelier hanging in a windowless room, chilled to a hair above absolute zero, protecting a fingernail-sized chip from the roar of the ordinary world so that a few dozen loops of superconducting metal could be held in a state of both, of maybe, of possibilities suspended and made to interfere like waves. It is worth returning to that image at the end, because it contains, in miniature, everything this book has been about — and because it is genuinely uncertain, in a way the reader is now equipped to feel, what that machine finally means.

It is uncertain in its physics, most literally. A quantum computer computes with probabilities and amplitudes, with superpositions that have no definite value until measured, with the collapse that Einstein could not accept and that we still cannot fully explain. Its every operation is a wager placed inside a cloud of possibility and cashed out by a measurement that destroys the cloud. The machine embodies the deepest and least resolved strangeness in all of physics — the measurement problem, the meaning of the wavefunction, the question of whether Deutsch's parallel worlds are real — and it works, flawlessly, whatever the answer to those questions turns out to be. We have built a device that runs on the one part of nature we understand least, and it does exactly what the equations say, and the equations remain a mystery. That is a strange thing to have accomplished.

It is uncertain in its engineering. The machine teeters on the edge of the possible, its qubits so fragile they must be shielded in the coldest place in the known universe, its computations racing against a decoherence that would dissolve them in a flicker, its scale a thousandfold or more short of what its greatest promises require. We have crossed, at last, the threshold that thirty years ago seemed perhaps uncrossable — error correction that improves as it grows, logical qubits that outlive their physical parts — but the distance from that hard-won proof of principle to a machine that can factor a real number or design a real drug remains vast, measured in millions of qubits not yet built, in codes not yet perfected, in years not yet counted.

And it is uncertain in its future. No one honestly knows whether the fault-tolerant machines will arrive in a decade or in three, whether the great applications will transform whole industries or settle into useful niches, whether the current wave of investment and excitement will carry through to fulfillment or break into a quantum winter of dashed expectations and withdrawn funding before the promise is kept. The optimists and the skeptics survey the same evidence and reach opposite forecasts, and the honest observer must admit that both could be right, that the destination is genuinely unknown. This is not a failure of the book to reach a verdict; it is the verdict. Quantum computing's ending is not yet written, and anyone who tells you otherwise — in either direction — is guessing or selling.

But uncertainty is not the same as emptiness, and the machine's unresolved future should not obscure what has already been achieved. In the span of a single human lifetime, a philosophical puzzle about the nature of reality became a rigorous science, then a global technological race, then a matter of national security and billions of dollars. A handful of theorists — Benioff, Feynman, Deutsch, Shor, Grover,

Preskill, Kitaev, and the many others whose names fill the pages behind this one — took the weirdness that had troubled physics for a century and turned it into a resource, a thing to build with. Engineers learned to trap single atoms in light, to cool circuits colder than interstellar space, to protect the most delicate states in nature from the noise of the world. And in the doing, we came to understand our own universe more deeply: what information is, what computation is, what the quantum world permits and forbids. The uncertain machine has already paid for itself in understanding, whatever it does next.

I do not know — no one knows — whether the reader who picks up this book in twenty years will do so on a device touched by quantum computation, or whether quantum computers will by then have broken the old codes and remade chemistry, or whether they will remain magnificent instruments of narrow use, their grandest promises still receding. What I do know is that the effort to build a computer out of the fabric of possibility itself — to compute with the deepest strangeness in nature, to fight the exponential wildness of the quantum world with a machine made of the same wild stuff — is one of the boldest things our species has attempted, and one of the most illuminating. The golden chandelier hums in its cold room, holding its impossible states against the pressing warmth of the ordinary world, computing with maybe. Whatever it finally becomes, it is already a wonder. And its story, which began with a glowing oven in Berlin in the last December of the nineteenth century, is very far from over.

A last word on how to hold all this

If you carry away one habit from this book, let it be a way of listening. When the next quantum headline arrives — and it will, breathless and confident — pause and ask a few quiet questions. Is this about *supremacy* (a useless benchmark) or *advantage* (a useful problem)? Are these *physical* qubits (noisy raw material) or *logical* qubits (the error-corrected kind that actually compute)? Was the classical comparison against the *best* known method, or a convenient one? Has the result been reproduced, and what do knowledgeable skeptics say? Is the problem one with real hidden structure that interference can exploit, or a generic task where quantum machines offer little? These questions are not cynicism; they are the courtesy of taking a remarkable field seriously enough to understand it truly.

And hold, alongside the skepticism, a genuine sense of wonder — for the story really is astonishing. In a single long arc, humanity took the most disturbing puzzles of twentieth-century physics — Einstein's spooky action, Schrödinger's cat suspended between life and death, Heisenberg's irreducible uncertainty — and, rather than merely arguing about what they meant, learned to *use* them. Entanglement became a fuel. Superposition became a register. Interference became an algorithm. The very fragility that makes quantum states so maddening to preserve became, in the sensor, a gift of unmatched precision, and, in the threshold theorem, a problem that mathematics proved could be tamed. Whether or not the great machines arrive on the timeline the optimists hope, that transformation — from paradox to engineering — is already one of the deepest intellectual achievements of our age.

The uncertain machine, then, is aptly named twice over: uncertain in the quantum sense, computing with the mathematics of maybe; and uncertain in its future, its grandest promises still poised between fulfillment and endless deferral. Both uncertainties are, in their way, the point. We are building a machine out of the deepest strangeness in nature, and in the building we are learning, as never before, what kind of universe we inhabit. That is reason enough to watch what comes next with patience, with rigor, and with awe.

Timeline

1900 — Max Planck introduces the quantum hypothesis to explain black-body radiation; energy comes in discrete grains. The quantum age begins.

1905 — Albert Einstein proposes that light itself is composed of quanta (photons), explaining the photoelectric effect; the seed of wave–particle duality.

1925–1927 — Quantum mechanics is formalized by Heisenberg (matrix mechanics), Schrödinger (wave mechanics), Born (the probabilistic interpretation), and Dirac; Heisenberg states the uncertainty principle.

1935 — Einstein, Podolsky, and Rosen publish the EPR paradox; Schrödinger names "entanglement."

1936 — Alan Turing defines the universal computing machine, founding the theory of computation.

1961 — Rolf Landauer proves that erasing information has a minimum energy cost: "information is physical."

1964 — John Bell proves Bell's theorem, making the reality of entanglement experimentally testable.

1973 — Charles Bennett proves that any computation can be done reversibly, without erasure.

1980 — Paul Benioff describes a quantum-mechanical model of a computer; Yuri Manin independently suggests quantum computing.

1981–82 — Richard Feynman argues that simulating quantum physics requires a quantum computer.

1982 — Alain Aspect's experiments confirm the violation of Bell's inequality, refuting local hidden variables.

1984 — Bennett and Brassard publish the BB84 protocol for quantum key distribution.

1985 — David Deutsch defines the universal quantum computer and the first quantum algorithm.

1991 — Artur Ekert proposes entanglement-based quantum key distribution (E91).

1994 — Peter Shor discovers his algorithm for factoring and discrete logarithms, threatening RSA; Daniel Simon demonstrates an exponential quantum speedup for a hidden-period problem.

1995 — Shor and, independently, Steane invent quantum error-correcting codes; Cirac and Zoller propose the trapped-ion quantum gate.

1996 — Lov Grover discovers his quantum search algorithm; Seth Lloyd proves quantum simulation is efficient; the threshold theorem for fault tolerance is established.

1998 — First two-qubit quantum computer (NMR) runs Grover's algorithm.

1999 — D-Wave Systems founded; NEC demonstrates a superconducting charge qubit.

2001 — IBM/Stanford factor 15 using Shor's algorithm on a seven-qubit NMR machine.

2007 — The transmon superconducting qubit is introduced at Yale.

2011 — D-Wave ships the first commercial quantum annealer.

2012 — John Preskill coins "quantum supremacy"; Wineland and Haroche win the Nobel Prize for quantum control.

2016 — China's Micius satellite demonstrates space-based quantum key distribution.

2018 — U.S. National Quantum Initiative and EU Quantum Flagship launched; Preskill coins "NISQ"; Ewin Tang's dequantization work begins deflating quantum machine-learning claims.

2019 — Google's Sycamore (53 qubits) claims quantum supremacy; IBM disputes the ten-thousand-year figure.

2020 — China's Jiuzhang demonstrates photonic quantum advantage via boson sampling.

2021 — IBM Eagle reaches 127 qubits; the Aspect–Clauser–Zeilinger Nobel Prize follows in 2022.

2022 — IBM Osprey reaches 433 qubits; Nobel Prize in Physics to Aspect, Clauser, and Zeilinger for entanglement experiments.

2023 — IBM Condor reaches 1,121 qubits; IBM reports quantum "utility" via error mitigation; neutral-atom teams demonstrate dozens of logical qubits.

August 2024 — NIST finalizes the first post-quantum cryptography standards (FIPS 203 ML-KEM, 204 ML-DSA, 205 SLH-DSA).

December 2024 — Google's Willow (105 qubits) demonstrates below-threshold quantum error correction — a turning point.

2025 — UN International Year of Quantum Science and Technology; Microsoft announces the (contested) Majorana 1 topological chip; Google claims verifiable advantage with "Quantum Echoes"; IBM unveils Nighthawk and Loon and targets verified quantum advantage for 2026; neutral-atom and trapped-ion systems report dozens of logical qubits.

2026 and beyond (projected) — Verified quantum advantage on a useful problem pursued; roadmaps point toward large-scale fault-tolerant machines (e.g., IBM's "Starling") by the end of the decade. The dates are targets, not certainties.

Who's Who

Scott Aaronson (b. 1981) — American computer scientist; leading quantum complexity theorist, co-proposer of boson sampling, and a prominent, skeptical public explainer of the field. Advised Ewin Tang.

Alain Aspect (b. 1947) — French physicist whose 1982 experiments confirmed the violation of Bell's inequality. Shared the 2022 Nobel Prize in Physics.

Paul Benioff (1930–2022) — American physicist who in 1980 described the first quantum-mechanical model of a computer.

Charles Bennett (b. 1943) — American physicist at IBM; proved reversible computation possible, co-invented quantum cryptography (BB84) and quantum teleportation.

John Bell (1928–1990) — Irish physicist; proved Bell's theorem (1964), making entanglement experimentally testable. Died before the Nobel recognizing work built on his.

Ethan Bernstein & Umesh Vazirani — computer scientists who defined the quantum complexity class BQP and the Bernstein–Vazirani problem, giving quantum computing its rigorous complexity theory.

Niels Bohr (1885–1962) — Danish physicist; architect of the Copenhagen interpretation and Einstein's great debating partner.

Max Born (1882–1970) — German physicist; gave the probabilistic (Born rule) interpretation of the wavefunction.

Gilles Brassard (b. 1955) — Canadian computer scientist; co-inventor, with Bennett, of the BB84 quantum- cryptography protocol.

Ignacio Cirac & Peter Zoller — physicists who proposed the trapped-ion quantum gate (1995), foundational to ion-trap quantum computing.

John Clauser (b. 1942) — American physicist; performed early Bell-inequality tests. Shared the 2022 Nobel Prize.

David Deutsch (b. 1953) — British physicist; defined the universal quantum computer (1985) and the first quantum algorithm; advocate of the many-worlds interpretation. Often called the father of quantum computing.

Paul Dirac (1902–1984) — British physicist; unified quantum mechanics and introduced bra–ket notation.

Albert Einstein (1879–1955) — proposed the photon (1905); later the leading critic of quantum mechanics via the EPR paradox and "spooky action at a distance."

Artur Ekert (b. 1961) — physicist who proposed entanglement-based quantum key distribution (E91, 1991).

Hugh Everett III (1930–1982) — American physicist; originated the many-worlds interpretation (1957).

Richard Feynman (1918–1988) — American physicist and Nobel laureate; argued in 1981–82 that simulating quantum physics requires a quantum computer, inspiring the field.

Daniel Gottesman — physicist who developed the stabilizer formalism unifying quantum error-correcting codes.

Lov Grover (b. 1961) — physicist at Bell Labs; discovered the quantum search algorithm (1996).

Serge Haroche (b. 1944) — French physicist; Nobel laureate (2012) for cavity-QED control of quantum systems.

Werner Heisenberg (1901–1976) — German physicist; created matrix mechanics and the uncertainty principle.

Alexei Kitaev (b. 1963) — Russian-American physicist at Caltech; invented topological quantum computation and the surface code, and contributed to the threshold theorem.

Rolf Landauer (1927–1999) — German-American physicist at IBM; proved that erasing information costs energy ("information is physical").

Seth Lloyd (b. 1960) — American physicist at MIT; proved quantum computers can efficiently simulate quantum systems (1996).

Mikhail Lukin (b. 1971) — Harvard physicist; leader in neutral-atom / Rydberg-array quantum computing.

Yuri Manin (1937–2023) — Russian mathematician who independently suggested quantum computing in 1980.

John Martinis (b. 1958) — American physicist who led Google's superconducting-qubit supremacy effort.

Christopher Monroe (b. 1965) — American physicist; trapped-ion pioneer and co-founder of IonQ.

Jian-Wei Pan (b. 1970) — Chinese physicist; leader of China's photonic quantum computing (Jiuzhang) and satellite quantum communication (Micius).

Max Planck (1858–1947) — German physicist; introduced the quantum in 1900.

John Preskill (b. 1953) — American theoretical physicist at Caltech; coined "quantum supremacy" (2012) and "NISQ" (2018).

Erwin Schrödinger (1887–1961) — Austrian physicist; wrote the wave equation and named "entanglement."

Peter Shor (b. 1959) — American mathematician; discovered the factoring algorithm (1994) and the first quantum error-correcting code (1995). The field's most consequential algorithmist.

Andrew Steane (b. 1965) — British physicist; devised the seven-qubit Steane error-correcting code (1996).

Ewin Tang (b. ~2000) — American computer scientist who, as an undergraduate, "dequantized" a leading quantum machine-learning algorithm (2018).

Alan Turing (1912–1954) — British mathematician; founded the theory of computation and the concept of the universal computer.

David Wineland (b. 1944) — American physicist; pioneer of trapped-ion control; Nobel laureate (2012).

Anton Zeilinger (b. 1945) — Austrian physicist; pioneer of experimental quantum information. Shared the 2022 Nobel Prize.

Ignacio Cirac (b. 1965) & **Peter Zoller** (b. 1952) — theorists whose 1995 proposal showed how trapped ions could be turned into a quantum computer, launching the first hardware platform.

Isaac Chuang (b. 1968) — American physicist; led the 2001 NMR experiment that factored 15 with Shor's algorithm and co-authored the field's standard textbook with Michael Nielsen.

John Clauser (b. 1942) — American physicist whose early-1970s experiments first tested Bell's inequality; shared the 2022 Nobel Prize.

Hugh Everett III (1930–1982) — American physicist who proposed the many-worlds interpretation (1957), then left academia for defense work, dismissed in his lifetime.

Daniel Gottesman — physicist who developed the stabilizer formalism (1997), turning quantum error correction into a systematic science.

Lov Grover (b. 1961) — Indian-American computer scientist at Bell Labs; invented the quantum search algorithm (1996) that bears his name.

Aram Harrow — physicist and co-author of the HHL algorithm (2009) for quantum linear systems.

Alexei Kitaev (b. 1963) — Russian-American theorist; invented the surface code and topological quantum computing, among the most influential practical contributions to the field.

Rolf Landauer (1927–1999) — German-American physicist at IBM; proved that information is physical (erasure costs energy) and a prominent early skeptic of quantum computing.

Seth Lloyd (b. 1960) — American physicist; proved that quantum computers can efficiently simulate quantum systems (1996), vindicating Feynman's dream.

Yuri Manin (1937–2023) — Russian mathematician who independently suggested the idea of quantum computing around 1980.

John Martinis (b. 1958) — American physicist who led the Google team that built Sycamore and achieved the first quantum-supremacy demonstration (2019).

Christopher Monroe (b. 1965) — American physicist; pioneer of trapped-ion quantum computing and co-founder of IonQ.

Michele Mosca — cryptographer whose "inequality" frames the urgency of migrating to post-quantum cryptography.

Jian-Wei Pan (b. 1970) — Chinese physicist; led the Jiuzhang photonic supremacy demonstration and the Micius quantum-communication satellite, making China a quantum leader.

John Preskill (b. 1953) — American theoretical physicist;

coined "quantum supremacy" and "NISQ," and a leading voice for rigor and honesty in the field.

Max Planck (1858–1947) — German physicist whose 1900 quantum hypothesis began the quantum age; a reluctant revolutionary.

Peter Shor (b. 1959) — American mathematician; devised the factoring algorithm (1994) that made quantum computing consequential, and the first quantum error-correcting code (1995).

Daniel Simon — computer scientist whose 1994 period-finding problem showed Shor the path to factoring.

Andrew Steane (b. 1965) — British physicist; devised the seven-qubit Steane code, importing classical coding theory into the quantum realm.

Ewin Tang (b. 1999) — American computer scientist who, as a teenager, "dequantized" several celebrated quantum machine-learning algorithms, puncturing inflated claims.

Glossary

Adiabatic quantum computation — A model of computation in which the answer is encoded in the ground state of a system reached by slow, continuous evolution; the basis of quantum annealing.

Amplitude (probability amplitude) — A complex number attached to each possibility in a quantum state; its squared magnitude gives the probability of that outcome. Amplitudes can be negative, enabling interference.

Amplitude estimation — A quantum technique, related to Grover's algorithm, giving a quadratic speedup for tasks such as Monte Carlo simulation.

Anyon — An exotic quasiparticle existing in two dimensions, with unusual exchange behavior; the basis of topological quantum computing.

BB84 — The first quantum key distribution protocol, proposed by Bennett and Brassard in 1984.

Bell state — A maximally entangled state of two qubits; the simplest and most fundamental entangled state.

Bell's theorem / inequality — A result (John Bell, 1964) showing that quantum correlations exceed what any local hidden-variable theory permits; confirmed experimentally, refuting local realism.

Bloch sphere — A geometric picture in which every state of a single qubit corresponds to a point on the surface of a sphere.

Born rule — The rule that the probability of a measurement outcome equals the squared magnitude of its amplitude.

BQP — "Bounded-error quantum polynomial time," the class of problems a quantum computer can solve efficiently; the formal definition of quantum computational power.

Boson sampling — A specialized sampling task, hard for classical computers, used in photonic demonstrations of quantum advantage.

Coherence / coherence time — The preservation of a qubit's quantum state and phase; the time over which it survives before decohering.

CNOT (controlled-NOT) — A two-qubit gate that flips a target qubit if a control qubit is |1>; a key entangling gate.

Decoherence — The loss of a quantum system's superposition and phase through interaction with its environment; the central obstacle to quantum computing.

Dequantization — Finding a classical algorithm that matches a claimed quantum speedup, showing the advantage was illusory (Ewin Tang, 2018).

Discrete logarithm problem — A hard mathematical problem underlying Diffie–Hellman and elliptic-curve cryptography; broken by Shor's algorithm.

Entanglement — A quantum correlation between systems whose joint state cannot be factored into independent parts; the key resource of quantum computing.

Error correction (quantum) — Protecting quantum information by encoding one logical qubit across many physical qubits and correcting errors via syndrome measurements.

Error mitigation — Statistically estimating a noise-free result from many noisy runs, without full error correction; a

NISQ-era stopgap.

Fault tolerance — Reliable computation despite component errors, achievable below the error threshold.

Gate (quantum) — A reversible (unitary) operation on one or more qubits; the building block of quantum circuits.

Grover's algorithm — A quantum algorithm giving a quadratic speedup for unstructured search (1996).

Hadamard gate (H) — A single-qubit gate that creates superposition; central to generating and harvesting interference.

Harvest now, decrypt later — The strategy of storing encrypted data today to decrypt it once a quantum computer exists.

Interference — The reinforcement or cancellation of quantum amplitudes; the true engine of quantum speedups.

Logical qubit — An error-corrected, reliable qubit emulated by many physical qubits.

Measurement — Reading a quantum system, collapsing its superposition to a definite outcome per the Born rule.

NISQ (Noisy Intermediate-Scale Quantum) — The current era of noisy, mid-sized, non-error-corrected quantum computers (Preskill, 2018).

No-cloning theorem — The impossibility of copying an unknown quantum state; foundational to quantum cryptography and shaping error correction.

Phase — The relative angle of a qubit's amplitudes (its "longitude" on the Bloch sphere); invisible to a naive measurement but essential to interference.

Photon — A quantum of light; a natural qubit and the carrier of quantum communication.

Physical qubit — An actual hardware qubit, noisy and error-prone.

Post-quantum cryptography (PQC) — Classical cryptography based on problems believed hard for quantum computers; standardized by NIST in 2024 (ML-KEM, ML-DSA, SLH-DSA).

Q-Day — The hypothetical day a quantum computer can break current public-key cryptography.

qLDPC code — Quantum low-density parity-check code; a low-overhead error-correcting code, a leading route to efficient fault tolerance.

Quantum annealing — A heuristic optimization method using adiabatic-like evolution; the basis of D-Wave's machines.

Quantum advantage / supremacy — A quantum computer outperforming any classical one on a task; "supremacy" for any task (even useless), "advantage" (strong sense) for a useful one.

Quantum Fourier transform (QFT) — A quantum subroutine that detects periodicity; the heart of Shor's algorithm.

Quantum key distribution (QKD) — Sharing secret keys with security guaranteed by physics (e.g., BB84, E91).

Qubit — The basic unit of quantum information; a two-level quantum system that can be in a superposition of 0 and 1.

Rydberg state — A highly excited atomic state used to mediate entangling gates between neutral atoms.

Shor's algorithm — A quantum algorithm for factoring and discrete logarithms (1994); breaks RSA and related public-key cryptography.

Stabilizer code / formalism — The general framework describing most quantum error-correcting codes.

Superconducting qubit — A qubit made from Josephson-junction circuits (e.g., the transmon), the leading industrial platform.

Superposition — A quantum state that is a linear combination of basis states; a qubit being "both 0 and 1."

Surface code — A leading two-dimensional error-correcting code requiring only nearest-neighbor interactions (Kitaev).

Syndrome measurement — A measurement that detects errors without revealing (and thus destroying) the encoded quantum information.

Threshold theorem — The result guaranteeing that fault-tolerant quantum computing is possible if physical error rates are below a critical threshold.

Topological quantum computing — Encoding information in the braiding of anyons, protected by topology; pursued by Microsoft, still unproven.

Transmon — The most common superconducting qubit design, robust against charge noise (2007).

Trapped-ion qubit — A qubit encoded in a single ion held by electromagnetic fields; the highest-fidelity platform.

Unitary — A reversible, norm-preserving operation; the mathematical form of all quantum gates.

Boson sampling — A specialized sampling problem, natural for photonic hardware and believed hard classically; used by the Jiuzhang machine to claim quantum computational supremacy.

BQP (bounded-error quantum polynomial time) — The class of problems a quantum computer can solve efficiently with high probability; the quantum analogue of the classical class P.

Cirac–Zoller gate — The 1995 proposal showing how a shared vibrational mode could entangle trapped ions, the blueprint for the first ion-trap quantum logic.

Clifford gates — A special set of quantum gates (including Hadamard, phase, and CNOT) that, by the Gottesman–

Knill theorem, can be simulated efficiently on a classical computer; quantum advantage requires non-Clifford "magic."

CNOT (controlled-NOT) — A two-qubit gate that flips a target qubit if a control qubit is 1; the canonical entangling gate.

Coherence time — How long a qubit preserves its quantum state before decoherence; a key measure of qubit quality (often split into T1, energy relaxation, and T2, dephasing).

Concatenation — Nesting an error-correcting code inside itself in layers, so that below-threshold hardware suppresses errors exponentially with each added layer.

Cooper pair — A bound pair of electrons that flows without resistance in a superconductor; the carrier of supercurrent underlying superconducting qubits.

Cryptographically relevant quantum computer (CRQC) — A machine large and reliable enough to run Shor's algorithm against real encryption keys; the definition of "Q-Day" capability.

Dequantization — Showing that a claimed quantum-algorithm speedup can be matched by a classical algorithm; associated with Ewin Tang's work on quantum machine learning.

Dilution refrigerator — The cryogenic apparatus that cools superconducting chips to millikelvin temperatures using a helium-3/helium-4 mixture; the iconic golden "chandelier."

DiVincenzo criteria — The five requirements (2000) a physical system must meet to be a viable quantum computer, from scalable qubits to reliable measurement.

Fault tolerance — The property of a computing scheme that continues to work correctly even when its own components (including the error-correction machinery) are faulty.

Fidelity — A measure of how accurately a quantum operation or state matches its ideal; "gate fidelity" of 99.9% means one error per thousand operations.

Gottesman–Knill theorem — The result that circuits built only from Clifford gates are efficiently classically simulable, locating the source of quantum advantage in non-Clifford resources.

HHL algorithm — A 2009 quantum algorithm (Harrow–Hassidim–Lloyd) for solving linear systems, offering an apparent exponential speedup hedged by severe input/output caveats.

Josephson junction — A thin insulating barrier between two superconductors through which paired electrons tunnel; the nonlinear element at the heart of superconducting qubits.

Learning with errors (LWE) — A lattice-based problem believed hard for quantum computers; the foundation of leading post-quantum encryption such as ML-KEM.

Logical qubit — An error-corrected qubit built from many physical qubits, reliable enough to compute with; the count that truly matters for capability.

Magic state — A special resource state that supplies the non-Clifford ingredient needed to lift a quantum computation beyond classical reach; produced by "magic-state distillation."

Mosca's inequality — The rule of thumb that if the required secrecy lifetime plus migration time exceeds the years until Q-Day, your long-lived secrets are already at risk.

Nitrogen-vacancy (NV) center — A defect in diamond whose electron spin serves as a room-temperature quantum sensor and qubit.

No-cloning theorem — The proof that an unknown quantum state cannot be copied; foundational to quantum cryptography and error correction.

NISQ (noisy intermediate-scale quantum) — Preskill's term for today's machines: dozens to hundreds of noisy qubits without full error correction.

Physical qubit — A single hardware qubit (one ion, one circuit, one atom); noisy and fragile, the raw material for logical qubits.

Post-quantum cryptography (PQC) — Classical encryption schemes designed to resist quantum attack, deployable on ordinary computers; standardized by NIST in 2024.

Quantum advantage — Solving a genuinely useful problem faster or better than any classical computer; distinct from (and harder than) supremacy.

Quantum annealing — A specialized analog method for optimization that exploits quantum tunneling; the approach of D-Wave's machines.

Quantum Fourier transform (QFT) — The quantum operation that detects periodicity, the engine at the heart of Shor's algorithm.

Quantum supremacy / computational advantage — Performing any task, however useless, that no classical computer can match in reasonable time; first claimed by Sycamore (2019).

Qubit — The quantum bit: a two-state quantum system that can occupy superpositions of 0 and 1.

Random circuit sampling — The contrived benchmark used by Sycamore and Willow to demonstrate supremacy; sampling from the output distribution of a random quantum circuit.

Rydberg blockade — The effect by which a neutral atom excited to a giant "Rydberg" state prevents a neighbor from being excited; the entangling mechanism of neutral-atom computers.

Shor's algorithm — The 1994 quantum algorithm that factors large numbers efficiently, threatening RSA and elliptic-curve cryptography.

Squeezed light — A quantum-engineered state of light with reduced noise in one property; used to enhance the sensitivity of gravitational-wave detectors.

Stabilizer code — A quantum error-correcting code specified by a set of commuting measurement operators (stabilizers); the unifying framework for most codes.

Superposition — A quantum state that is a weighted blend of multiple possibilities at once, with amplitudes that can interfere.

Surface code — Kitaev's topological error-correcting code needing only nearest-neighbor interactions and tolerating ~1% error; the workhorse of superconducting quantum computing.

Threshold theorem — The proof that if physical error rates fall below a critical value, arbitrarily reliable quantum computation is possible; the theoretical foundation of fault tolerance.

Toffoli gate — A reversible three-bit (controlled-controlled-NOT) gate, universal for reversible classical logic and a building block in quantum circuits.

Topological quantum computing — A hardware approach that stores information in global, topological properties (e.g., braided anyons) so that it is intrinsically protected from local noise.

Trotterization — Slicing a continuous quantum evolution into a sequence of small discrete steps so it can be run on a gate-based simulator; central to Lloyd's simulation algorithm.

Variational quantum eigensolver (VQE) — A hybrid quantum-classical algorithm for estimating molecular ground-state energies on near-term noisy machines.

qLDPC codes — Quantum low-density parity-check codes: error-correcting codes that promise far lower overhead than the surface code, at the cost of requiring longer-range connectivity.

Further Reading

For readers who wish to go deeper, the following books, articles, and resources offer paths into the subject, from accessible introductions to rigorous textbooks.

Accessible introductions

- Chris Bernhardt, *Quantum Computing for Everyone* (MIT Press, 2019). The leading light-math introduction to qubits, gates, entanglement, and algorithms; the best next step for a reader of this book.
- John Gribbin, *Computing with Quantum Cats: From Colossus to Qubits* (2013). A narrative popular-science history of the people and ideas, requiring no mathematics.
- George Johnson, *A Shortcut Through Time: The Path to the Quantum Computer* (2003). An early, readable popular account.

Foundational and philosophical

- David Deutsch, *The Fabric of Reality* (1997) and *The Beginning of Infinity* (2011). The founder of quantum computing on the multiverse, computation, and the nature of knowledge.

- Scott Aaronson, *Quantum Computing Since Democritus* (2013). A witty, rigorous tour of complexity theory and quantum foundations by a leading theorist; also his blog, *Shtetl-Optimized,* for sharp commentary on the field.

Technical texts (for readers with mathematics)

- Michael A. Nielsen & Isaac L. Chuang, *Quantum Computation and Quantum Information* (Cambridge, 2000; 10th- anniversary edition, 2010). "Mike & Ike," the canonical graduate textbook and the standard reference.
- John Preskill, lecture notes for Physics 219, *Quantum Computation* (Caltech, freely available online). Deep, authoritative, and widely used.
- Jack D. Hidary, *Quantum Computing: An Applied Approach* (2019/2021). Bridges theory and hands-on programming.

Landmark papers (for the historically curious)

- R. P. Feynman, "Simulating Physics with Computers," *International Journal of Theoretical Physics* (1982).
- D. Deutsch, "Quantum theory, the Church–Turing principle and the universal quantum computer," *Proceedings of the Royal Society* (1985).
- P. W. Shor, "Polynomial-Time Algorithms for Prime Factorization and Discrete Logarithms on a Quantum Computer," *SIAM Journal on Computing* (1997).
- L. K. Grover, "A fast quantum mechanical algorithm for database search," *STOC* (1996).
- J. Preskill, "Quantum Computing in the NISQ era and beyond," *Quantum* (2018).

- Google Quantum AI et al., "Quantum supremacy using a programmable superconducting processor," *Nature* (2019); and "Quantum error correction below the surface code threshold," *Nature* (2024/2025).

Current landscape and reference

- Olivier Ezratty, *Understanding Quantum Technologies* (updated annually, freely available online). An encyclopedic, continually revised survey of the entire field — hardware, algorithms, cryptography, and the industry landscape.
- NIST Post-Quantum Cryptography project (csrc.nist.gov/projects/post-quantum-cryptography) for the standardized quantum-resistant algorithms.
- Industry news trackers such as *The Quantum Insider* and *Quantum Computing Report* for the fast-moving frontier — read with the skeptical eye this book has recommended.

A guided path deeper

For readers who want to go further, the following offer well-trodden next steps, arranged roughly from most accessible to most demanding.

For the curious general reader. Start with accessible popular accounts that convey the ideas without equations, then move to the many excellent long-form explainer articles and lecture videos produced by university groups and science journalists. Scott Aaronson's blog and writings are an unusually honest, technically grounded antidote to hype, and are warmly recommended for anyone who wants to tell the real from the oversold.

For the mathematically inclined. The standard graduate textbook, Nielsen and Chuang's *Quantum Computation and*

Quantum Information, remains the field's bible and is surprisingly readable for a determined reader with some linear algebra; it repays slow study. Shorter lecture-note sets by leading researchers (widely available online) cover the core algorithms — Deutsch–Jozsa, Simon, Shor, Grover — in full detail.

On cryptography and the quantum threat. The published post-quantum standards and the accompanying explanatory reports from national standards bodies are the authoritative source on how the world is defending itself; they are more readable than one might fear and convey the urgency of the migration.

On the physics foundations. For the deep strangeness underlying it all — superposition, entanglement, Bell's theorem — the accessible writings of the physicists who did the work, and good histories of the quantum revolution, ground the computing story in the century of physics that made it possible.

Keeping up with a fast-moving field. Because quantum computing advances quickly and is heavily hyped, the best habit is to follow a few trusted, skeptical sources rather than chase headlines. Read announcements critically using the questions in the appendix "How to Read Quantum News." Remember that press releases are marketing; peer-reviewed papers, independent replications, and the reactions of knowledgeable skeptics are what separate durable results from noise.

A Note on Sources and Method

This book is a work of narrative science history, written for the general reader and drawn from the public record and the standard published scholarship rather than from original research. Its purpose is synthesis and explanation — to tell the story of quantum computing whole, from the birth of quantum mechanics to the frontier of the mid-2020s, in language a non-specialist can follow — and it stands on the shoulders of the physicists, computer scientists, historians, and science writers who created and chronicled the field.

For the physics and the foundational theory, I have relied on the canonical accounts: Nielsen and Chuang's textbook, Preskill's lecture notes and essays, and the accessible expositions of Bernhardt, Aaronson, and Deutsch listed in the Further Reading. For the history of quantum mechanics itself — Planck, Einstein, Bohr, Heisenberg, Schrödinger, Born, Bell — I have drawn on the general body of physics history and on the primary papers where they bear directly on the computing story. For the milestones of quantum computing, I have followed the landmark papers (Feynman 1982, Deutsch 1985, Shor 1994, Grover 1996, the threshold-theorem papers of 1996–98, Preskill's NISQ essay of 2018, and the Google *Nature* papers of 2019 and 2024) and the reporting of the scientific and technical press.

Throughout, I have tried to observe a discipline appropriate to a fast-moving field: to distinguish settled science from provisional claim, to date the recent milestones, and to flag vendor announcements — quantum-advantage, logical-qubit, and speedup claims from companies about their own machines — as the dated, often pre-peer-review claims they are, rather than as established facts. The history of quantum computing is, as this book has stressed, littered with claims that did not survive independent scrutiny: eroded supremacy figures, dequantized speedups, retracted results. Where I report such claims, especially those from 2023 onward, the reader should understand them as snapshots of a contested and rapidly shifting frontier, accurate to the best of my knowledge as of the time of writing but subject to revision as the science advances and the classical competition responds.

Two general observations about method deserve stating. First, on the mathematics: this book contains almost none, by deliberate choice, and where I have used metaphors — the wave, the noise-cancelling headphones, the golden chandelier — I have tried to correct them where they mislead, above all the pervasive and wrong image of a quantum computer "trying all answers at once." Any metaphor for quantum mechanics is imperfect; I have preferred honest imperfection to false precision. Second, on judgment: quantum computing invites both breathless enthusiasm and weary dismissal, and I have tried throughout to occupy the harder middle ground — genuine wonder at what has been achieved, honest skepticism about what has merely been promised — trusting the reader to weigh the evidence rather than accept a verdict. Where I have erred, I hope it is on the side of clarity and candor.

The subject changes monthly, and any book about it is a photograph of a moving object. For the current state of the field beyond this book's horizon, the reader is directed to the continually updated resources in the Further Reading — Ezratty's annual survey above all — and, always, to the skeptical questions this book has tried to make second nature: Is the task useful or contrived? Has the classical competition genuinely been bested? Are the qubits physical or logical? Has the claim survived independent replication? Those questions, more than any particular fact that may soon be out of date, are what I hope the reader carries away.

Landmark papers and further sources

The following foundational works are cited throughout this book and repay direct study by readers with the relevant background; full details are readily found through any scholarly search.

- M. Planck (1900), on the law of black-body radiation — the birth of the quantum.
- A. Einstein (1905), on the photoelectric effect and light quanta.
- W. Heisenberg (1925); E. Schrödinger (1926); M. Born (1926) — the founding papers of quantum mechanics.
- A. Einstein, B. Podolsky, N. Rosen (1935), the EPR paper; E. Schrödinger (1935), naming entanglement.
- A. Turing (1936), on computable numbers — the universal machine.
- R. Landauer (1961), on the thermodynamics of information; C. Bennett (1973), on reversible computation.
- J. S. Bell (1964), on the Einstein–Podolsky–Rosen paradox; A. Aspect et al. (1982), experimental tests.
- P. Benioff (1980); R. Feynman (1982), "Simulating Physics with Computers"; D. Deutsch (1985), the universal quantum computer.
- D. Deutsch and R. Jozsa (1992); E. Bernstein and U. Vazirani (1993); D. Simon (1994).
- P. Shor (1994), factoring and discrete logarithms; L. Grover (1996), quantum search.
- P. Shor (1995), the first quantum error-correcting code; A. Steane (1996), the Steane code; D. Gottesman

(1997), stabilizer codes.

- A. Aharonov and M. Ben-Or; D. Aharonov; A. Kitaev; E. Knill, R. Laflamme, W. Zurek (c. 1996–98), the threshold theorem.
- A. Kitaev (1997/2003), the surface and toric codes; topological quantum computation.
- S. Lloyd (1996), universal quantum simulators.
- C. Bennett and G. Brassard (1984), the BB84 quantum-key-distribution protocol; C. Bennett et al. (1993), quantum teleportation.
- E. Knill, R. Laflamme, G. Milburn (2001), linear-optics quantum computing (the KLM scheme).
- F. Arute et al. (Google, 2019), quantum supremacy using a programmable superconducting processor.
- H.-S. Zhong et al. (2020), quantum computational advantage using photons (Jiuzhang).
- A. Harrow, A. Hassidim, S. Lloyd (2009), quantum linear systems (HHL); E. Tang (2019), dequantization.
- NIST (2024), the finalized post-quantum cryptography standards (FIPS 203/204/205).
- Google Quantum AI (2024), quantum error correction below the surface-code threshold (Willow).

Standard reference text: M. A. Nielsen and I. L. Chuang, *Quantum Computation and Quantum Information.*

Appendix: Frequently Asked Questions

Over the years, certain questions about quantum computing come up again and again — from curious newcomers, from journalists, from worried executives, from students. This appendix gathers the most common of them and answers each briefly and honestly, distilling the arguments made at length in the body of the book. If you read nothing else, read this.

Will a quantum computer replace the computer on my desk? No. Quantum computers are not faster classical computers, and they never will be for ordinary tasks — email, spreadsheets, web browsing, video, gaming. They are specialized instruments that outperform classical machines only on a narrow class of problems with special mathematical structure. Your laptop is safe. The quantum computer, if it fulfills its promise, will be a remote, refrigerated, expensive machine accessed over the cloud for particular scientific and cryptographic tasks — a supercomputer for special problems, not a personal device.

Does a quantum computer try all answers at once? Not usefully. It can be placed in a superposition of all possible inputs, but measuring that superposition yields just one random answer and destroys the rest. The entire art of quantum algorithms is arranging interference so that the wrong answers cancel and the right one dominates before you look. "Tries everything at once" is the single most misleading sentence about quantum computing; the reality is subtler, harder, and confined to specially structured problems.

When will quantum computers break encryption? No one knows. Breaking real encryption with Shor's algorithm requires a large, error-corrected machine — likely thousands of logical qubits, translating to perhaps a million or more physical qubits — that does not exist and is, by most estimates, a decade or more away, if it arrives on schedule at all. But because of "harvest now, decrypt later," the threat is already urgent for long-lived secrets, which is why the world is migrating to post-quantum cryptography now.

Should I be worried about my data today? For most everyday purposes, no immediate action is needed on your part; the migration to quantum-resistant encryption is being handled by the companies and standards bodies that build the internet's infrastructure. But if you hold secrets that must remain secret for ten or twenty years, assume they could be harvested today and decrypted after Q-Day, and ensure your systems adopt post-quantum cryptography as it becomes available. Symmetric encryption (like AES) survives the quantum era with longer keys; it is public-key cryptography (RSA, elliptic curves) that faces total breakage.

How many qubits does a "real" quantum computer need? It depends entirely on the task and on qubit quality. Useful chemistry simulations may need hundreds to thousands of logical qubits; breaking RSA needs a few thousand logical qubits. Because each logical (error-corrected) qubit may require hundreds or thousands of physical qubits, the physical-qubit counts required run to the hundreds of thousands or millions. Today's largest machines have a few hundred noisy physical qubits and a handful of logical ones. The gap is enormous, and closing it is the central challenge of the field.

What is the difference between physical and logical qubits, and why does it matter? A physical qubit is one piece of hardware — noisy and fragile. A logical qubit is an error-corrected qubit built from many physical ones, reliable enough to compute with. Headlines count physical qubits; capability depends on logical qubits, of which there can be a thousand times fewer. When you read a qubit-count claim, ask whether it refers to physical or logical qubits — the difference is the difference between hype and substance.

Is quantum supremacy the same as a useful quantum computer? No. Quantum supremacy means doing some task — any task, however useless — that no classical computer can match. It was achieved in 2019 using a deliberately pointless benchmark. Useful quantum advantage — solving a valuable real-world problem better than any classical computer — has not been achieved and is a much harder, later milestone. Conflating the two is a chronic source of confusion.

What will quantum computers actually be good for? The most solidly grounded application is simulating quantum systems — molecules, materials, chemical reactions — which could transform chemistry, drug discovery, energy, and materials science. This was the field's founding purpose. Cryptography (breaking old codes) is the most dramatic threat. Search and optimization offer real but modest (quadratic) speedups that may or may not prove practically worthwhile. Claims about revolutionizing artificial intelligence rest on shaky foundations and deserve deep skepticism.

Which company or country is winning? No one is clearly "winning," partly because "winning" has several meanings achieved at different times. No single hardware technology has prevailed — superconducting circuits, trapped ions, neutral atoms, photonics, and silicon spins are all in serious contention. The United States and China lead the global competition, with Europe and others significant. Beware anyone claiming a clear leader; ask which finish line they mean.

Is quantum computing just hype? No — but it is heavily hyped. The achievements are real: supremacy demonstrated, error correction crossed below threshold, logical qubits growing in number, cryptography genuinely threatened, sensing already useful. The physics is sound and the long-term potential is real. But useful large-scale machines remain years or decades away and are not guaranteed on any timeline, and the promises have been persistently oversold. The mature view holds both truths: genuine substance and genuine distance.

Could quantum computers become conscious, or break the laws of physics, or enable time travel? No, no, and no. A quantum computer computes exactly the same set of functions a classical computer can — not one more; it merely does a few of them faster. It cannot solve uncomputable problems, cannot signal faster than light (entanglement carries no usable message on its own), and has nothing to do with consciousness or time travel. Much science-fiction-flavored quantum hype is simply false. The real story is strange enough without invention.

Might quantum computing simply never work at scale? It is possible, though it looks increasingly unlikely after the below-threshold error-correction demonstrations. The obstacles — the crushing overhead of error correction, the wiring and refrigeration bottlenecks, the sheer engineering of a million-qubit machine — are formidable, and no law of physics guarantees they will all be overcome affordably. Responsible experts hold open the possibility that useful fault-tolerant machines take much longer than hoped, or arrive in a more limited form than the grandest visions promise. Uncertainty about the timeline is honest; the science itself is not in doubt.

Appendix: How to Read Quantum News

Quantum computing generates a relentless stream of headlines, and most of them are, to put it gently, unreliable — inflated by corporate marketing, garbled by well-meaning journalists, or simply misunderstanding the science. This short appendix is a practical toolkit: a set of questions and warning signs to apply to any quantum computing story so that you can tell the durable results from the noise. It is, in a sense, the whole book compressed into a checklist.

Seven questions to ask of any quantum claim

1. **Supremacy or advantage?** Is the story about doing a *useless* task no classical computer can match (supremacy — already achieved, of limited practical meaning), or about solving a *useful* problem better than classical machines (advantage — much harder, largely unproven)? Marketing loves to blur these. If a headline implies a useful breakthrough, check whether the underlying task is actually useful or merely a contrived benchmark.

2. **Physical or logical qubits?** A claim of "1,000 qubits" almost always means physical qubits — noisy raw mate-

rial. The number that determines real capability is the count of logical, error-corrected qubits, which can be a thousand times smaller. A machine with fewer, better qubits can outperform one with more, worse qubits. Always ask which kind of qubit is being counted.

3. Compared against the best classical method? A quantum "speedup" is only meaningful relative to the *best* classical algorithm, not a convenient or naive one. Many quantum advantage claims evaporate when a cleverer classical method — often invented specifically in response — matches them. Ask: has this been benchmarked against state-of-the-art classical computing, and what happened when the classical experts responded?

4. Peer-reviewed and reproduced? Press releases are marketing; peer-reviewed papers are evidence; independent replication is confirmation. A company blog post announcing a milestone is not the same as a vetted, reproduced scientific result. Weight your belief accordingly, and wait for the skeptics to weigh in before concluding a revolution has occurred.

5. What do knowledgeable skeptics say? The quantum field has a healthy culture of expert skeptics who scrutinize bold claims. If a result is real and important, it will survive their examination; if it is hype, they will say so. Seek out the reactions of independent researchers, not just the announcing company's spokespeople.

6. Does the problem have exploitable structure? Quantum computers help dramatically only with problems that have special hidden mathematical structure (like the periodicity Shor exploits) or, more modestly, with unstructured search (Grover's quadratic gain). For generic problems with no such structure, quantum machines offer little or nothing. Ask whether the touted application is actually the kind of problem quantum computers are good at.

7. What is the timeline, really? Is the claimed capability available now, or is it a projection — a roadmap target for

2029, a "path toward," an "in principle"? Roadmaps are aspirations, not deliveries, and the field's history is littered with slipped timelines. Distinguish sharply between what a machine does today and what its makers hope it will do years hence.

Warning signs of overhyped coverage

Certain phrases and framings should trigger heightened skepticism. Be wary of stories claiming quantum computers will "solve problems impossible for classical computers" without specifying which problems; of "millions of times faster" or "instantly" without a careful classical comparison; of quantum computing "revolutionizing AI" (a claim resting on shaky, often-dequantized foundations); of any suggestion that quantum computers "try all possibilities simultaneously" and read off the answer; and of breathless language about breaking "all encryption tomorrow" (the threat is real but not imminent, and the defense already exists). Be especially cautious when the source is a company announcing its own product, an investor promoting a holding, or a journalist who appears to be paraphrasing a press release without independent expertise.

Signs of trustworthy coverage

Conversely, good quantum journalism and honest claims share recognizable features. They distinguish physical from logical qubits and supremacy from advantage. They specify the exact task and compare against the best classical methods, and they report what happened when classical experts responded. They cite peer-reviewed work and quote independent skeptics. They are precise about timelines, separating present achievement from future projection. They convey both the genuine progress and the genuine remaining distance, resisting both breathless hype and reflexive dismissal. Above all, they leave you better able to reason about the field, not merely more excited or more alarmed.

A closing rule of thumb

When in doubt, apply the physicist's discipline: an extraordinary claim requires extraordinary evidence, and the more a story promises, the more scrutiny it deserves. Quantum computing is a genuinely important and genuinely difficult technology, and it is precisely because it matters that it is worth understanding truly rather than through the distorting lens of hype. The reader who applies these questions — patiently, curiously, skeptically — will, over time, develop something rarer and more valuable than any single fact about quantum computing: the judgment to know what to believe.

Appendix: Landmark Experiments and What They Proved

Theory maps the possible; experiment decides the actual. This appendix gathers the pivotal experiments of quantum computing's history in one place, and for each states plainly what it demonstrated — and, just as importantly, what it did not. Read together, they trace the field's passage from philosophical dispute to laboratory reality.

Aspect's Bell tests (1982, and the loophole-free tests of 2015). Alain Aspect's Paris experiments measured pairs of entangled photons while switching the measurement settings so rapidly that no light-speed signal could pass between the detectors in time, and found Bell's inequality decisively violated. *What it proved:* the world is not governed by local hidden variables — reality is genuinely non-classical, and entanglement is real, not a statistical illusion. Later "loophole-free" experiments (2015, in Delft, Vienna, and Boulder) closed the last technical escape routes. *What it did not prove:* which interpretation of quantum mechanics is correct — all interpretations agree on the results. These tests underpin the 2022 Nobel Prize.

The first trapped-ion gate (NIST, 1995). Following the

Cirac–Zoller proposal, David Wineland's group demonstrated a controlled quantum logic gate between the internal and motional states of a single trapped ion. *What it proved:* quantum logic gates could be built in real hardware; the machine had left the page. *What it did not prove:* that the approach could scale — a single gate is a long way from a computer.

The NMR factoring of 15 (IBM/Stanford, 2001). Isaac Chuang's team used a seven-spin molecule in a test tube to run Shor's algorithm and factor 15 into 3 x 5. *What it proved:* Shor's algorithm could run on physical hardware; a symbolic milestone. *What it did not prove:* genuine scalable quantum computation — the room-temperature ensemble may have contained no true entanglement, and the approach proved a dead end. A cautionary tale that a headline result can mask a technology with no future.

Quantum teleportation (Innsbruck and Rome, 1997). Building on the 1993 protocol of Bennett and colleagues, experimenters transmitted an unknown quantum state from one location to another using shared entanglement and classical communication. *What it proved:* entanglement is a usable resource that can be spent to accomplish otherwise-impossible tasks. *What it did not prove:* faster-than-light communication — teleportation requires an ordinary classical message to complete, and carries no signal on its own.

Micius satellite quantum communication (China, 2016–2017). Jian-Wei Pan's team used a dedicated satellite to distribute entangled photons and quantum keys between ground stations separated by more than a thousand kilometers. *What it proved:* quantum-secured communication can work over intercontinental distances by routing photons through space, sidestepping fiber losses; a major step toward a quantum internet and a marker of China's leadership. *What it did not prove:* a full quantum internet — practical long-distance quantum repeaters remain unbuilt.

Google Sycamore's quantum supremacy (2019). A 53-

qubit superconducting processor sampled from a random quantum circuit in about 200 seconds, a task Google estimated would take a supercomputer 10,000 years. *What it proved:* a programmable quantum machine could enter a regime beyond easy classical simulation — the first experimental proof of quantum computational supremacy. *What it did not prove:* any useful capability (the task is useless), nor an unassailable gap (IBM argued a clever classical method could do it in days, and later classical algorithms narrowed the margin further).

Jiuzhang's photonic advantage (China, 2020). A fixed optical apparatus performed boson sampling far beyond classical reach. *What it proved:* quantum supremacy is reproducible on a completely different platform and task, strengthening the case that it is real; and that China is at the forefront. *What it did not prove:* programmable, universal computation (Jiuzhang did one fixed task), and its margins too were later challenged by improved classical methods.

IBM's "quantum utility" simulation (2023). A 127-qubit processor with error mitigation simulated a complex quantum magnetic system in a regime where straightforward classical simulation struggled. *What it proved:* noisy machines can serve as useful scientific tools at meaningful scale. *What it did not prove:* a clear advantage — classical researchers soon reproduced the results with clever methods, in the now-familiar pattern.

Neutral-atom logical qubits (Harvard/QuEra, 2023). A reconfigurable array of neutral atoms created dozens of logical qubits and performed entangling operations between them, exploiting the atoms' mobility. *What it proved:* error-corrected logical qubits can be built and operated at a scale of dozens on a fast-rising platform. *What it did not prove:* fault-tolerant computation at useful scale — the numbers, though impressive, remain far short of what applications require.

Google Willow's below-threshold error correction (2024).

A 105-qubit chip ran surface-code error correction at increasing sizes and showed the logical error rate dropping — roughly halving — as the code grew. *What it proved:* the single most important thing the field had waited three decades to see — that scaling up error correction suppresses errors, crossing below the threshold and vindicating the threshold theorem. The path to reliable, arbitrarily long quantum computation is demonstrably open. *What it did not prove:* a large machine — Willow protected only one logical qubit, and the road from one to the thousands needed remains long and steep.

Taken together, these experiments tell a coherent story: the strangeness is real (Bell tests), it can be controlled (ion gates, teleportation), it can outpace classical computation (Sycamore, Jiuzhang), it can be put to scientific use (utility experiments), and — most crucially — its fatal fragility can be overcome (Willow). Each was a genuine milestone; none was the finish line. The pattern of the field is written in this list: real progress, real remaining distance, and a classical community that keeps the quantum claims honest.

Appendix: The Mathematics, Gently

This book has told the story of quantum computing almost entirely in words and pictures, avoiding equations so as to remain welcoming to every reader. But some readers will want to peek, just a little, behind the curtain — to see the actual mathematical objects that the words describe, without being buried in formalism. This appendix offers exactly that: the gentlest possible glimpse of the mathematics, enough to make the ideas concrete for the curious, and entirely skippable for everyone else. No prior mathematics beyond arithmetic and a willingness to think carefully is assumed.

A qubit is an arrow

The state of a single qubit can be written as a combination of the two basic states, which physicists write as |0> and |1> — those angular brackets, called "kets," are just a notation for "the quantum state labeled 0" and "the quantum state labeled 1." A general qubit state looks like this: a|0> + b|1>, where a and b are two numbers called amplitudes. You can picture the state as an arrow, and a and b as telling you how much of it points along the "0 direction" and how much along the "1 direction."

The amplitudes obey one rule: the squares of their sizes must add up to one, |a|-squared + |b|-squared = 1. This is because |a|-squared is the probability of getting the answer 0 if you measure, and |b|-squared is the probability of getting 1, and the probabilities of the only two possible outcomes must total one. So if a qubit is in the state where a and b are equal, each equal to about 0.707 (which is one divided by the square root of two), then |a|-squared = |b|-squared = 0.5, meaning a fifty-fifty chance of each outcome — a perfectly balanced superposition, an arrow pointing at the equator of the Bloch sphere.

The crucial feature, the one that makes everything quantum, is that a and b can be *negative* (and, in full generality, complex numbers). This is what separates a qubit from a mere probabilistic coin. A classical coin has probabilities, always positive, that can only add up when combined. Amplitudes can be positive or negative, and so they can *cancel*. That cancellation is interference, and it is the whole source of quantum power.

Interference in one line

Here is the simplest possible glimpse of interference. The Hadamard gate, applied to |0>, produces the balanced superposition (|0> + |1>) divided by the square root of two — an equal blend, both amplitudes positive. Applied to |1>, it produces (|0> - |1>) divided by the square root of two — an equal blend, but now with a *minus* sign on the |1> part.

Now watch what happens if you apply the Hadamard gate *twice* to |0>. The first Hadamard makes (|0> + |1>) over root two. The second Hadamard acts on each piece: it turns |0> into (|0> + |1>) and |1> into (|0> - |1>), and when you add these together, the |1> parts — one plus, one minus — *cancel exactly*, while the |0> parts reinforce. The result is simply |0> again, with certainty. Two Hadamards return you to the start, and they do so because of interference:

the two paths to the |1> outcome had opposite amplitudes and annihilated each other. This tiny calculation — amplitudes cancelling on the unwanted outcome, reinforcing on the wanted one — is, in miniature, exactly the mechanism behind Deutsch's algorithm, Shor's algorithm, and every quantum speedup there is. Everything else is this, scaled up and made clever.

Why two qubits hold four numbers

A single qubit needs two amplitudes, a and b. Two qubits together have four basic states — |00>, |01>, |10>, |11> — and a general two-qubit state is a combination of all four, needing four amplitudes. Three qubits need eight; n qubits need 2-to-the-n amplitudes. This doubling is the exponential resource. For 300 qubits, 2-to-the-300 is a number larger than the count of atoms in the observable universe — which is why no classical computer can write down the state of a few hundred entangled qubits, and why a quantum computer, which simply *is* such a state, holds something no classical machine can.

An entangled state is one that *cannot* be separated into a description of each qubit on its own. For example, the two-qubit state (|00> + |11>) divided by the square root of two — one of the "Bell states" — cannot be written as (something for the first qubit) combined with (something for the second qubit). The qubits have no individual states, only a shared one: measure the first and get 0, and the second is instantly 0 too; get 1, and the second is 1. This inseparability is entanglement expressed in symbols, and it is the connective tissue that lets a register of qubits behave as a single indivisible whole.

Measurement, in symbols

When you measure the qubit a|0> + b|1>, you get the outcome 0 with probability |a|-squared and the outcome 1 with probability |b|-squared, and the state then "collapses" to whichever result you got — becoming simply |0> or |1>, the superposition gone. This is why a qubit yields only one bit when measured, no matter how rich its amplitudes were: you sample once, according to the probabilities, and the rest is lost. The art of quantum algorithms is to arrange, through interference, for the amplitude of the desired answer to be large — close to one — so that the single sample you draw is, with high probability, the answer you want.

The exponential, felt as time

One last piece of arithmetic conveys why the stakes are so high. The best known classical methods for factoring a number grow roughly exponentially with the number's length, while Shor's quantum algorithm grows only as a small power (a polynomial) of the length. Concretely: factoring a 2,048-bit number classically would take longer than the age of the universe, while a large enough quantum computer running Shor's algorithm might do it in hours. The difference between exponential and polynomial growth is the difference between "never" and "an afternoon," and it is the whole reason a machine built of maybe could shatter the locks of the internet.

A closing reassurance

If none of this appendix was necessary for you to appreciate the story, that is by design — the ideas of quantum computing can be genuinely understood in words and pictures, and the equations, while beautiful, are servants of the ideas rather than the other way around. But if these few symbols made something click — if you saw, in the cancelling amplitudes of two Hadamard gates, the very engine of Shor's assault on cryptography — then you have glimpsed the mathematical heart of the strangest and most consequential machine humanity has yet attempted to build. The rest, as they say, is detail: profound, difficult, and endlessly unfolding detail, but detail all the same, radiating outward from the simple, radical idea that information itself is quantum.

www.ingramcontent.com/pod-product-compliance
Lightning Source LLC
LaVergne TN
LVHW100132061126
851857LV00010B/63
* 9 7 9 8 2 3 5 2 8 4 8 4 5 *